ADVANCES IN
NUMERICAL HEAT TRANSFER

Volume 4

T0225398

Series in Computational and Physical Processes in Mechanics and Thermal Sciences

Series Editors

W. J. Minkowycz

Mechanical and Industrial Engineering
University of Illinois at Chicago
Chicago, Illinois

E. M. Sparrow

Mechanical Engineering
University of Minnesota
Minneapolis, Minnesota

ADVANCES IN NUMERICAL HEAT TRANSFER

Volume 4

NANOPARTICLE HEAT TRANSFER AND FLUID FLOW

Edited by

W. J. Minkowycz
Mechanical and Industrial Engineering
University of Illinois at Chicago
Chicago, Illinois

E. M. Sparrow
Mechanical Engineering
University of Minnesota, Twin Cities
Minneapolis, Minnesota

J. P. Abraham
School of Engineering
University of St. Thomas
St. Paul, Minnesota

CRC Press
Taylor & Francis Group
Boca Raton London New York

CRC Press is an imprint of the
Taylor & Francis Group, an **informa** business

CRC Press
Taylor & Francis Group
6000 Broken Sound Parkway NW, Suite 300
Boca Raton, FL 33487-2742

First issued in paperback 2017

Version Date: 20121023

ISBN 13: 978-1-138-07654-9 (pbk)
ISBN 13: 978-1-4398-6192-9 (hbk)

Library of Congress Cataloging-in-Publication Data

Nanoparticle heat transfer and fluid flow / [edited by] W.J. Minkowycz, E.M. Sparrow, J.P. Abraham.
  pages cm
  Includes bibliographical references and index.
  ISBN 978-1-4398-6192-9 (hardback)
  1. Heat--Transmission. 2. Heat exchangers--Thermodynamics. 3. Nanoparticles--Fluid dynamics. I. Minkowycz, W. J. II. Sparrow, E. M. (Ephraim M.) III. Abraham, J. P. (John P.)

QC320.N36 2013
620.1'15--dc23   2012030319

Contents

Preface

The day of nanoparticles and nanofluids has arrived, and the applications of these media are legion. Here, attention is focused on such disparate applications as biomedical, energy conversion, material properties, and fluid flow and heat transfer. The common denominator of the chapters that set forth these applications here is numerical quantification, modeling, simulation, and presentation.

The first chapter of this volume conveys a broad overview of nanofluid applications, while the second chapter presents a critical synthesis of the variants within the thermophysical properties of nanofluids and then narrows the focus to the applications of nanoparticles in the biomedical field. Chapters 3 and 4 deepen the biomedical emphasis. Equally reflective of current technological and societal themes is energy conversion from dispersed forms to more concentrated and utilizable forms, and these issues are treated in Chapters 5 and 6.

Basic to the numerical modeling and simulation of any thermofluid process are material properties. Nanofluid properties have been shown to be less predictable and less repeatable than are those of other media that participate in fluid flow and heat transfer. Property issues for nanofluids are set forth in Chapters 6 and 7.

The last three chapters each focus on a specific topic in nanofluid flow and heat transfer. Chapter 8 deals with filtration. Microchannel heat transfer has been identified as the preferred means for the thermal management of electronic equipment, and the role of nanofluids as a coolant is discussed in Chapter 9. Natural convection is conventionally regarded as a low heat-transfer coefficient form of convective heat transfer. Potential enhancement of natural convection due to nanoparticles is the focus of Chapter 10.

Editors

W. J. Minkowycz is the James P. Hartnett Professor of Mechanical Engineering at the University of Illinois at Chicago. He joined the faculty at UIC in 1966. His primary research interests lie in the numerical modeling of fluid flow and heat transfer problems. Professor Minkowycz is currently the editor-in-chief of the *International Journal of Heat and Mass Transfer, Numerical Heat Transfer,* and *International Communications in Heat and Mass Transfer.* He has won numerous awards for his excellence in teaching, research, and service to the heat transfer community.

E. M. Sparrow is a professor of mechanical engineering at the University of Minnesota. He has taught and performed research there since 1955. He is a member of the National Academy of Engineering, a Max Jakob awardee, and is a Morse Alumni Distinguished Teaching Professor and Institute Professor. He has published over 600 peer-reviewed articles on a wide variety of topics in heat transfer and fluid flow.

J. P. Abraham has worked in the area of thermal sciences for approximately 20 years. His research areas include nanoscale thermal processes, energy production and distribution, climate monitoring, and medical device development. He has about 150 journal publications, conference presentations, book chapters, and patents. He teaches courses in the undergraduate and graduate mechanical engineering programs at the University of St. Thomas, St. Paul, Minnesota.

Contributors

Ronald J. Adrian
Department of Mechanical and
    Aerospace Engineering
Arizona State University
Tempe, Arizona

Corinne S. Baresich
Department of Mechanical and
    Aerospace Engineering
Rutgers, The State University of
    New Jersey
Piscataway, New Jersey

John C. Bischof
Department of Mechanical Engineering
and
Department of Biomedical Engineering
and
Department of Urologic Surgery
University of Minnesota
Minneapolis, Minnesota

Anna S. Cherkasova
Department of Mechanical and
    Aerospace Engineering
Rutgers, The State University of
    New Jersey
Piscataway, New Jersey

Massimo Corcione
Dipartimento di Ingegneria
    Astronautica, Elettrica ed Energetica
Sapienza Università di Roma
Rome, Italy

Omar De Leon
Department of Mechanical and
    Aerospace Engineering
University of Miami
Coral Gables, Florida

Richard E. DeVor (Deceased)
Department of Mechanical
    Science and Engineering
University of Illinois
Urbana, Illinios

Michael L. Etheridge
Department of Mechanical
    Engineering
and
Department of Biomedical
    Engineering
University of Minnesota
Minneapolis, Minnesota

Yu Feng
Department of Mechanical
    and Aerospace
    Engineering
North Carolina State
    University
Raleigh, North Carolina

Rhonda R. Franklin
Department of Electrical and
    Computer Engineering
University of Minnesota
Minneapolis, Minnesota

Shiv G. Kapoor
Department of Mechanical
    Science and Engineering
University of Illinois
Urbana, Illinois

Khalil M. Khanafer
Department of Biomedical
    Engineering
University of Michigan
Ann Arbor, Michigan

Clement Kleinstreuer
Department of Mechanical and
    Aerospace Engineering
and
Department of Biomedical
    Engineering
North Carolina State
    University
Raleigh, North Carolina

Jie Li
Department of Mechanical and
    Aerospace Engineering
North Carolina State University
Raleigh, North Carolina

Chen Lin
Department of Mechanical and
    Aerospace Engineering
Rutgers, The State University of
    New Jersey
Piscataway, New Jersey

Ronghui Ma
Department of Mechanical
    Engineering
University of Maryland Baltimore
    County
Baltimore, Maryland

Navid Manuchehrabadi
Department of Mechanical
    Engineering
University of Maryland Baltimore
    County
Baltimore, Maryland

Todd P. Otanicar
Department of Mechanical
    Engineering
University of Tulsa
Tulsa, Oklahoma

Patrick E. Phelan
Department of Mechanical and
    Aerospace Engineering
Arizona State University
Tempe, Arizona

Ravi S. Prasher
Department of Mechanical and
    Aerospace Engineering
Arizona State University
Tempe, Arizona

Jerry W. Shan
Department of Mechanical and
    Aerospace Engineering
Rutgers, The State University of
    New Jersey
Piscataway, New Jersey

Di Su
Department of Mechanical
    Engineering
University of Maryland
    Baltimore County
Baltimore, Maryland

Robert Taylor
School of Mechanical and
    Manufacturing Engineering
and
School of Photovoltaic and
    Renewable Energy Engineering
University of New South Wales
Sydney, Australia

Kambiz Vafai
Department of Mechanical Engineering
University of California
Riverside, California

John E. Wentz
School of Engineering
University of St. Thomas
St. Paul, Minnesota

Kaufui V. Wong
Department of Mechanical and
    Aerospace Engineering
University of Miami
Coral Gables, Florida

Zhuomin M. Zhang
G.W. Woodruff School of Mechanical
    Engineering
Georgia Institute of Technology
Atlanta, Georgia

Liang Zhu
Department of Mechanical Engineering
University of Maryland Baltimore
    County
Baltimore, Maryland

Qunzhi Zhu
School of Energy and Environment
    Engineering
Shanghai University of Electric Power
Shanghai, China

1 Review of Nanofluid Applications

Kaufui V. Wong and Omar De Leon

CONTENTS

1.1 INTRODUCTION

Nanofluids are dilute liquid suspensions of nanoparticles with at least one of their principal dimensions smaller than 100 nm. From the literature, nanofluids have been found to possess enhanced thermophysical properties such as thermal conductivity, thermal diffusivity, viscosity, and convective heat-transfer coefficients compared with those of base fluids like oil or water [1–6].

1

From this review, it will be seen that nanofluids clearly display enhanced thermal conductivity, which goes up with increasing volumetric fraction of nanoparticles. The current review does concentrate on this relatively new class of fluids and not on colloids, which are nanofluids because the latter have been used for a long time. Review of experimental studies clearly showed a lack of consistency in the reported results of different research groups regarding thermal properties [7,8]. The effects of several important factors such as particle size and shapes, clustering of particles, temperature of the fluid, and relationship between the surfactant and effective thermal conductivity of nanofluids have not been studied adequately. It is important to conduct more research so as to ascertain the effects of these factors on the thermal conductivity of wide range of nanofluids.

Classical models have not been successful in explaining the observed enhanced thermal conductivity of nanofluids. Recently developed models only include one or two postulated mechanisms of nanofluids heat transfer. For instance, there has not been much fundamental work reported on the determination of the effective thermal diffusivity of nanofluids nor heat-transfer coefficients for nanofluids in natural convection [9].

There is a growth in the use of colloids, which are nanofluids in the biomedical industry for sensing and imaging purposes. This is directly related to the ability to design novel materials at the nanoscale level alongside recent innovations in analytical and imaging technologies for measuring and manipulating nanomaterials. This has led to the fast development of commercial applications, which use a wide variety of manufactured nanoparticles. The production, use, and disposal of manufactured nanoparticles will lead to discharges of nanoparticles to the air, soils, and water systems. Negative effects are likely and quantification and minimization of these effects on environmental health is obligatory. Actual knowledge of concentrations and physico-chemical properties of manufactured nanoparticles under realistic conditions is important to predicting their fate, behavior, and toxicity in the natural aquatic environment. The aquatic colloid and atmospheric ultrafine particle literature both offer evidence as to the presumptive behavior and impacts of manufactured nanoparticles [10], and there is no pretense that a review duplicating similar literature about the use of colloids which are also nanofluids is attempted in the current review, which is an update of Ref. [11].

As they have enhanced properties such as for thermal transfer, nanofluids can be used in a wide range of engineering applications from the automotive industry to the medical arena to applications in power plant cooling systems as well as computers.

1.2 HEAT-TRANSFER APPLICATIONS

1.2.1 INDUSTRIAL COOLING APPLICATIONS

Routbort et al. [12] started a project in 2008 that employed nanofluids for industrial cooling that could result in great energy savings and resulting emissions reductions. For U.S. industry, the replacement of cooling and heating water with nanofluids has the potential to conserve 1 trillion Btu of energy. For the U.S. electric power industry, using nanofluids in closed-loop cooling cycles could save about 10–30 trillion Btu per year (equivalent to the annual energy consumption of about 50,000–150,000 households). The associated emissions reductions would be approximately 5.6 million metric tons of carbon dioxide; 8600 metric tons of nitrogen oxides; and 21,000 metric tons of sulfur dioxide.

For Michelin North America tire plants, the productivity of numerous industrial processes is constrained by the lack of facility to cool the rubber efficiently as it is being processed. This requires the use of over 2 million gallons of heat-transfer fluids for Michelin's North American plants. It is Michelin's goal in this project to obtain a 10% productivity increase in its rubber processing plants if suitable water-based nanofluids can be developed and commercially produced in a cost-effective manner.

Han et al. [13] have used phase change materials as nanoparticles in nanofluids to simultaneously enhance the effective thermal conductivity and specific heat of the fluids. As an example, a suspension of indium nanoparticles (melting temperature, 157°C) in polyalphaolefin has been synthesized using a one-step, nanoemulsification method. The fluid's thermo-physical properties, that is, thermal conductivity, viscosity, and specific heat, and their temperature dependence were measured experimentally. The observed melting–freezing phase transition of the indium nanoparticles significantly augmented the fluid's effective specific heat.

This study is one of the first few to address thermal diffusivity; however, similar studies allow industrial cooling applications to continue without a thorough understanding of all the heat-transfer mechanisms in nanofluids.

1.2.2 Smart Fluids

In this new age of energy consciousness, the lack of abundant sources of clean energy and the widespread use of battery-operated devices, such as cell phones and laptops, have highlighted the necessity for a smart technological handling of energetic resources. Nanofluids have been shown to be able to handle this role in some instances as a smart fluid. Vailati et al. [14] showed that a fluid transported more heat from the bottom plate to the top plate in the convecting state than in the static, conducting state that resulted when the particles were initially allowed to sink toward the heat source. There are practical uses for a fluid with two different possible rates of heat transfer. Most standard heat exchange systems try to transfer as much heat as possible, but heat transfer with a "smart" fluid could provide controlled heating or cooling of industrial processes or electronics as needed. To leap the chasm to heating and cooling technologies, the researchers will have to show more evidence of a stable operating system that responds to a larger range of heat flux inputs. This warrants further research and developmental work.

In a paper published in the March 2009 issue of *the Physical Review Letters*, Donzelli et al. [15] showed that a particular class of nanofluids can be used as a smart material working as a heat valve to control the flow of heat. The nanofluid can be readily configured either in a "low" state, where it conducts heat poorly, or in a "high" state, where the dissipation is more efficient. This versatility presents a more efficient approach to mitigating heat, or conducing heat, within heat-transfer processes or unintentional heat-transfer processes such as the heating of an electronic device.

One goal of the oil and gas research consortium is to develop subsurface micro- and nano-sensors that can be injected into oil and gas wellbores [16]. Owing to their very small size, these sensors would migrate out of the wellbores and into and through pores of the surrounding geological structure to obtain data about the physical and chemical characteristics of hydrocarbon reservoirs, thereby helping to map and key these reservoirs in terms of additional information. The data collected could enable a more efficient recovery of hydrocarbon resources. This could be particularly beneficial for enhanced oil recovery applications.

1.2.3 NUCLEAR REACTORS

Kim et al. [17] at the Nuclear Science and Engineering Department of the Massachusetts Institute of Technology (MIT), performed a study to assess the feasibility of nanofluids in nuclear applications by improving the performance of any water-cooled nuclear system that is limited by heat-removal processes. Possible applications include pressurized water reactor (PWR) primary coolant, standby safety systems, accelerator targets, plasma divertors, and so on [18].

In a PWR nuclear power plant system, the limiting process in the generation of steam is critical heat flux (CHF) between the fuel rods and the water when vapor bubbles cover the surface of the fuel rods and significantly reduce the heat-transfer coefficient. Using nanofluids instead of water, the fuel rods become coated with nanoparticles such as alumina, which actually push newly formed bubbles away, preventing the formation of a layer of vapor around the rod and subsequently increasing the CHF significantly.

After testing in MIT's Nuclear Research Reactor, preliminary experiments have shown promising success where it is seen that PWR is significantly more productive. Nanofluids as a coolant could also be used in emergency cooling systems, where they could cool down overheated surfaces more quickly, leading to an improvement in power plant safety.

Some issues regarding the use of nanofluids in a power plant system include the unpredictability of the amount of nanoparticles that are carried away by the boiling vapor. One other concern is what extra safety measures have to be taken in the disposal of the nanofluid. The application of nanofluid coolant to boiling water reactors is predicted to be minimal because nanoparticle carryover to the turbine and condenser would raise erosion and fouling concerns.

From Jackson's study [19], it was observed that considerable enhancement in the CHF can be achieved by creating a structured surface from the deposition of nanoparticles. If the deposition film characteristics such as the structure and thickness can be controlled, it may be possible to increase the CHF with little decrease in the heat transfer. Whereas the nanoparticles themselves cause no significant difference in the pool-boiling characteristics of water, the boiling of nanofluids shows promise as a simple way to create an enhanced surface.

The use of nanofluids in nuclear power plants seems like a potential future application [18]. Several significant gaps in knowledge are evident at this time including demonstration of the nanofluid thermal-hydraulic performance at proto-typical reactor conditions and the compatibility of the nanofluid chemistry with the reactor materials.

Another possible application of nanofluids in nuclear systems is the alleviation of postulated severe accidents during which the core melts and relocates to the bottom of the reactor vessel. If such accidents were to occur, it is desirable to retain the molten fuel within the vessel by removing the decay heat through the vessel wall. This process is limited by the occurrence of CHF on the vessel outer surface, but analysis indicates that the use of nanofluid can increase the in-vessel retention capabilities of nuclear reactors by as much as 40% [20].

Many water-cooled nuclear power systems are CHF limited, but the application of nanofluids can greatly increase the CHF of the coolant so that there is a bottom-line economic benefit while also raising the safety standard of the power plant system.

1.2.4 EXTRACTION OF GEOTHERMAL POWER AND OTHER ENERGY SOURCES

The world's total geothermal energy resources were calculated to be over 13,000 ZJ in a report from MIT [21]. Currently only 200 ZJ would be extractable; however, with technological improvements, over 2000 ZJ could be extracted and could supply the world's energy needs for a long time. When extracting energy from the earth's crust that varies in thickness between 5 and 10 km and temperature between 500°C and 1000°C, nanofluids can be used to cool the pipes exposed to such high temperatures. When drilling, nanofluids can serve in cooling the machinery and equipment working in high friction and high temperature environments. As a "fluid superconductor," nanofluids could be used as a working fluid to extract energy from the earth's core and aid in the process of creating energy within a PWR power plant system producing large amounts of work energy.

In the sub-area of drilling technology, which is fundamental to geothermal power, improved sensors and electronics cooled by nanofluids capable of operating at higher temperature in down-hole tools, and revolutionary improvements utilizing new methods of rock penetration cooled and lubricated by nanofluids will lower production costs. Such improvements will enable access to deeper, hotter regions in high-grade formations or to economically acceptable temperatures in lower-grade formations. One of the patents reviewed later in this chapter addressed this issue.

In the sub-area of power conversion technology, improving heat-transfer performance for lower-temperature nanofluids, and developing plant designs for higher resource temperatures to the supercritical water region would lead to an order of magnitude (or more) gain in both energy reservoir performance and heat-to-power conversion efficiency.

Tran et al. [22], funded by the United States Department of Energy, performed research targeted at developing a new class of highly specialized drilling fluids that may have superior performance in high-temperature drilling. This research is applicable to high-pressure high-temperature drilling, which may be pivotal in opening up large quantities of previously unrecoverable domestic fuel resources. Commercialization could be the bottleneck of progress in this sub-field.

1.3   AUTOMOTIVE APPLICATIONS

Engine oils, automatic transmission fluids, coolants, lubricants, and other synthetic high-temperature heat-transfer fluids found in conventional truck thermal systems—radiators, engines, heating, ventilation, and air-conditioning—have much room for improvement in terms of heat-transfer properties. These could benefit from the high thermal conductivity offered by nanofluids that resulted from addition of nanoparticles [23,24].

1.3.1   NANOFLUID COOLANT

In searching for ways to improve the aerodynamic designs of vehicles and, subsequently, the fuel economy, manufacturers must reduce the amount of energy needed to overcome wind resistance on the road. At high speeds, approximately 65% of the total energy utilization from a truck is expended in overcoming the aerodynamic drag. This fact is partly due to the large radiator in front of the engine positioned to maximize the cooling effect of oncoming air.

The use of nanofluids as coolants would allow for smaller size and better positioning of the radiators. Owing to the fact that there would be less fluid due to the higher efficiency, coolant pumps could be shrunk and truck engines could be operated at higher temperatures allowing for more horsepower while still meeting stringent emission standards.

Argonne researchers [25] have determined that the use of high-thermal conductive nanofluids in radiators can lead to a reduction in the frontal area of the radiator to as much as 10%. This reduction in aerodynamic drag can lead to a fuel savings of up to 5%. The application of nanofluid also contributed to a reduction of friction and wear, reducing parasitic energy losses, better operation of components such as pumps and compressors, and consequently leading to more than 6% fuel savings. It is possible that greater improvement of savings could be obtained in the future.

In order to determine whether nanofluids degrade radiator materials, Singh et al. [25] have built and calibrated an apparatus that can imitate the coolant flow in a radiator and are currently testing and measuring material loss of typical radiator materials by various nanofluids. Erosion of radiator material is determined by weight-loss measurements as a function of fluid velocity and impact angle of the fluid.

In their tests, they observed no erosion using nanofluids made from base fluids ethylene and tri-chloroethylene glycols with velocities as high as 9 m/s and at 30°–90° impact angles. There was erosion observed with copper nanofluid at a velocity of 9.6 m/s and impact angle of 90°. The corresponding recession rate was calculated to be 0.065 mils/year of vehicle operation.

Through preliminary investigation, it was determined that copper nanofluids produce a higher wear rate than the base fluid and this is possibly due to oxidation of copper nanoparticles. A lower wear and friction rate was seen for alumina nanofluids in comparison to the base fluid. Some interesting erosion test results from Singh et al. [25] are shown in Tables 1.1 and 1.2.

TABLE 1.1
Erosion Test Results for 50% Ethlyene Glycol, 50% H₂O Aluminum 3003—50°C Rig

Impact Angle (°)	Velocity (m/s)	Time (h)	Weight Loss (mg)
90	8.0	236	0 ± 0.2
90	10.5	211	0 ± 0.2
50	6.0	264	0 ± 0.2
50	10.0	244	0 ± 0.2
30	8.0	283	0 ± 0.2
30	10.5	293	0 ± 0.2

Source: Adapted from D. Singh et al. 2006. Heavy Vehicle Systems Optimization Merit Review and Peer Evaluation, Argonne Nat. Lab., Annual Report, 2006.

TABLE 1.2
Erosion Test Results for Cu Nanoparticles in Tri-Chloroethylene Glycol on Al 3003—50°C Rig

Impact Angle (°)	Velocity (m/s)	Time (h)	Weight Loss (mg)
90	4.0	217	0 ± 0.2
30	4.0	311	0 ± 0.2
90	7.6	341	0 ± 0.2
30	7.6	335	0 ± 0.2
30	9.6	336	0 ± 0.2

Source: Adapted from D. Singh et al. 2006. Heavy Vehicle Systems Optimization Merit Review and Peer Evaluation, Argonne Nat. Lab., Annual Report, 2006.

Shen et al. [26] researched the wheel wear and tribological characteristics in wet, dry, and minimum quantity lubrication (MQL) grinding of cast iron. Water-based alumina and diamond nanofluids were applied in the MQL-grinding process, and the grinding results were compared with those of pure water. Nanofluids demonstrated the benefits of reducing grinding forces, improving surface roughness, and preventing burning of the work piece. Contrasted to dry grinding, MQL grinding could considerably lower the grinding temperature, which would prevent burning of the work piece.

More research must be conducted on the tribological properties using nanofluids of a wider range of particle loadings as well as on the erosion rate of radiator material to help develop predictive models for nanofluid wear and erosion in engine systems. Future research initiatives involve nanoparticles materials containing aluminum and oxide-coated metal nanoparticles. Additional research and testing in this area will assist in the design of engine cooling and other thermal management systems that involve nanofluids.

Future engines that are designed using nanofluids' cooling properties would be able to run at more optimal temperatures allowing for increased power output. With a nanofluids engine, components would be smaller and weigh less allowing for better gas mileage, saving consumers money and resulting in fewer emissions for a cleaner environment.

1.3.2 NANOFLUIDS IN FUEL

The aluminum nanoparticles, produced using a plasma arc system, are covered with thin layers of aluminum oxide owing to the high oxidation activity of pure aluminum, thus creating a larger contact surface area with water and allowing for increased decomposition of hydrogen from water during the combustion process. During this combustion process, the alumina acts as a catalyst and the aluminum nanoparticles then serve to decompose the water to yield more hydrogen. It was shown that the combustion of diesel fuel mixed with an aqueous aluminum nanofluid increased the total combustion heat while decreasing the concentration of smoke and nitrous oxide in the exhaust emission from the diesel engine [27].

1.3.3 BRAKE AND OTHER VEHICULAR NANOFLUIDS

As vehicle aerodynamics are improved and drag forces are reduced, there is a higher demand for braking systems with higher and more efficient heat-dissipation mechanisms and properties such as brake nanofluid.

A vehicle's kinetic energy is dispersed through the heat produced during the process of braking, and this is transmitted throughout the brake fluid in the hydraulic braking system. If the heat causes the brake fluid to reach its boiling point, a vapor lock is created that retards the hydraulic system from dispersing the heat caused from braking. Such an occurrence will in turn cause a brake malfunction and create a safety hazard in vehicles. As the heat generated from braking easily affects brake oil, nanofluids with enhanced characteristics maximize performance in heat transfer as well as in removing any safety concerns.

Copper-oxide brake nanofluid (CBN) is manufactured using the method of arc-submerged nanoparticle synthesis system. Essentially this is done by melting bulk copper metal used as the electrode which is submerged in dielectric liquid within a vacuum-operating environment, and the vaporized metals are condensed in the dielectric liquid [27].

Aluminum-oxide brake nanofluid (AOBN) is made using the plasma charging arc system. This is performed in a very similar manner to that of the arc-submerged nanoparticle synthesis system method. The aluminum metal is vaporized by the plasma electric arc at a high temperature and mixed thoroughly with the dielectric liquid [27].

CBN has a thermal conductivity 1.6 times higher than that of the brake fluid designated DOT3, whereas AOBN's thermal conductivity is only 1.5 times higher than DOT3. This enhanced thermal conductivity optimizes heat transmission and lubrication.

CBN and AOBN both have enhanced properties such as a higher boiling point, higher viscosity, and a higher conductivity than that of traditional brake fluid (DOT3). By yielding a higher boiling point, conductivity, and viscosity, CBN and AOBN reduce the occurrence of vapor lock and offer increased safety while driving. Important findings of Kao et al. [27] are shown in Figure 1.1 and Table 1.3.

In the nanofluid research applied to the cooling of automatic transmissions, Tzeng et al. [28] dispersed CuO and Al_2O_3 nanoparticles into engine transmission oil. The experimental setup was the transmission of a four-wheel-drive vehicle. The transmission had an advanced rotary blade coupling, where high local temperatures occurred

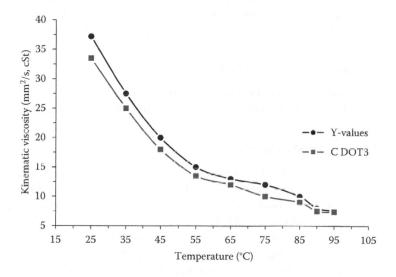

FIGURE 1.1 CBN temperature and viscosity fluctuations. (Adapted from M.J. Kao et al. 2007. *Journal of the Chinese Society of Mechanical Engineers*, 28(2), 123–131.)

TABLE 1.3
CBN and AOBN Boiling Point and Thermal Conductivity Values

	DOT3[a]	CBN 2 wt% (CuO + DOT3)	DOT3[a]	AOBN 2 wt% (Al_2O_3 + DOT3)
Boiling point (°C)	270	278	240	248
Conductivity (25°C) (W/m°C)	0.03	0.05	0.13	0.19

Source: Adapted from M.J. Kao et al. 2007. *Journal of the Chinese Society of Mechanical Engineers*, 28(2), 123–131.

[a] Different DOT3 brake fluids were used.

at high rotating speeds. Temperature measurements were taken on the exterior of the rotary-blade-coupling transmission at four engine-operating speeds (range from 400 to 1600 rpm), and the optimum composition of nanofluids with regard to heat-transfer performance was studied. The results indicated that CuO nanofluids resulted in the lowest transmission temperatures both at high and low rotating speeds. Therefore, the use of nanofluids in the transmission has a clear advantage from the thermal performance viewpoint. As in all nanofluid applications, however, consideration must be given to such factors as particle settling, particle agglomeration, and surface erosion.

In automotive lubrication applications, Zhang [29] reported that surface-modified nanoparticles stably dispersed in mineral oils are effective in reducing wear and enhancing load-carrying capacity. Results from a research project involving industry and academia point to the use of nanoparticles in lubricants to enhance tribological properties such as load-carrying capacity, wear resistance, and friction reduction between moving mechanical components. Such results are promising for enhancing heat-transfer rates in automotive systems through the use of nanofluids.

1.4   ELECTRONIC APPLICATIONS

Nanofluids are used for cooling of microchips in computers and elsewhere. They are also used in other electronic applications that use microfluidic applications.

1.4.1   COOLING OF MICROCHIPS

A principal limitation on developing smaller microchips is the high heat-flux density. Nanofluids can be used for liquid cooling of computer processors due to their high thermal conductivity. It is predicted that the next generation of computer chips will produce localized heat flux over 10 MW/m^2, with the total power exceeding 300 W. In combination with thin-film evaporation, the nanofluid oscillating heat pipe (OHP) cooling system, in which the working fluid oscillates between the evaporator and condenser sections, will be able to remove heat fluxes over 10 MW/m^2 and serve as the next-generation cooling device that will be able to handle the heat dissipation coming from new technologies [30].

In order to visually observe the oscillation, researchers had to modify the metal pipe system of the OHP to use glass or plastic for visibility. However, as OHP systems are usually made of copper, the use of glass or plastic changes the thermal-transfer properties of the system and subsequently altering the performance of the system and the legitimacy of the experimental data [30].

So as to obtain experimental data while maintaining the integrity of the OHP system, Arif [31] employed neutron imaging to study the liquid flow in a 12-turn nanofluid OHP, where the 12 indicates how many times the fluid oscillates between the condenser and evaporator. As a consequence of the high-intensity neutron beam from an amorphous silicon imaging system, they were able to capture dynamic images at 1/30th of a second. The nanofluid used was composed of diamond nanoparticles suspended in water.

Even though nanofluids and OHPs are not new discoveries, combining their unique features allows for the nanoparticles to be completely suspended in the base liquid, increasing their heat-transport capability. As nanofluids have a strong temperature-dependent thermal conductivity and they show a nonlinear relationship between thermal conductivity and concentration, they are high-performance conductors with an increased CHF. The OHP takes intense heat from a high-power device and converts it into kinetic energy of fluids while not allowing the liquid and vapor phases to interfere with each other as they flow in the same direction.

In their experiment, Ma et al. [32] introduced diamond nanoparticles into high-performance liquid chromatography water. The movement of the OHP keeps the nanoparticles from settling and thus improved the efficiency of the cooling device. At an input power of 80 W, the diamond nanofluid decreased the temperature difference between the evaporator and the condenser from 40.9°C to 24.3°C (Figure 1.2).

However, for the water, as the heat input increases, the oscillating motion increases and the resultant temperature difference between the evaporator and condenser does not continue to increase after a certain power input. This phenonmenon inhibits the effective thermal conductivity of the nanofluid from continuously increasing. However, at its maximum power level of 336 W, the temperature difference for the nanofluid OHP was still less than that of the OHP with pure water. Hence, it has been shown that the nanofluid can significantly increase the heat-transport capability of the OHP.

Lin et al. [33] investigated nanofluids in pulsating heat pipes by using silver nanoparticles, and discovered encouraging results. The silver nanofluid improved heat-transfer characteristics of the heat pipes compared with that of pure water.

Nguyen et al. [34] investigated the heat-transfer enhancement and behavior of Al₂O₃ water nanofluid with the intention of using it in a closed cooling system designed for microprocessors or other electronic devices. The experimental data support that the inclusion of nanoparticles into distilled water produce a significant

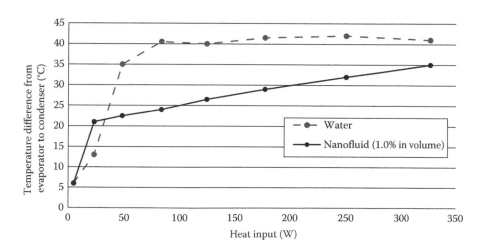

FIGURE 1.2 Effect of nanofluid on heat transport capability in an OHP. (Adapted from H.B. Ma et al. 2006. *Journal of Heat Transfer*, 128, 1213–1216.)

increase of the cooling convective heat-transfer coefficient. At a given particle concentration of 6.8%, the heat-transfer coefficient increased as much as 40% compared with the base fluid of water. Smaller Al_2O_3 nanoparticles likewise showed higher convective heat-transfer coefficients than the larger ones.

Further research of nanofluids in electronic cooling applications will lead to the development of the next generation of cooling devices that incorporate nanofluids for ultrahigh-heat-flux electronic systems.

1.4.2 MICROSCALE FLUIDIC APPLICATIONS

The manipulation of small volumes of liquid is necessary in fluidic digital display devices, optical devices, and microelectromechanical systems (MEMSs) such as lab-on-chip analysis systems. This can be done by electrowetting or reducing the contact angle by an applied voltage, the small volumes of liquid. Electrowetting on dielectric actuation is one very useful method of microscale liquid manipulation.

Vafaei et al. [35] discovered that nanofluids are effective in engineering the wettability of the surface and possibly of surface tension. Using a goniometer, it was found that even the addition of a very low concentration of bismuth telluride nanofluid dramatically changed the wetting characteristics of the surface. Concentrations as low as 3×10^{-6} increased the contact angle to over 40°, distinctly indicating that the nanoparticles change the force balance in the vicinity of the triple line. The contact angle, θ_o, seen in Figure 1.3, rises with the concentration of the nanofluid, reaches a maximum, and then decreases (Figure 1.4) [36].

The droplet contact angle was observed to change depending on the size of the nanoparticles as well as concentration. Smaller nanoparticles are more effective in increasing the contact angle. The reason for this effect is that smaller particles would provide more surface-to-volume area for the same concentration.

Dash et al. [36] used the electrowetting on dielectrics effect to demonstrate that nanofluids display increased performance and stability when exposed to electric fields. The experiment consisted of placing droplets of water-based solutions containing bismuth telluride nanoparticles onto a Teflon-coated silicon wafer. A strong change in the angle at which the droplet contacted the wafer was observed when an electric field was applied to the droplet. The change noticed with the nanofluids in place was significantly greater than when not using nanofluids. The bismuth telluride nanofluid also displayed enhanced droplet stability and absence of the contact angle saturation effect compared with solutions of 0.01 N Na_2SO_4 and thioglycolic acid in deionized water.

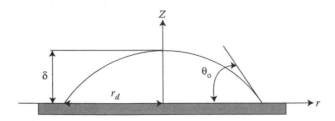

FIGURE 1.3   Schematic diagram of droplet shape.

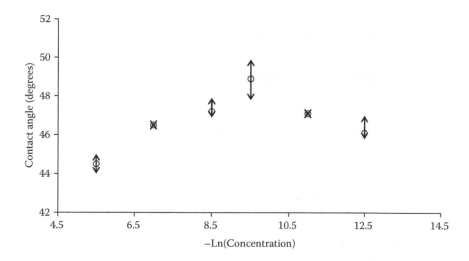

FIGURE 1.4 Variation of contact angle for 10 nm bismuth telluride nanoparticles concentration on a glass substrate. (Adapted from R.K. Dash et al. 2007 *IOP Publishing: Nanotechnology*, 18, 1–6.)

That the contact angle of droplets of nanofluids can be changed as potential applications for efficiently moving liquids in microsystems, allowing for new methods for focusing lenses in miniature cameras and for cooling computer chips.

1.4.3 MICROREACTORS

Microreactors, as their name implies, are miniscule and it is due to their small size that they hold advantageous properties. This makes them ideal for safe storage and transport of explosive, toxic, and/or harmful materials. Some microreactors are barely larger than 1 cm. In microreactors, nanofluids could be used where the particles act as nanocatalysts in the reactor. The narrow channels and thin walls are better suited for testing potential explosive reactions [37]. Using nanofluids increases the inherently high heat- and mass-transfer characteristics of microreactors. A higher level of control and reactant manipulation in the microreactor is achievable because of the very high heat- and mass-transfer rates.

It is extremely important for microreactors to control reactions because any potentially dangerous reactions should be kept under control. Microchannels are more efficient because of the big length-to-depth ratios that they have. This results in a higher number of molecular collisions per unit volume than in a conventional reactor. The result is desirable because the efficiency of reactions increases when the frequency of molecular collisions increases.

Microreactors find use in oil refineries that utilize gas-to-liquid (GTL) technologies. GTL technology permits gas that would usually be a waste product to be converted into liquid form either directly or by converting it to synthetic gas (hydrogen and carbon monoxide) first. In its liquid form, it can readily be used as fuel. The World Bank [38] estimates that more than 150 billion m³ of natural gas is wasted annually by way of venting and flaring; this amount of gas adds 400 million tons of carbon dioxide to the annual world greenhouse gas emission budget. Microreactors could be used for storage and transportation of GTL products.

1.5 BIOMEDICAL APPLICATIONS

1.5.1 NANO-DRUG DELIVERY

Most bio-MEMS studies were carried out in academia in the 1990s, while recently commercialization of such devices have started. Examples include an electronically activated drug-delivery microchip [39] that uses a controlled delivery system via integration of silicon and electro-active polymer technologies, along with MEMS-based DNA sequencer developed by Cepheid [40] and arrays of in-plane and out-of-plane hollow micro-needles for dermal/transdermal drug delivery [41,42]. Others include nanomedicine applications of nanogels or gold-coated nanoparticles [43]. An objective of the advanced endeavors in developing integrated micro- or nano-drug-delivery systems is the interest in easily monitoring and controlling target-cell responses to pharmaceutical stimuli to understand biological cell activities or to enable drug-development processes.

Conventional drug delivery is characterized by the "high-and-low" phenomenon, which is essentially distributing the dosage of medicine to either an overly large region or not distributing enough to a single region. Micro-devices, in contrast, facilitate precise drug delivery by both implanted and transdermal techniques. This means that when a drug is dispensed conventionally, drug concentration in the blood will increase, peak, and then drop as the drug is metabolized. The cycle is repeated for each drug dose. Employing nano-drug delivery systems, controlled drug release takes place over an extended period of time. Thus, the desired drug concentration will be sustained within the therapeutic window as required.

A nano-drug-supply system, that is, a bio-MEMS, was introduced by Kleinstreuer et al. [44]. Their main concern was the conditions for delivering uniform concentrations of the supplied nanodrugs at the microchannel exit. A heat flux, which depends on the levels of nano-fluid and purging fluid velocity, was added to ascertain that drug delivery to the living cells occurs at an optimal temperature, that is, 37°C. The added wall heat flux had also a positive influence on drug-concentration uniformity. In general, channel length, particle diameter, and the Reynolds number of both the nanofluid supply and main microchannels affect the nano-drug concentration uniformity. As the transport mechanisms are dependent on convection, diffusion, longer channels, smaller particle diameters as well as lower Reynolds numbers are desirable for the best, that is, uniform drug delivery.

1.5.2 CANCER THERAPEUTICS

There is a new initiative that takes advantage of several properties of certain nanofluids to use in cancer imaging and drug delivery. This initiative involves the use of iron-based nanoparticles as delivery vehicles for drugs or radiation in cancer patients. Magnetic nanofluids are to be used with magnets to guide the particles up the bloodstream to a tumor. This will allow doctors to deliver high local doses of drugs or radiation without damaging nearby healthy tissue, which may be a significant side effect of traditional cancer treatment methods. In addition, magnetic nanoparticles are more adhesive to tumor cells than to nonmalignant cells, and they absorb much more power than microparticles in alternating current magnetic fields tolerable in humans; they make excellent candidates for cancer therapy.

Magnetic nanoparticles are used because as compared with other metal-type nanoparticles, they provide a means for handling and manipulation of the nanofluid by magnetic force [45]. This combination of targeted delivery and controlled release will also decrease the likelihood of systemic toxicity as the drug is encapsulated and biologically unavailable during transit in systemic circulation. The nanofluid containing magnetic nanoparticles also acts as a super-paramagnetic fluid that in an alternating electromagnetic field absorbs energy producing a controllable hyperthermia. By enhancing the chemotherapeutic efficacy, the hyperthermia is able to produce a preferential radiation effect on malignant cells [46].

There are numerous biomedical applications that involve nanofluids such as magnetic cell separation, drug delivery, hyperthermia, and contrast enhancement in magnetic resonance imaging. Depending on the specific application, there are different chemical syntheses developed for various types of magnetic nanofluids that allow for the careful tailoring of their properties for different requirements in applications. Surface coating of nanoparticles and the colloidal stability of biocompatible water-based magnetic fluids, which are fluids with magnetic properties that can be used to deliver medicine to specific locations, are the two particularly important factors that affect successful application [47,48].

Nanofluids could be applied to almost any disease treatment techniques by reengineering the nanoparticles' properties. In Vekas' study, the nanoparticles were laced with the drug docetaxel to be dissolved in the cells' internal fluids, releasing the anticancer drug at a predetermined rate [47,48]. The nanoparticles contain targeting molecules called aptamers that recognize the surface molecules on cancer cells, preventing the nanoparticles from attacking other cells. In order to prevent the nanoparticles from being destroyed by macrophages—cells that guard against foreign substances entering our bodies—the nanoparticles also have polyethylene glycol molecules. The nanoparticles are excellent drug-delivery vehicles because they are so small that living cells absorb them when they arrive at the cell surface.

For most biomedical uses, the magnetic nanoparticles should be below 15 nm in size and stably dispersed in water. A potential magnetic nanofluid that could be used for biomedical applications is one composed of FePt nanoparticles. This FePt nanofluid possesses an intrinsic chemical stability and a higher saturation magnetization making it ideal for biomedical applications. However, before magnetic nanofluids can be used as drug-delivery systems, more research must be conducted on the nanoparticles containing the actual drugs and the release mechanism.

1.5.3 CRYOPRESERVATION

Conventional cryopreservation protocols for slow freezing or vitrification involve cell injury due to ice formation/cell dehydration or toxicity of high cryoprotectant concentrations, respectively. In a study, He et al. [49] developed a novel cryopreservation technique to achieve ultra-fast cooling rates using a quartz microcapillary. The quartz microcapillary enabled vitrification of murine embryonic stem (ES) cells using an intracellular cryoprotectant concentration in the range used for slowing freezing (molarity of 1–2). More than 70% of the murine ES cells post-vitrification attached with respect to nonfrozen control cells, and the proliferation rates of the two groups were alike. Preservation of undifferentiated properties of the pluripotent murine ES cells post-vitrification cryopreservation was verified using three different types of assays. These results indicate that vitrification at a low concentration (molarity of 2) of intracellular cryoprotectants is a viable and effective approach for the cryopreservation of murine ES cells.

1.5.4 NANOCRYOSURGERY

Cryosurgery is a procedure that uses freezing to destroy undesired tissues. This therapy is becoming popular because of its important clinical advantages. Although it still cannot be regarded as a routine method of cancer treatment, cryosurgery is quickly becoming as an alternative to traditional therapies.

Simulations were performed by Yan and Liu [50] on combined phase change bioheat-transfer problems in a single-cell level and its surrounding tissues to show the difference of transient temperature response between conventional cryosurgery and nanocryosurgery. According to theoretical findings and existing experimental measurements, intentional loading of nanoparticles with high thermal conductivity into the target tissues can reduce the final temperature, increase the maximum freezing rate, and enlarge the ice volume obtained in the absence of nanoparticles. In addition, introduction of nanoparticle-enhanced freezing could also make conventional cryosurgery more flexible in many aspects such as artificially interfering in the size, shape, image, and direction of iceball formation. The concepts of nanocryosurgery may offer new opportunities for future tumor treatment.

With respect to the choice of particles for enhancing freezing, magnetite (Fe_3O_4) and diamond are perhaps the most popular and appropriate because of their good biological compatibility. Particle sizes less than 10 μm are sufficiently small to permit effective delivery to the site of the tumor, either via encapsulation in a larger molarity or suspension in a carrier fluid. Introduction of nanoparticles into the target via a nanofluid would effectively increase the nucleation rate at a high-temperature threshold.

1.5.5 SENSING AND IMAGING

Colloidal gold has been used for several centuries, be it as colorant of glass ("purple of Cassius") and silk, in medieval medicine for the diagnosis of syphilis or, more recently, in chemical catalysis, nonlinear optics, supramolecular chemistry, molecular recognition, and the biosciences. Colloidal gold is often referred to as the most stable of all colloids. Its history, properties, and applications have been reviewed extensively. For a thorough and up-to-date overview, a study by Daniel and Astruc [51] and the references cited therein may be consulted. As stated in the introduction, no attempt is made here to review the use of colloids, which are also nanofluids. An increase in the use of colloids is expected in this category.

1.6 OTHER APPLICATIONS

1.6.1 NANOFLUID DETERGENT

Nanofluids do not behave in the same manner as simple liquids with classical concepts of spreading and adhesion on solid surfaces [52–54]. This fact opens up the possibility of nanofluids being superior prospects in the processes of soil remediation, oil recovery, lubrication, and detergency. Future engineering applications could abound in such processes.

Wasan and Nikolov [55] were able to use reflected-light digital video microscopy to ascertain the mechanism of spreading dynamics in liquids containing nanosized polystyrene particles. They were able to exhibit the two-dimensional crystal-like formation of the polystyrene spheres in water and how this enhances the spreading dynamics of a micellar fluid, which is an aqueous solution in which surfactant molecules self assemble in the presence of certain organic salts, at the three-phase region [55].

When encountering an oil drop, the polystyrene nanoparticles gather and rearrange around the drop creating a wedge-like region between the surface and the oil drop, as illustrated in Figure 1.5. The nanoparticles then diffuse into the wedge film and cause an increase in concentration and subsequently an increase in disjoining pressure around the film region. Owing to the increase in pressure, the oil–solution interface

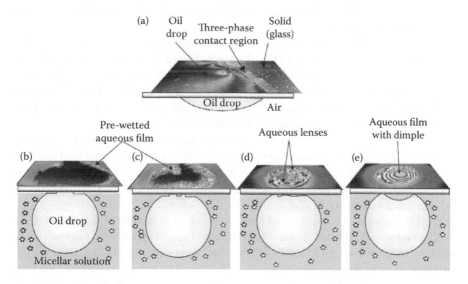

FIGURE 1.5 (a) Photomicrograph showing the oil drop placed on a glass surface and differential interference patterns formed at the three-phase contact region. (b) Photomicrographs taken after addition of the nanofluid at (c) 30 s; 2 min; (d) 4 min; (e) 6 min region. (Adapted from D.T. Wasan and A.D. Nikolov. 2003. *Nature*, 423, 156–159.)

moves forward allowing the polystyrene nanoparticles to spread along the surface. It is this mechanism that causes the oil drop to detach completely from the surface.

Wasan and Nikolov [55] did an additional experiment where they introduced an electrolyte into the process in order to decrease the interfacial tension at the interface of the oil and the nanofluid but found that the drop did not become detached from the surface. They actually observed a diminished disjoining pressure contrary to logical prediction. Additional research must be carried out in this area to better understand such behavior of the nanofluid.

Generally, the occurrence that involves the increased spreading of detergent surfactants, which are not only limited to polystyrene nanoparticles and enhanced oil-removal processes, but offers a new way of removing stains and grease from surfaces. This type of nanofluid also has potential in the commercial extraction of oil from the ground as well as the remediation of oil spills.

1.7 NANOFLUID PATENTS

Arguably, the first patent regarding nanofluids was granted to Choi and Eastman [56] in 2001 for enhancing the heat transfer in fluids such as deionized water, ethylene glycol, and oil by dispersing nanocrystalline particles of substances such as copper, copper oxide, aluminum oxide, or alumina. Their open literature publications justify their patent claims, and indeed created a significant and important sub-area of nanotechnology both in scientific research as well as in the patent world.

There were many patents concerning nanofluids granted both in the United States and internationally between 2002 and 2007, all of which have been adequately reviewed by Nosfor [57]. The current review will outline the patents that came after 2007 until the present time.

Ohira et al. [58] created a technique for reducing the reactivity or toxicity of a liquid. Nanoparticles of a metal or the like are dispersed in a liquid fluid. In addition, the flow resistance of the liquid fluid can be raised, and the leakage of the liquid fluid from minute cracks can be reduced. By using the liquid as a heat-transfer medium is a heat exchanger, heat-transfer performance equivalent to or higher than the heat-transfer performance of the original heat exchanger can be obtained.

Davidson and Bradshaw [59] came up with compositions with nanoparticle size conductive material powder and methods of using them for transferring heat between a heat source and a heat sink. In particular, they used nanoparticle size conductive material such as nanoparticle size diamond powders to enhance the thermal capacity and thermal conductivity of heat-transfer media such as transformer oil.

Wu [60] designed a heat-conducting water solution nanofluid for use in a car cooling system. It is formed by mixing aluminum oxide solution with titanium oxide solution, which is then mixed with diluents and dispersing agents and an emulsifying agent so as to disperse the solutions uniformly. The titanium oxide cleans lime scale, and the emulsifying agent adheres to wall surfaces allowing the alumina to release energy continuously. The nanometer-scale materials speed up the micro-explosion of the cooling water so as to optimize the cooling effect and increase the heat-dispersing efficacy significantly.

Hoi et al. [61] introduced an oil-based nanofluid with high thermal efficiency. The oil-based nanofluid was created to improve the energy efficiency, service life, and reliability of a potential transformer by increasing the heat-transfer characteristics, such as conductivity, of the insulating oil by 10% or more.

Earlier that year, Lockwood [62] enhanced the thermal conductivity of fluids with graphite nanoparticles. The graphite is dispersed in the fluid by various methods, including ultrasonication, milling, and chemical dispersion. Carbon nanotubes with graphitic structure are another preferred source of carbon nanomaterial, although other carbon nanomaterials are useable. To render long-term stability, the use of one or more chemical dispersants is preferred. The thermal conductivity enhancement, compared with the fluid without carbon nanomaterial, is proportional to the amount of carbon nanomaterials (carbon nanotubes and/or graphite) added. Fluid compositions have enhanced thermal conductivity up to 250% greater than their conventional analogs. The petroleum liquid medium can be any petroleum distillates or synthetic petroleum oils, greases, gels, or oil-soluble polymer composition.

Zhang et al. [63] also invented a special gear oil that contained nanomaterials. They used nanomaterials as a viscosity modifier and thermal conductivity improver for gear oil and other lubricating oil compositions. Their novel gear oils have a higher viscosity index; higher shear stability, and improved thermal conductivity compared to currently available gear oils. The preferred nanoparticles also reduce the coefficient of friction, including reduced friction in the boundary lubrication regime. These properties are obtained by replacing part or all of the polymer thickener or viscosity-index improver or some other part of the composition normally used in gear oils with nanomaterials of suitable shape, size, and composition.

Zhang et al. [64] introduced nanostructures in a liquid that provided a means for changing the physical and/or chemical properties of the liquid. Improvements in heat transfer, electrical properties, viscosity, and lubricity can be realized on dispersion of nanotubes in liquids. Some preferred surfactants are identified which can disperse carbon nanotubes in petroleum liquid media utilizing selected dispersants and mixing methods to form stable carbon nanostructure fluid dispersions. These nanofluids are applicable to vehicle engineering, in particular to suspension shock-absorbing devices and can be used in shock absorbers for cars, and trucks, and other vehicles.

Zhong et al. [65] was awarded a patent for an aluminum oxide organic nanofluid for engine high-temperature cooling technology. The magnitude of the flow resistance is smaller; hence, the requirements of the engine high-temperature cooling technology can be met. The alumina organic nanofluid is conducive to the development of a new generation of microchannels of a vehicle thermal management system and a compact and light heat exchanger.

Liu et al. [66] created a low-temperature, phase-change cold storage nanofluid, which improves the heat-transfer characteristic of a traditional cold storage working medium. This nanofluid can effectively reduce the temperature difference between a refrigerant and a cold storage working medium, improve the degree of nucleation supercooling, increase the cold storage efficiency, and reduce the volume of a heat exchanger. This same nanofluid has significant influence on the air conditioning and cold storage industry, and a positive impact on effective energy utilization. Further developmental work and implementation of this nanofluid is indicated.

Griffo and Keshavan [67] patented a high-performance rock bit grease. The lubricant is for lubricating journal bearings in a rock bit for drilling earth formations. It can be used for high temperature (environments exceeding 121°C). For additives to prove beneficial in grease used in drilling applications, it is necessary to balance thermal performance, the load-carrying capacity, and seal/gland wear. Generally, lubricants that reduce seal and gland wear typically lack sufficient film strength, that is, load carrying capacity, and lubricants with sufficient film strength tend to show excessive seal and glad wear and are used as a drill bit lubricant. Hence, there exists a need for a lubricant that exhibits improved thermal performance, a tight seal, and good load-carrying capacity with reduced seal and gland wear.

Jeffcoate et al. [68] were awarded a patent for a fluid composition having enhanced heat-transfer efficiency. The fluid composition includes a coolant and a plurality of nanoparticles dispersed within the coolant. The plurality of nanoparticles includes glass, silica, pumices, metal compounds adapted to react with chloride in the coolant, and/or mixtures thereof. The plurality of nanoparticles substantially increases the heat capacity of the coolant and enhances heat-transfer efficiency of the fluid composition. Research into the correct mix and concentrations of nanoparticles to obtain the correct enhancements in both thermal conductivity and thermal capacity would be a natural extension to research in this sub-area.

Jung et al. [69] disclosed a blanket using lithium nanofluid and a fusion reactor using the same, which may remarkably reduce reactivity with water to improve stability and reliability. In this case, the lithium nanofluid may be used to breed the tritium and cool the fusion reactor. According to them, a fusion reactor has been made using a fusion reaction and utilizes the lithium nanofluid as a coolant. In this case, the lithium nanofluid may be used to multiply the fuel of the fusion reactor. In addition, the multiplying agent and the coolant may be the same. As the multiplying agent and coolant, the lithium nanofluid may be preferably used. In this case, the lithium nanofluid may be obtained by dispersing a metal or a metal oxide nanoparticles in liquid lithium.

Farmer [70] created a nanofluid that includes nanoparticles plus a dielectric or ionic fluid. This can be used in an energy conversion and storage device. Electrochemical energy storage is required for grid storage, wireless communications, portable computing, future fleets of electric and hybrid electric vehicles, and many other applications.

1.8 CONCLUSION

Nanofluids are important because they can be used in numerous applications involving heat transfer as well as industrial processes, nuclear reactors, transportation, electronics, biomedicine, food, and detergents. Colloids, which are also nanofluids, have been used in the biomedical field for a long time, and their use will continue to grow. Nanofluids have also been demonstrated for use as smart fluids. Problems of nanoparticle agglomeration, settling, and erosion potential all need to be examined in greater detail in the applications. Nanofluids employed in experimental research have to be well characterized with respect to particle size, size distribution, shape, and clustering so as to render the results most widely applicable. Even though the science of nanofluids is not fully understood, many of them have been reproduced and used in applications. Commercialization has happened for many of the promising ones, especially in the automotive industry. Colloids, which are also nanofluids, will see an increase in use in biomedical engineering and the biosciences.

Nanofluid patents have been reviewed until the present time. Further research still has to be done on the synthesis and applications of many nanofluids so that they may be applied as predicted. Nevertheless, there have been many discoveries and improvements identified about the characteristics of nanofluids in the surveyed applications. There is great potential for developing systems that are more efficient and/or smaller by employing nanofluids.

REFERENCES

1. S.U.S. Choi. 2009, Nanofluids: From vision to reality through research, *Journal of Heat Transfer*, 131(3), 9.
2. W. Yu, D. M. France, J.L. Routbort, and S.U.S. Choi. 2008, Review and comparison of nanofluid thermal conductivity and heat transfer enhancements, *Heat Transfer Engineering*, 29(5 May 2008), 432–460.
3. T. Tyler, O. Shenderova, G. Cunningham, J. Walsh, J. Drobnik, and G. McGuire. 2006, Thermal transport properties of diamond-based nanofluids and nanocomposites, *Diamond and Related Materials*, 15, 2078–2081.
4. S.K. Das, S.U.S. Choi, and H.E. Patel. 2006, Heat transfer in nanofluids—a review, *Heat Transfer Engineering*, 27(10), 3–19.
5. M.S. Liu, M.C.C. Lin, I.T. Huang, and C.C. Wang. 2005, Enhancement of thermal conductivity with carbon nanotube for nanofluids, *International Communications in Heat and Mass Transfer*, 32, 1202–1210.
6. S.U.S. Choi, Z.G. Zhang, P. Keblinski, and H.S. Nalwa (ed.) 2004, *Nanofluids. Encyclopedia of Nanoscience and Nanotechnology*, 6, 757–737. American Scientific Publishers, Los Angeles, CA, USA.
7. S.M.S. Murshed, S.H. Tan, and N.T. Nguyen. 2008, Temperature dependence of interfacial properties and viscosity of nanofluids for droplet-based microfluidics, *Journal of Physics D: Applied Physics,* 41, 085502, 1–5.
8. K.V. Wong and T. Kurma. 2008, Transport properties of alumina nanofluids, *Nanotechnology*, 19, 345702, 8.
9. K.V. Wong, B. Bonn, S. Vu, and S. Samedi. 2007, Study of nanofluid natural convection phenomena in rectangular enclosures, *Proceedings of IMECE 2007*, Nov. 2007, Seattle, WA.
10. Y. Ju-Nam and J.R. Lead. 2008, Manufactured nanoparticles: An overview of their chemistry, interactions and potential environmental implications, *Science of the Total Environment*, 400, 396–414.
11. K.V. Wong and O. De Leon. 2010, Applications of nanofluids: Current and future, *J. Advances in Mech. Engineering*, 10, 1–11. Accessed online at http://www.hindawi.com/journals/ame/2010/519659/.
12. J. Routbort et al. 2009, Argonne National Lab, Michellin North America, St. Gobain Corp., http://www1.eere.energy.gov/industry/nanomanufacturing/pdfs/nanofluids_industrial_cooling.pdf.
13. Z.H. Han, F.Y. Cao, and B. Yang. 2008, Synthesis and thermal characterization of phase-changeable Indium/Polyalphaolefin nanofluids, *Applied Physics Letters*, 92, 243104. Accessed online at http://dx.doi.org/10.1063/1.2944914.

14. A. Vailati, R. Cerbino, S. Mazzoni, M. Giglio, G. Nikolaenko, C.J. Takacs, D.S. Cannell, W.V. Meyer, and A.E. Smart. 2006, Gradient-driven fluctuations experiment: Fluid fluctuations in microgravity, *Applied Optics*, 45(10), 2155–2165.
15. G. Donzelli, R. Cerbino, and A. Vailati. 2009, Bistable heat transfer in a nanofluid, *Physics Review Letters*, 102(10), 1–4.
16. S.L. Durham. 2009, Researchers are Thinking Small. Explorer (online). August 2009.
17. S.J. Kim, I.C. Bang, J. Buongiorno, and L.W. Hu. 2007, Surface wettability change during pool boiling of nanofluids and its effect on critical heat flux, *International Journal of Heat and Mass Transfer*, 50, 4105–4116.
18. J. Buongiorno, L. Hu, S.J. Kim, R. Hannink, B. Truong, and E. Forrest. 2008, Nanofluids for enhanced economics and safety of nuclear reactors: An evaluation of the potential features, issues and research gaps, *Nuclear Technology*, 162(1), 80–91.
19. J.E. Jackson. 2007, Investigation into the pool-boiling characteristics of gold nanofluids, M.S. Thesis, U. Missouri-Columbia.
20. J. Buongiorno, L.W. Hu, G. Apostolakis, R. Hannink, T. Lucas, and A. Chupin. 2009, A feasibility assessment of the use of nanofluids to enhance the in-vessel retention capability in light-water reactors, *Nuclear Engineering and Design* 239, 941–948.
21. M.I.T. 2007, The Future of Geothermal Energy, Sponsored by U.S.D.O.E., p. 372.
22. P.X. Tran, D.K. Lyons et al. 2007, Nanofluids for Use as Ultra-Deep Drilling Fluids, U.S.D.O.E., http://www.netl.doe.gov/publications/factsheets/rd/R&D108.pdf.
23. W.Yu, D.M. France, J.L. Routbort, and S.U.S. Choi (2008), Review and comparison of nanofluid thermal conductivity and heat transfer enhancements, *Heat Transfer Engineering*, 29, 432–460.
24. M. Chopkar, P.K. Das, and I. Manna. 2006, Synthesis and characterization of nanofluid for advanced heat transfer applications, *Scripta Materialia*, 55, 549–552.
25. D. Singh, J. Toutbort, G. Chen, J. Hull, R. Smith, O. Ajayi, and W. Yu. 2006, Heavy Vehicle Systems Optimization Merit Review and Peer Evaluation, Argonne Nat. Lab., Annual Report, 2006.
26. B. Shen, A.J. Shih, S.C. Tung, and M. Hunter. 2009, Application of nanofluids in minimum quantity lubrication grinding, *Tribology and Lubrication Technology*, 51(6), 730–737.
27. M.J. Kao, H. Chang, Y.Y. Wu, T.T. Tsung, and H.M. Lin. 2007, Producing aluminum-oxide brake nanofluids using plasma charging system, *Journal of the Chinese Society of Mechanical Engineers*, 28(2), 123–131.
28. S.C. Tzeng, C.W. Lin, and K.D. Huang. 2005, Heat transfer enhancement of nanofluids in rotary blade coupling of four-wheel-drive vehicles. *Acta Mechanica* 179, 11–23.
29. Z. Zhang and Q. Que. 1997, Synthesis, structure and lubricating properties of dialkyl-dithiophosphate-modified Mo-S compound nanoclusters. *Wear*, 209, 8–12.
30. H.B. Ma, C. Wilson, B. Borgmeyer, K. Park, Q. Yu, U.S. Choi, and M. Tirumala. 2006, Effect of nanofluid on the heat transport capability in an oscillation g heat pipe, *Applied Physics Letters*, 88, 1–3.
31. M. Arif. 2006, Neutron Imaging for Fuel Cell Research, Imaging and Neutron Workshop, Oak Ridge, TN, October 2006.
32. H.B. Ma, C. Wilson, Q. Yu, K. Park, U.S. Choi, and M. Tirumala. 2006, An experimental investigation of heat transport capability in a nanofluid oscillating heat pipe, *Journal of Heat Transfer*, 128, 1213–1216.
33. Y.H. Lin, S.W. Kang, and H.L. Chen. 2008, Effect of silver nano-fluid on pulsating heat pipe thermal performance, *Applied Thermal Engineering*, 28(11–12), 1312–1317.
34. C.T. Nguyen, G. Roy, C. Gautheir, and N. Galanis. 2007, Heat transfer enhancement using Al_2O_3- water nanofluid for an electronic liquid cooling system, *Applied Thermal Engineering*, 27, 1501–1506.

35. S. Vafaei, T.B. Tasciuc, M.Z. Podowski, A. Purkayastha, G. Ramanath, and P.M. Ajayan. 2006, Effect of nanoparticles on sessile droplet contact angle, *Institute of Physics Publishing: Nanotechnology*, 17, 2523–2527.

36. R.K. Dash, T.B. Tasciuc, A. Purkayastha, and G. Ramanath. 2007, Electrowetting on dielectric-actuation of microdroplets of aqueous bismuth telluride nanoparticles suspensions, *IOP Publishing: Nanotechnology*, 18, 1–6.

37. C.S. Kumar (ed.), 2010, *Microfluidic Devices in Nanotechnology: Fundamental Concepts*, John Wiley, Hoboken, NJ.

38. World Bank, GGFR partners unlock value of wasted gas, *World Bank*, 14 December 2009. Retrieved 28 December 2010. http://web.worldbank.org/WBSITE/EXTERNAL/TOPICS/EXTSDNET/0,,contentMDK:22416844~menuPK:64885113~pagePK:64885161~piPK:64884432~theSitePK:5929282,00.html

39. R.S. Shawgo, A.C.R. Grayson, Y. Li, and M.J. Cima. 2002, Bio-MEMS for drug delivery, *Curr. Opin. Solid State Mater. Sci.* 6, 329–334.

40. Cepheid. 2009, http://www.Cepheid.Com.

41. A. Ovsianikov, B. Chichkov, P. Mente, N.A. Monteiro-Riviere, A. Doraiswamy, and R. J. Narayan. 2007, Two photon polymerization of polymer–ceramic hybrid materials for transdermal drug delivery, *Int. J. Ceramic Technol.*, 4, 22–29.

42. K. Kim, and J.B. Lee. 2007, High aspect ratio tapered hollow metallic microneedle arrays with microfluidic interconnector, *Microsyst. Technol.*, 13, 231–235.

43. V. Labhasetwar and D.L. Leslie-Pelecky. 2007, *Biomedical Applications of Nanotechnology*, Wiley-Interscience, A John Wiley & Son, Inc., Publication, Hoboken, New Jersey 2007.

44. C. Kleinstreuer, J. Li, and J. Koo. 2008, Microfluidics of nano-drug delivery, *International Journal of Heat and Mass Transfer,* 51, 5590–5597.

45. D. Bica, L. Vekas, M V Avdeev, O Marinica, V. Socoliuc, M. Balasoiu, and V.M. Garamus. 2007, Sterically stable water based magnetic fluids: Synthesis, structure and properties, *Journal of Magnetism and Magnetic Materials*, 311, 17–21.

46. P.C. Chiang, D.S. Hung, J.W. Wang, C.S. Ho, and Y.D. Yao. 2007, Engineering water dispersible Fept nanoparticles for biomedical applications, *IEEE Tranaction on Magnetics*, 43(6), 2445–2447.

47. L. Vekas, D. Bica, and O. Marinica. 2006, Magnetic nanofluids stabilized with various chain length surfactants, *Romanian Reports in Physics*, 58(3), 257–267.

48. L. Vekas, D. Bica, and M.V. Avdeev. 2007, Magnetic nanoparticles and concentrated magnetic nanofluids: Synthesis, properties and some applications, *China Particuology*, 5, 43–49.

49. X. He, E.Y.H. Park, A. Fowler, M. L. Yarmush, and M. Toner. 2008, Vitrification by ultra-fast cooling at a low concentration of cryoprotectants in a quartz micro-capillary: A study using murine embryonic stem cells, *Cryobiology*, 56, 223–232.

50. J. Yan and J. Liu. 2008, Nanocryosurgery and its mechanisms for enhancing freezing efficiency of tumor tissues, *Nanomedicine: Nanotechnology, Biology, and Medicine* 4, 79–87.

51. M.C. Daniel and D. Astruc. 2004, Gold nanoparticles: Assembly, supramolecular chemistry, quantum-size-related properties, and applications toward biology, catalysis, and nanotechnology, *Chemical Reviews*, 104, 293–346.

52. K. Sefiane, J. Skilling, and J. Macgillivray. 2008, Contact line motion and dynamic wetting of nanofluid solutions, *Advances in Colloid and Interface Science*, 138(2), 101–120.

53. S.M.S. Murshed, S.H. Tan, and N.T. Nguyen. 2008, Temperature dependence of interfacial properties and viscosity of nanofluids for droplet-based microfluidics, *Journal of Physics D: Applied Physics,* 41, 1–5.

54. Y.H. Jeong, W.J. Chang, and S.H. Chang. 2007, Wettability of heated surfaces under pool boiling using surfactant solutions and nano-fluids, *Int. Journal of Heat and Mass Transfer*, 51(11–12), 3025–3031.

55. D.T. Wasan and A.D. Nikolov. 2003, Spreading of nanofluids on solids, *Nature*, 423, 156–159.
56. S.U.S. Choi Eastman, J.A. 2001: US20016221275.
57. E.C. Nosfor. 2008, Recent Patents on Nanofluids (Nanoparticles In Liquids) Heat Transfer.
58. H. Ohira, K. Ara, and M. Konomura, (Feb 2008), US Patent 7,326,368. Reducing the Reactivity or Toxicity of a Liquid.
59. J.L. Davidson, D.T. Bradshaw, (June 2008), US Patent 7,348,298. Conductive Material Powder.
60. C. Wu, (May 2008), US Patent 7,374,698. Nanometer Heat-Conducting Water Solution for Use in a Car Cooling System.
61. C. Hoi, J.M. Oh, H.S. Yoo, (May 2008), Korean Patent KR2008008625. Oil-Based Nanofluid with High Thermal Efficiency.
62. F.E. Lockwood, (March 2008), US Patent 7,348,298. Enhancing Thermal Conductivity of Fluids with Graphite Nanoparticles.
63. Z. Zhang, G. Wu, F.E. Lockwood, T.R. Smith, (Nov 2008), US Patent 7,449,432. Gear Oil Composition Containing Nanomaterial.
64. Z. Zhang, G. Wu, F.E. Lockwood, D.J. Dotson, (Dec 2008), US Patent 7,470,650. Shock Absorber Fluid Composition Containing Nanostructures.
65. X. Zhong, X. Yu, X. Peng, L. Xia, (Dec 2008), Chinese Patent CN101323777. Aluminum Oxide Organic Nanofluid for Engine High Temperature Cooling Technology.
66. H. Liu, Q. He, S. Yin, (Sep 2009), Chinese Patent CN101525530. Low Temperature Phase Change Cold Storage Nanofluid.
67. A. Griffo, M. Keshavan, (July 2010), US Patent 7,749,947. High Performance Rock Bit Grease.
68. C.S. Jeffcoate, F.J. Marinho, A.V. Gershun, (Oct 2010), US Patent 7,820,066. Fluid Composition with Enhanced Heat Transfer Efficiency.
69. K.S. Jung, B.G. Hong, D.W. Lee, (June 2010), Korean Patent Kr20100068676. Blanket Using Lithium Nanofluid and a Fusion Reactor Using the Same.
70. J.C. Farmer, (Nov 2010), US Patent Application 2010291429. Electrochemical Nanofluid.

2 The Role of Nanoparticle Suspensions in Thermo/Fluid and Biomedical Applications

Khalil M. Khanafer and Kambiz Vafai

CONTENTS

2.1 INTRODUCTION

Recent advances in nanomaterials and nanotechnology have led to the development of new class of heat-transfer fluids containing nanometer-sized particles called nanoparticles typically made of carbon nanotubes, metals, or oxides. Nanofluids are engineered by suspending nanoparticles with average sizes below 100 nm in a base fluid such as water, ethylene glycol, and oil [1]. Compared with the base fluid, nanofluids have distinctive properties that make them attractive in many applications such as pharmaceutical processes, transportation industry, thermal management of electronics, fuel cells, boiler flue gas temperature reduction, heat exchangers, etc. [2]. Extensive research studies on heat-transfer enhancement using nanofluids were conducted both experimentally and theoretically in the literature [1–15]. Conflicting results on the heat transfer enhancement using nanofluids in forced and natural convection are reported in the literature. Pak and Cho [16] illustrated that the Nusselt number for Al_2O_3–water and TiO_2–water nanofluids increased with increasing Reynolds number and volume fraction of nanoparticles. Nevertheless, the convective heat-transfer coefficient for nanofluids at a volume fraction of 3% was found to be 12% smaller than that of the base fluid when assuming a constant average velocity [16]. Yang et al. [17] studied experimentally the convective heat-transfer coefficients of graphite–water nanofluids under laminar flow in a horizontal tube heat exchanger. Their experimental heat-transfer coefficients showed that the nanoparticles increased the heat-transfer coefficient of the fluid system in laminar flow, but the increase was much less than that predicted by the existing correlation based on static thermal conductivity measurements. However, many other researchers have reported forced convective heat-transfer enhancement using nanofluids [18–22].

Not many studies are found in the literature on the application of nanofluids in natural convective heat transfer. Khanafer et al. [6] analyzed numerically natural convection heat transfer of nanofluids in an enclosure under various physical parameters. Their results showed that the average Nusselt number increases with an increase in the nanoparticles volume fraction for different Grashof numbers. Kim et al. [23] introduced a factor to explain the effect of nanoparticle addition on the convective instability and heat-transfer characteristics of a base fluid. The new factor included the effect of the ratio of the thermal conductivity of nanoparticles to that of the base fluid, the shape factor of the nanoparticles, the volume fraction of nanoparticles, and the heat capacity ratio. Their results indicate that the heat-transfer coefficient in the presence of nanofluids increases with an increase in the volume fraction of nanoparticles. Ghasemi and Aminossadati [24] numerically studied natural convection heat transfer in an inclined enclosure filled with a CuO–water nanofluid for various pertinent parameters such as Rayleigh number, inclination angle, and solid volume fraction. Their results indicated that the addition of nanoparticles improves heat-transfer performance. In addition, they showed that there is an optimum solid volume fraction that maximizes heat-transfer rate. Natural convective heat-transfer enhancement using nanofluids was also demonstrated experimentally by Nnanna et al. [25] and Nnanna and Routhu [26].

Conversely, Putra et al. [13] illustrated experimentally that the presence of nanoparticles (Al_2O_3 and CuO) in water-based nanofluids inside a horizontal cylinder decreased natural convective heat-transfer coefficient with an increase in the volume fraction of nanoparticles, particle density as well as the aspect ratio of the cylinder. Ding et al. [27] have also reported experimentally that the natural convective heat-transfer coefficient decreases systematically with an increase in nanoparticle concentration, and the deterioration was partially related to the higher viscosity of nanofluids. Chang et al. [28] considered natural convection experiments with Al_2O_3 micro-particle (approximately 250 nm) aqueous suspensions in thin enclosures. Their results illustrated that nanoparticles have insignificant effect on the Nusselt number values for a vertical enclosure. Nevertheless, for horizontal enclosure, there was a decrease in Nusselt number compared with pure water at lower Rayleigh numbers and higher particle concentrations. The researchers attributed this anomalous behavior to sedimentation.

Presently, there are no robust theoretical models to determine the anomalous thermal conductivity enhancement of nanofluids. Many researchers have attributed the thermal conductivity enhancement of nanofluids to thermal conductivities of fluid and nanoparticles, shape and surface area of nanoparticles, volume fraction, and temperature [29]. Keblinski et al. [29] and Eastman et al. [30] proposed four main mechanisms for thermal conductivity enhancement of nanofluids. These consist of Brownian motion of nanoparticles, molecular-level layering of the liquid at the liquid/particle interface, heat transport within the nanoparticles, and the effects of nanoparticle clustering. On the basis of molecular dynamics simulations and the simple kinetic theory, Evans et al. [31] demonstrated that the hydrodynamics effects associated with Brownian motion have a small effect on the thermal conductivity of the nanofluid. Conflicting results were reported in the literature associated with the effect of solid/liquid interfacial layer (i.e., the interface between the nanoparticle and the fluid) on the thermal conductivity enhancement of nanofluids [32–37].

Yu and Choi [32,33] and Xue and Xu [34] suggested a theoretical model for the effect of a solid/liquid interface based on the Hamilton–Crosser model for suspensions of nano-spherical particles. Their results showed that the solid/liquid interfacial layers play an important role in enhancing the thermal conductivity of nanofluids. Considering the interface effect between the solid particles and the base fluid in nanofluids, Xue [35] developed a model for the effective thermal conductivity of nanofluids based on Maxwell theory and average polarization theory. Xue [35] suggested that the developed model can interpret the anomalous enhancement of the effective thermal conductivity of the nanofluid. On the basis of molecular dynamic simulations and simple liquid–solid interfaces, Xue et al. [37] illustrated that the layering of the liquid atoms at the liquid–solid interface does not have any significant effect on the thermal-transport properties of nanofluids.

Although many possible mechanisms were proposed in the literature, there are no robust theoretical and experimental studies that explain the basis for possible heat-transfer enhancement when using nanofluids. As such, it is still unclear as to what are the best models to use for the thermal conductivity and viscosity of nanofluids. Therefore, the aim of this study is to analyze the variants within the thermophysical characteristics of nanofluids especially with respect to the thermal conductivity and viscosity models and propose possible physical reasons for the deviations between experimental and analytical studies.

2.2 ANALYTICAL MODELS FOR PHYSICAL PROPERTIES OF NANOFLUIDS

2.2.1 DENSITY

The density of nanofluid can be determined analytically based on the physical principle of the mixture rule as

$$\rho_{eff} = \left(\frac{m}{V}\right)_{eff} = \frac{m_f + m_p}{V_f + V_p} = \frac{\rho_f V_f + \rho_p V_p}{V_f + V_p} = (1 - \phi_p)\rho_f + \phi_p\rho_p \qquad (2.1)$$

where f and p refer to the fluid and nanoparticle, respectively, and $\phi_p = (V_p/V_f + V_p)$ is the volume fraction of the nanoparticles. To test the validity of Equation 2.1, Pak and Cho [16] and Ho et al. [38] conducted experimental studies to measure the density of Al_2O_3–water nanofluids at room temperature as depicted in Figure 2.1a. Excellent agreement was found between the experimental results and the predictions using Equation 2.1 as shown in Figure 2.1a. Ho et al. [38] also measured the density of Al_2O_3–water nanofluid at different temperatures and nanoparticle volume fractions. Khanafer and Vafai [39] developed a correlation for the density of Al_2O_3–water nanofluid using the experimental data of Ho et al.'s study [38] as a function of temperature and volume fraction of nanoparticles as follows:

$$\rho_{eff} = 1001.064 + 2738.6191\phi_p - 0.2095T; \quad 0 \le \phi_p \le 0.04, 5 \le T\,(^\circ C) \le 40 \quad (2.2)$$

The R^2 of the regression is 99.97% and the maximum relative error is 0.22%. It is clear from Figure 2.1b that the present regression (Equation 2.2) is in excellent agreement with the density measurements of Ho et al. [38].

2.2.2 HEAT CAPACITY OF NANOFLUIDS

The vast majority of studies on nanofluids have used an analytical model for the specific heat by assuming thermal equilibrium between the nanoparticles and the base fluid phase as follows:

$$(\rho c)_{eff} = \rho_{eff}\left(\frac{Q}{m\Delta T}\right)_{eff} = \rho_{eff}\frac{Q_f + Q_p}{(m_f + m_p)\Delta T} = \rho_{eff}\frac{(mc)_f \Delta T + (mc)_p \Delta T}{(m_f + m_p)\Delta T}$$

$$\rightarrow (\rho c)_{eff} = \rho_{eff}\frac{(\rho c)_f V_f + (\rho c)_p V_p}{\rho_f V_f + \rho_p V_p} \qquad (2.3)$$

$$\Rightarrow c_{eff} = \frac{(1 - \phi_p)\rho_f c_f + \phi_p\rho_p c_p}{\rho_{eff}}$$

where ρ_p is the density of the nanoparticle, ρ_f is the density of the base fluid, ρ_{eff} is the density of the nanofluid, and c_p and c_f are the heat capacities of the nanoparticle

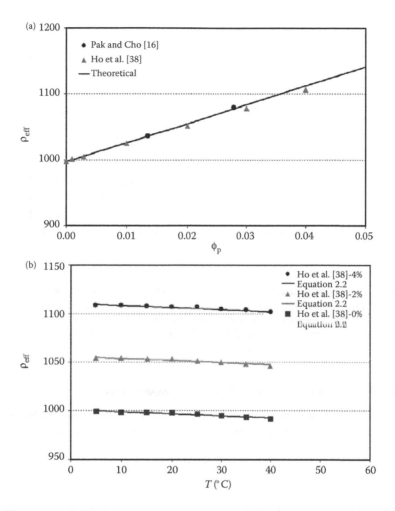

FIGURE 2.1 Effect of the volume fraction on the density of the Al$_2$O$_3$–water nanofluid: (a) room temperature; (b) various temperatures. (Reprinted from K. Khanafer and K. Vafai. 2011, A critical synthesis of thermophysical characteristics of nanofluids, *International Journal of Heat and Mass Transfer* 54, 4410–4428, Copyright 2011, with permission from Elsevier.)

and the base fluid, respectively. In contrast, some researchers [16,40–42] suggest a simpler expression given by

$$c_{eff} = (1 - \phi_p)c_f + \phi_p c_p \tag{2.4}$$

The experimental data of Zhou and Ni [43] were used to evaluate the validity of Equations 2.3 and 2.4, Figure 2.2 shows a comparison of the specific heat of Al$_2$O$_3$–water nanofluid at room temperature using both equations with the experimental data of Zhou and Ni [43] for various volume fractions ($\phi_p = 0$–21.7%). Figure 2.2 shows that model I given in Equation 2.3 compares very well with the experimental data of Zhou and Ni [43].

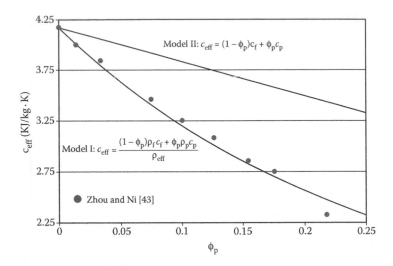

FIGURE 2.2 Comparison of the heat capacity of Al_2O_3–water nanofluid obtained by models I and II given in Equations 2.3 and 2.4 and the experimental data of Zhou and Ni [43]. (Reprinted from K. Khanafer and K. Vafai. 2011, A critical synthesis of thermophysical characteristics of nanofluids, *International Journal of Heat and Mass Transfer* 54, 4410–4428, Copyright 2011, with permission from Elsevier.)

2.2.3 THERMAL EXPANSION COEFFICIENT OF NANOFLUIDS

The thermal expansion coefficient of nanofluids can be approximated by utilizing the volume fraction of the nanoparticles on a weight basis as follows [6]:

$$\beta_{eff} = \frac{(1 - \phi_p)(\rho\beta)_f + \phi_p(\rho\beta)_p}{\rho_{eff}} \tag{2.5}$$

where β_f and β_p are the thermal expansion coefficients of the base fluid and the nanoparticle, respectively. However, some investigators give a simpler model for the thermal expansion coefficient of the nanofluid as [44,45]:

$$\beta_{eff} = (1 - \phi_p)\beta_f + \phi_p\beta_p \tag{2.6}$$

Ho et al. [38] conducted an experimental study to estimate the thermal expansion of Al_2O_3–water nanofluid at various volume fractions of nanoparticles. The values of the thermal expansion of Al_2O_3–water nanofluid predicted by Equations 2.5 and 2.6 were compared with the experimental data of Ho et al. [38] at a temperature of 26°C. Figure 2.3a shows that neither Equation 2.5 nor Equation 2.6 can be utilized to correctly determine the thermal expansion of nanofluid as compared with the experimental data of Ho et al. [38]. The effect of varying the temperature and volume fraction of nanoparticles on the thermal expansion coefficient of Al_2O_3–water nanofluid was investigated by Ho et al. [38]. Khanafer and Vafai [39] developed a correlation for the thermal expansion coefficient of Al_2O_3–water nanofluid was developed based

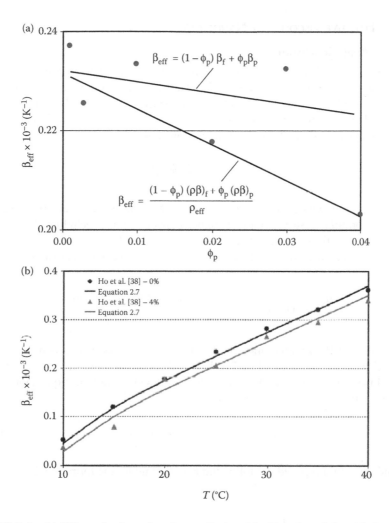

FIGURE 2.3 (a) Effect of volume fraction as displayed by Equations 2.5 and 2.6 at room temperature; (b) temperature effect as displayed by a comparison between Equation 2.7 and experimental data of Ho et al. [38]. (Reprinted from K. Khanafer and K. Vafai. 2011, A critical synthesis of thermophysical characteristics of nanofluids, *International Journal of Heat and Mass Transfer* 54, 4410–4428, Copyright 2011, with permission from Elsevier.)

on the experimental data of Ho et al. [38] as a function of temperature and volume fraction of nanoparticles. This correlation (Al_2O_3–water) can be expressed as [39]:

$$\beta_{eff} = (-0.479\phi_p + 9.3158 \times 10^{-3}T - \frac{4.7211}{T^2}) \times 10^{-3};$$

$$0 \le \phi_p \le 0.04, 10°C \le T \le 40°C \tag{2.7}$$

Figure 2.3b shows the validity of the correlation given by Equation 2.7 compared with the experimental data [38].

2.2.4   EFFECTIVE VISCOSITY OF NANOFLUIDS

2.2.4.1   Analytical Studies

Different analytical models of viscosity have been developed in the literature to model the effective viscosity of nanofluid as a function of volume fraction. Einstein [46] determined the effective viscosity of a suspension of spherical solids as a function of volume fraction (volume concentration <5%) using the phenomenological hydrodynamic equations. This equation was presented as

$$\mu_{eff} = (1 + 2.5\phi_p)\mu_f \tag{2.8}$$

Since Einstein's model, several equations have been developed in an effort to extend Einstein's formula to suspensions of higher concentrations, including the effect of non-spherical particle concentrations [47–51]. For example, Brinkman [47] presented a viscosity model that extended Einstein's equation to concentrated suspensions:

$$\mu_{eff} = \frac{1}{(1 - \phi_p)^{2.5}} = (1 + 2.5\phi_p + 4.375\phi_p^2 + \cdots)\mu_f \tag{2.9}$$

The effect of Brownian motion on the effective viscosity in a suspension of rigid spherical particles was studied by Batchelor [48]. For isotropic structure of suspension and based on reciprocal theorem in Stokes flow to obtain an expression for the bulk stress, the effective viscosity was given by

$$\mu_{eff} = (1 + 2.5\phi_p + 6.2\phi_p^2)\mu_f \tag{2.10}$$

Lundgren [49] proposed the following equation under the form of a Taylor series in ϕ_p:

$$\mu_{eff} = \frac{1}{1 - 2.5\phi_p}\mu_f = (1 + 2.5\phi_p + 6.25\phi_p^2 + O(\phi_p^3))\mu_f \tag{2.11}$$

It is noticeable that if the terms $O(\phi_p^2)$ and higher are neglected, the above correlation reduces to that of Einstein's model. Table 2.1 summarizes the most common analytical expressions for the viscosity of nanofluids as a function of the volume fraction of the nanoparticles [39].

2.2.4.2   Experimental Studies

A number of experimental studies have been carried out in the literature to determine the dynamic viscosity of nanofluids [16,56–65]. Masuda et al. [57] were the first to measure the dynamic viscosity of several water-based nanofluids for temperatures ranging from room condition to 67°C. Wang et al. [56] obtained some data for the dynamic viscosity of Al_2O_3–water and Al_2O_3–ethylene glycol mixtures at various temperatures.

TABLE 2.1
Summary of Significant Number of Models Found in the Literature

Models	Effective Viscosity	Physical Model	Remarks
Einstein [46]	$\mu_{eff} = (1 + 2.5\phi_p)\mu_f$	Based on the phenomenological hydrodynamic equations	Infinitely dilute suspension of spheres (no interaction between the spheres)
Brinkman [47]	$\mu_{eff} = \dfrac{1}{(1-\phi_p)^{2.5}}$ $= (1 + 2.5\phi_p + 4.375\phi_p^2 + \cdots)\mu_f$	Considered a suspension containing n solute particles in a total volume V Based on Einstein model Derived by considering the effect of the addition of one solute molecule to an existing solution	Valid for relatively low particle volume fraction ($\phi_p \leq 2\%$) Spherical particles Valid for high-moderate particle concentrations Used Einstein's factor: $(1 + 2.5\phi_p)$
Batchelor [48]	$\mu_{eff} = (1 + \eta\phi_p + k_H\phi_p^2)\mu_f$ $= (1 + 2.5\phi_p + 6.2\phi_p^2)\mu_f$	Based on reciprocal theorem in Stokes flow problem to obtain an expression for the bulk stress due to the thermodynamic forces Incorporated both effects: hydrodynamic effects and Brownian motion	Rigid and spherical particles Brownian motion Isotropic structure Huggins coefficient: $k_H = 6.2$ (5.2 from hydrodynamic effects and 1.0 from Brownian motion)
Lundgren [49]	$\mu_{eff} = \dfrac{1}{1-2.5\phi_p}\mu_f$ $= (1 + 2.5\phi_p + 6.25\phi_p^2 + \cdots)\mu_f$	Based on a Taylor series expansion in terms of ϕ_p	Dilute concentration of spheres Random bed of spheres
Graham [50]	$\mu_{eff} = (1 + 2.5\phi_p)\mu_f$ $+ \left[\dfrac{4.5}{(h/r_p)(2 + h/r_p)(1 + h/r_p)^2}\right]\mu_f$	A cell theory was used to derive the dependence of the zero-shear-rate viscosity on volume concentration for a suspension of uniform, solid, neutrally buoyant spheres	Agrees well with Einstein's for small ϕ_p r_p is the particle radius and h is the inter-particle spacing

continued

TABLE 2.1 (continued)
Summary of Significant Number of Models Found in the Literature

Models	Effective Viscosity	Physical Model	Remarks
Simha [51]	$$\mu_{\text{eff}} = \left[1 + 2.5\phi_p + \left(\frac{125}{64\phi_{p\max}}\right)\phi_p^2 + \cdots\right]\mu_f$$	Based on Cage model of liquids and solutions	Spherical particles
Mooney [52]	$$\mu_{\text{eff}} = \exp\left(\frac{2.5\phi_p}{1 - k\phi_p}\right)\mu_f$$ $$= \{1 + 2.5\phi_p + [3.125 + (2.5k)]\phi_p^2 + \cdots\}\mu_f$$ $$1.35 < k < 1.91$$	Einstein's viscosity equation for an infinitely dilute suspension of spheres was extended to apply to a suspension of finite concentration	Rigid spherical spheres. Monodisperse suspension of finite concentration. Not valid at high concentrations. Considered the volume fraction of a suspension to be divided into two portions
Eilers [53]	$$\mu_{\text{eff}} = \mu_f\left[1 + \frac{1.25\phi_p}{1 - \phi_p/0.78}\right]$$ $$= (1 + 2.5\phi_p + 4.75\phi_p^2 + \cdots)\mu_f$$	Based on first-order interaction between particles (crowding effect). Based on experimental data	Suspensions of bitumen spheres. Curve fitting of the experimental data
Saito [55]	$$\mu_{\text{eff}} = \left(1 + \frac{2.5}{1 - \phi_p}\phi_p\right)\mu_f$$ $$= (1 + 2.5\phi_p + 2.5\phi_p^2 + \cdots)\mu_f$$	Developed based on a theory for spherical solute-molecules in which a single solute-molecule is placed in the field of flow, obtained by averaging over all the possible positions of a second solute-molecule	Spherical rigid particles. Brownian motion. Very small particles
Frankel and Acrivos [55]	$$\mu_{\text{eff}} = \left(\frac{9}{8}\frac{(\phi_p/\phi_{p\max})^{1/3}}{1 - (\phi_p/\phi_{p\max})^{1/3}}\right)\mu_f$$	An asymptotic technique was used to derive the functional dependence of effective viscosity on concentration for a suspension of uniform solid spheres, in the limit as concentration approaches its maximum value	Uniform solid particles

Source: Reprinted from K. Khanafer and K. Vafai. 2011, A critical synthesis of thermophysical characteristics of nanofluids, *International Journal of Heat and Mass Transfer* 54, 4410–4428, Copyright 2011, with permission from Elsevier.

Because the formulas such as the one proposed by Einstein [46] and later improved by Brinkman [47] and Batchelor [48] underestimate the viscosity of the nanofluids when compared with the measured data, Maiga et al. [58,59] performed a least-square curve fitting of some experimental data of Wang et al. [56] including Al_2O_3 in water and Al_2O_3 in ethylene glycol. Table 2.2 demonstrates a summary of various dynamic viscosity models at room temperature based on the experimental data.

TABLE 2.2
Summary of Viscosity Models at Room Temperature Based on Experimental Data

Models	Effective Viscosity (Regression)	Remarks
Maiga et al. [58]	$\mu_{eff} = \left(1 + 7.3\phi_p + 123\phi_p^2\right)\mu_f$	Least-square curve fitting of Wang et al. [56] data Al_2O_3–water, $d_p = 28$ nm
Maiga et al. [58]	$\mu_{eff} = \left(1 - 0.19\phi_p + 306\phi_p^2\right)\mu_f$ $d_p = 28$ nm	Least-square curve fitting of experimental data [56,57] Al_2O_3–ethylene glycol
Khanafer and Vafai [39]	$\mu_{eff} = \left(1 + 0.164\phi_p + 302.34\phi_p^2\right)\mu_f$ $d_p = 28$ nm	Least-square curve fitting of experimental data [55,57] Al_2O_3–ethylene glycol
Buongiorno [66]	$\mu_{eff} = \left(1 + 39.11\phi_p + 533.9\phi_p^2\right)\mu_f$	Curve fitting of Pak and Cho [16] data Al_2O_3–water, d_p – 13 nm
Buongiorno [66]	$\mu_{eff} = \left(1 + 5.45\phi_p + 108.2\phi_p^2\right)\mu_f$	Curve fitting of Pak and Cho [16] data TiO_2–water, $d_p = 27$ nm
Khanafer and Vafai [39]	$\mu_{eff} = \left(1 + 23.09\phi_p + 1525.3\phi_p^2\right)\mu_f$ $0 \leq \phi_p \leq 0.04$	Curve fitting of Pak and Cho [16] data Al_2O_3–water, $d_p = 13$ nm
Khanafer and Vafai [39]	$\mu_{eff} = \left(1 + 3.544\phi_p + 169.46\phi_p^2\right)\mu_f$ $0 \leq \phi_p \leq 0.1$	Curve fitting of Pak and Cho [16] data TiO_2–water, $d_p = 27$ nm
Nguyen et al. [61]	$\mu_{eff} = \mu_f \times 0.904e\ 0.148\phi_p;\ d_p = 47$ nm $\mu_{eff} = \left(1 + 0.025\phi_p + 0.015\phi_p^2\right)\mu_f;\ d_p = 36$ nm	Curve fitting of the experimental data Al_2O_3–water
Nguyen et al. [61]	$\mu_{eff} = \left(1.475 - 0.319\phi_p + 0.051\phi_p^2 + 0.009\phi_p^3\right)\mu_f$	Curve fitting of the experimental data CuO–water, $d_p = 29$ nm
Tseng and Lin [62]	$\mu_{eff} = 13.47 \exp(35.98\phi_p)\quad \mu_f;\ 0.05 \leq \phi_p \leq 0.12$	TiO_2–water Shear rate = 100 s^{-1}

Source: Reprinted from K. Khanafer and K. Vafai. 2011, A critical synthesis of thermophysical characteristics of nanofluids, *International Journal of Heat and Mass Transfer* 54, 4410–4428, Copyright 2011, with permission from Elsevier.

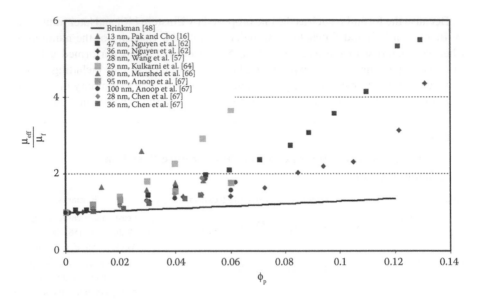

FIGURE 2.4 Relative viscosity measurement as a function of the volume fraction, ϕ_p, at ambient temperature (Al_2O_3–water nanofluid). (Reprinted from K. Khanafer and K. Vafai. 2011, A critical synthesis of thermophysical characteristics of nanofluids, *International Journal of Heat and Mass Transfer* 54, 4410–4428, Copyright 2011, with permission from Elsevier.)

Moreover, Figure 2.4 shows a comparison of the relative dynamic viscosity of Al_2O_3–water nanofluid from various research studies in the literature at room temperature. Figure 2.4 shows that Brinkman model [47], which was derived for two-phase mixture, is to some extent sufficient to estimate the viscosity for relatively low-volume fraction of particles (i.e., $\phi_p \leq 2\%$). Although it noticeably underestimates the nanofluid viscosity when compared with experimental data at high particle concentrations. The differences in the relative viscosity among the experimental data as shown in Figure 2.4 may be due to the difference in the size of the particle clusters, dispersion techniques, and the methods of measurements. This clearly illustrates the discrepancy between the researchers in measuring the dynamic viscosity of nanofluids.

2.3 EFFECT OF TEMPERATURE ON THE DYNAMIC VISCOSITY OF NANOFLUIDS

It should be noted that all the above-mentioned correlations (Tables 2.1 and 2.2) were developed to relate the dynamic viscosity as a function of volume fraction only, without temperature-dependence considerations. Few studies were conducted in the literature on the effect of temperature on the dynamic viscosity of nanofluids [60,61,66–70]. Nguyen et al. [61] analyzed experimentally the effect of temperature on the dynamic viscosities of two water-based nanofluids, namely

Al_2O_3–water ($d_p = 47$ nm, 36 nm) and CuO–water ($d_p = 29$ nm) mixtures. The following correlations were developed by Nguyen et al. [61] for estimating the dynamic viscosity for all nanofluids tested at particle concentrations of 1% and 4%, respectively:

$$\mu_{eff}(cP) = (1.125 - 0.0007 \times T)\mu_f; \; \phi_p = 1\% \tag{2.12}$$

$$\mu_{eff}(cP) = (2.1275 - 0.0215 \times T + 0.0002 \times T^2)\mu_f; \; \phi_p = 4\% \tag{2.13}$$

where T is the temperature in °C. It can be noticed from Equations 2.12 and 2.13 that Nguyen et al. [61] did not explicitly express the dynamic viscosity as a function of temperature and volume fraction. Palm et al. [67] proposed equations for the dynamic viscosity (Pa. s) by means of the polynomial curve fitting based on the data reported by Putra et al. [13]. The resulting equations as a function of temperature, expressed in Kelvin, for Al_2O_3–water are:

$$\mu_{eff} = 0.034 - 2 \times 10^{-4}T + 2.9 \times 10^{-7}T^2, \phi_p = 1\% \tag{2.14}$$

$$\mu_{eff} = 0.039 - 2.3 \times 10^{-4}T + 3.4 \times 10^{-7}T^2, \phi_p = 4\% \tag{2.15}$$

Tables 2.3 and 2.4 present a summary of different models of the dynamic viscosity of nanofluids as a function of temperature and volume fraction of nanoparticles. Khanafer and Vafai [39] developed a general correlation (Equation 2.16) for the effective viscosity of Al_2O_3–water using various experimental data found in the literature (Figure 2.5a) as a function of volume fraction, nanoparticles diameter, and temperature as follows:

$$\mu_{eff}(cP) = -0.4491 + \frac{28.837}{T} + 0.574\phi_p - 0.1634\phi_p^2 + 23.053\frac{\phi_p^2}{T^2}$$
$$+ 0.0132\phi_p^3 - 2354.735\frac{\phi_p}{T^3} + 23.498\frac{\phi_p^2}{d_p^2} - 3.0185\frac{\phi_p^3}{d_p^2}; \tag{2.16}$$
$$1\% \le \phi_p \le 9\%, 20 \le T(°C) \le 70, 13\,\text{nm} \le d_p \le 131\,\text{nm}$$

The validity of the above correlation (Equation 2.16) is shown in Figure 2.5b. As can be noticed in Figure 2.5a, the viscosity of the nanofluid decreases with an increase in the temperature. Moreover, there is no agreement between researchers about the experimentally measured values of the nanofluid's viscosity. Published results indicate a surprising range of variation of the results.

TABLE 2.3
Effect of Temperature and Volume Fraction on the Dynamic Viscosity of Nanofluids (Al₂O₃–Water)

Reference	Model (Regression)	Remarks
Khanafer and Vafai [39]	$\mu_{eff} = 0.444 - 0.254\phi_p + 0.0368\phi_p^2 + 26.333\dfrac{\phi_p}{T} - 59.311\dfrac{\phi_p^2}{T^2}$ $20 \le T(^\circ C) \le 70;\ \phi_p = 1.34\%,\ 2.78\%$	Curve fitting of Pak and Cho [16] data $d_p = 13$ nm Units: mPa.s
Palm et al. [67]	$\mu_{eff} = 0.034 - 2\times10^{-4}T(K) + 2.9\times10^{-7}T^2(K),\ \phi_p = 1\%$ $\mu_{eff} = 0.039 - 2.3\times10^{-4}T(K) + 3.4\times10^{-7}T^2(K),\ \phi_p = 4\%$	Curve fitting of the experimental data, Putra et al. [13] $d_p = 131.2$ nm Units: Pa.s Units: mPa.s
Nguyen et al. [61]	$\mu_{eff} = (1.125 - 0.0007\times T(^\circ C))\mu_f;\ \phi_p = 1\%$ $\mu_{eff} = (2.1275 - 0.0215\times T(^\circ C) + 0.0002\times T^2(^\circ C))\mu_f;\ \phi_p = 4\%$	
Khanafer and Vafai [39]	$\mu_{eff} = -0.4892 + \dfrac{26.9036}{T} + 0.6837\phi_p + \dfrac{24.1141}{T^2} - 0.1785\phi_p^2 + 0.1818\dfrac{\phi_p}{T} + 27.015\dfrac{\phi_p^2}{T^2}$ $+\ 0.0132\phi_p^3 - 2940.1775\dfrac{\phi_p}{T^3};\quad 1\% \le \phi_p \le 9.4\%,\ 20 \le T(^\circ C) \le 70$	Curve fitting of Nguyen et al. [61] data $d_p = 47$ nm Units: mPa.s
Khanafer and Vafai [39]	$\mu_{eff} = -0.1011 + \dfrac{18.0162}{T} + 0.3619\phi_p + \dfrac{164.0837}{T^2} - 0.0966\phi_p^2 + 0.1609\dfrac{\phi_p}{T} + 22.4901\dfrac{\phi_p^2}{T^2}$ $+\ 0.0078089\phi_p^3 - 2316.3754\dfrac{\phi_p}{T^3};\quad 1\% \le \phi_p \le 9.1\%,\ 20 \le T(^\circ C) \le 70$	Curve fitting of Nguyen et al. [61] data $d_p = 36$ nm Units: mPa.s
Khanafer and Vafai [39]	$\mu_{eff} = -0.4491 + \dfrac{28.837}{T} + 0.574\phi_p - 0.1634\phi_p^2 + 23.053\dfrac{\phi_p^2}{T^2} + 0.0132\phi_p^3 - 2354.735\dfrac{\phi_p}{T^3}$ $+\ 23.498\dfrac{\phi_p^2}{d_p^2} - 3.0185\dfrac{\phi_p^3}{d_p^2};\quad 1\% \le \phi_p \le 9\%,\ 20 \le T(^\circ C) \le 70, 13\,nm \le d_p \le 131\,nm$	Curve fitting of various experimental data available in the literature [13,16,61,71] Units: mPa.s
Namburu et al. [69,60]	$Log(\mu_{eff}) = Ae^{-BT}$, in mmPa.s $A = -0.29956\phi_p^3 + 6.7388\phi_p^2 - 55.444\phi_p + 236.11$ $B = (-6.4745\phi_p^3 + 140.03\phi_p^2 - 1478.5\phi_p + 20341)/10^6$	Experimental Al₂O₃–ethylene glycol and water mixture $1\% \le \phi_p \le 10\%,\ d_p = 53$ nm $238 < T < 323\ K$

Source: Reprinted from K. Khanafer and K. Vafai. 2011, A critical synthesis of thermophysical characteristics of nanofluids, *International Journal of Heat and Mass Transfer* 54, 4410–4428, Copyright 2011, with permission from Elsevier.

TABLE 2.4
Effect of Temperature and Volume Fraction on the Dynamic Viscosity of Nanofluids (TiO$_2$–Water, CuO–Water)

Models	Effective Viscosity (Regression)	Remarks
Duangthongsuk and Wongwises [68]	$\dfrac{\mu_{\text{eff}}}{\mu_f} = 1.0226 + 0.0477\phi_p - 0.0112\phi_p^2;\ T = 15^\circ C$ $\dfrac{\mu_{\text{eff}}}{\mu_f} = 1.013 + 0.092\phi_p - 0.015\phi_p^2;\ T = 25^\circ C$ $\dfrac{\mu_{\text{eff}}}{\mu_f} = 1.018 + 0.112\phi_p - 0.0177\phi_p^2;\ T = 35^\circ C$	Experimental data TiO$_2$–Water, $0.2 \le \phi_p \le 2\%$ $d_p = 21$ nm
Khanafer and Vafai [39]	$\dfrac{\mu_{\text{eff}}}{\mu_f} = 1.0538 + 0.1448\phi_p - 3.363 \times 10^{-3}T - 0.0147\phi_p + 6.735 \times 10^{-5}T^2 - 1.337\dfrac{\phi_p}{T}$ $15^\circ C \le T \le 35^\circ C, 0.2\% \le \phi_p \le 2\%$	Curve fitting of the experimental data [68] TiO$_2$–water $d_p = 21$ nm
Khanafer and Vafai [39]	$\mu_{\text{eff}} = 0.6002 - 0.569\phi_p + 0.0823\phi_p^2 + 28.8763\dfrac{\phi_p}{T} - 20.2202\dfrac{\phi_p^2}{T^2} + 561.3175\dfrac{\phi_p^3}{T^3}$ $20 \le T(^\circ C) \le 70;\quad \phi_p = 0.99\%, 2.04\%, 3.16\%$	Curve fitting of Pak and Cho [16] data TiO$_2$–water $d_p = 27$ nm Units: mmPa.s
Khanafer and Vafai [39]	$\mu_{\text{eff}} = -0.4262 + \dfrac{8.4312}{T} + 0.898\phi_p + \dfrac{524.7147}{T^2} - 0.2217\phi_p^2 - 4.7329\dfrac{\phi_p}{T} + 70.3105\dfrac{\phi_p^2}{T^2}$ $+ 0.0176\phi_p^3 - 5559.4641\dfrac{\phi_p}{T^3};\quad 1\% \le \phi_p \le 9\%,\ 20 \le T(^\circ C) \le 70$	Curve fitting of Nguyen et al. [59] data CuO–water $d_p = 29$ nm Units: mmPs.s
Namburu et al. [60]	$\text{Log}(\mu_{\text{eff}}) = Ae^{-BT},\ \text{in mm Pa.s}$ $A = 1.8375\phi_p^2 - 29.643\phi_p + 165.56$ $B = 4 \times 10^{-6}\phi_p^2 - 0.001\phi_p + 0.0186$	CuO–ethylene glycol and water mixture $1 \le \phi_p \le 6\%, d_p = 29$ nm $238 < T < 323$ K

continued

TABLE 2.4 (continued)

Effect of Temperature and Volume Fraction on the Dynamic Viscosity of Nanofluids (TiO₂–Water, CuO–Water)

Models	Effective Viscosity (Regression)	Remarks
Kulkarni et al. [63,64]	$\ln \mu_{eff} = A\left(\dfrac{1}{T}\right) - B$, in mm Pa.s $A = 20587\phi_p^2 + 15857\phi_p + 1078.3$ $B = -107.12\phi_p^2 + 53.54\phi_p + 2.8715$	CuO–water. $0.05 \le \phi_p \le 0.15$ $d_p = 29$ nm. $278 \le T \le 323$ K Shear rate = 100 1/s
Koo and Kleinstreuer [70]	$\mu_{eff} = \mu_{static} + \mu_{Brownian}$ $\mu_{Brownian} = 5 \times 10^4 \beta \rho_f \phi_p \sqrt{\dfrac{\kappa T}{\rho_p d_p}} f(\phi_p, T)$ $f(\phi_p, T) = (-6.04\phi + 0.4705)T + (1722.3\phi_p - 134.63)$ $\beta = \begin{cases} 0.0137(100\phi_p)^{-0.8229}, & \phi_p < 0.01 \\ 0.0011(100\phi_p)^{-0.7272}, & \phi_p > 0.01 \end{cases}$ $1\% < \phi_p < 4\%, \ 300 < T < 325\text{K}$	CuO–water

Source: Reprinted from K. Khanafer and K. Vafai. 2011, A critical synthesis of thermophysical characteristics of nanofluids, *International Journal of Heat and Mass Transfer* 54, 4410–4428, Copyright 2011, with permission from Elsevier.

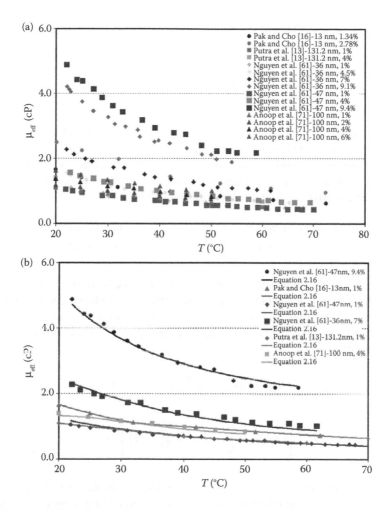

FIGURE 2.5 Effect of the volume fraction and temperature on the effective viscosity of Al$_2$O$_3$–water nanofluid: (a) experimental measurements; (b) comparison of Equation 2.16 developed in the current work with the experimental data. (Reprinted from K. Khanafer and K. Vafai. 2011, A critical synthesis of thermophysical characteristics of nanofluids, *International Journal of Heat and Mass Transfer* 54, 4410–4428, Copyright 2011, with permission from Elsevier.)

2.4 THERMAL CONDUCTIVITY OF NANOFLUIDS

Several experimental and theoretical studies were reported in the literature with respect to modeling thermal conductivity of nanofluids. The published results are in disagreement regarding the mechanisms for heat-transfer enhancement as well as a cohesive possible clarification with respect to the large discrepancies in the results even for the same base fluid and nanoparticles size. At present, there are no theoretical results available in the literature that can accurately determine the thermal conductivity of nanofluids. The existing results were generally based on the definition of the effective thermal conductivity of a two-component mixture as follows [72]:

$$k_{eff} = \frac{k_f(1 - \phi_p)(dT/dx)_f + k_p\phi_p(dT/dx)_p}{\phi_p(dT/dx)_p + (1 - \phi_p)(dT/dx)_f} \tag{2.17}$$

where $(dT/dx)_f$ is the temperature gradient within the fluid and $(dT/dx)_p$ is the temperature gradient through the particle. The Maxwell model [73] was one the first models developed for solid–liquid mixture with relatively large particles based on the solution of heat-conduction equation through a stationary random suspension of spheres. The effective thermal conductivity is given by

$$k_{eff} = \frac{k_p + 2k_f + 2\phi_p(k_p - k_f)}{k_p + 2k_f - \phi_p(k_p - k_f)} k_f = k_f + \frac{3\phi_p(k_p - k_f)}{k_p + 2k_f - \phi_p(kp - k_f)} \tag{2.18}$$

where k_p is the thermal conductivity of the particles, k_f is the fluid thermal conductivity, and ϕ_p is the volume fraction of the suspended particles. The Maxwell model is precise to the order of $\frac{1}{p}$ and applicable for the range of $\phi_p \ll 1$ or $|(k_p/k_f) - 1| \ll 1$, Bruggeman [74] developed a model to study the interactions between randomly dispersed spherical particles as follows:

$$\frac{k_{eff}}{k_f} = \frac{(3\phi_p - 1)(k_p/k_f) + \{3(1 - \phi_p) - 1\} + \sqrt{\Delta}}{4};$$

$$\Delta = \left[(3\phi_p - 1)\frac{k_p}{k_f} + \{3(1 - \phi_p) - 1\}\right]^2 + 8\frac{k_p}{k_f} \tag{2.19}$$

The Bruggeman model [74] is applicable for large volume fraction of spherical particles. For low-volume fractions, the Bruggeman model [74] results reduce to the Maxwell model [73]. For non-spherical particles, Hamilton and Crosser [72] proposed a model for the effective thermal conductivity of two-component mixtures as a function of the thermal conductivity of both the base fluid and the particle, volume fraction of the particles, and the shape of the particles. For the thermal conductivity ratio of two phases larger than 100 ($k_p/k_f > 100$), the thermal conductivity of two-component mixtures can be expressed as follows [72]:

$$k_{eff} = \frac{k_p + (n - 1)k_f + (n - 1)\phi_p(k_p - k_f)}{k_p + (n - 1)k_f - \phi_p(k_p - k_f)} k_f \tag{2.20}$$

where n is the empirical shape factor given by $n = 3/\psi$, and ψ is the particle sphericity, defined by the ratio of the surface area of a sphere with volume equal to that of the particle, to the surface area of the particle. Tables 2.5 through 2.7 review some relevant models for the effective thermal conductivity of nanofluids including the effects of Brownian motion and the nano-layer.

TABLE 2.5

Summary of Theoretical Models for the Effective Thermal Conductivity of Nanofluids

Models	Expressions	Physical Model	Remarks
Maxwell [73]	$$\frac{k_{\text{eff}}}{k_f} = \frac{k_p + 2k_f + 2\phi_p(k_p - k_f)}{k_p + 2k_f - \phi_p(k_p - k_f)}$$	Based on the conduction solution through a stationary random suspension of spheres	Spherical particles. Accurate to order ϕ_p^1.
Bruggeman [74]	$$\frac{k_{\text{eff}}}{k_f} = \frac{(3\phi_p - 1)k_p/k_f + \{3(1 - \phi_p) - 1\} + \sqrt{\Delta}}{4}$$ $$\Delta = \left[(3\phi_p - 1)\frac{k_p}{k_f} + \{3(1 - \phi_p) - 1\}\right]^2 + 8\frac{k_p}{k_f}$$	Based on the differential effective medium theory to estimate the effective thermal conductivity of composites at high particle concentrations. It consists in building up the composite medium through a process of incremental homogenization	Applicable to high-volume fraction of spherical particles. Suspension with spherical inclusions. No shape factor
Hamilton and Crosser [72]	$$\frac{k_{\text{eff}}}{k_f} = \frac{k_p + (n-1)k_f + (n-1)\phi_p(k_p - k_f)}{k_p + (n-1)k_f - \phi_p(k_p - k_f)}$$	Based on the effective thermal conductivity of a two-component mixture	Spherical and non-spherical particles. $n = 3$ (spheres), $n = 6$ (cylinders)
Wasp [75]	$$\frac{k_{\text{eff}}}{k_f} = \frac{k_p + 2k_f + 2\phi_p(k_p - k_f)}{k_p + 2k_f - \phi_p(k_p - k_f)}$$	Based on effective thermal conductivity of a two-component mixture	Special case of Hamilton and Crosser's model with $n = 3$
Jeffery [76]	$$\frac{k_{\text{eff}}}{k_f} = 1 + 3\eta\phi_p + \phi_p^2\left(3\eta^2 + \frac{3\eta^2}{4} + \frac{9\eta^3}{16}\frac{\kappa + 2}{2\kappa + 3} + \cdots\right)$$ $$\kappa = \frac{k_p}{k_f},\ \eta = \frac{\kappa - 1}{\kappa + 2}$$	Based on the conduction solution through a stationary random suspension of spheres	High order terms represent pair interactions of randomly dispersed spherical particles. Accurate to order ϕ_p^2

continued

TABLE 2.5　(continued)
Summary of Theoretical Models for the Effective Thermal Conductivity of Nanofluids

Models	Expressions	Physical Model	Remarks
Davis [77]	$\dfrac{k_{\text{eff}}}{k_f} = 1 + \dfrac{3(\kappa-1)}{(\kappa+2)-\phi_p(\kappa-1)}\left[\phi_p + f(\kappa)\phi_p^2 + O(\phi_p^3)\right]$ $\kappa = \dfrac{k_p}{k_f}$	Green's theorem was applied to the space occupied by the matrix material (spherical inclusions) Decaying temperature field was used	Accurate to order ϕ_p^2 High-order terms represent pair interactions of randomly dispersed particles $f(\kappa) = 2.5$ for $\kappa = 10$ $f(\kappa) = 0.5$ for $\kappa = \infty$
Lu and Lin [78]	$\dfrac{k_{\text{eff}}}{k_f} = 1 + a\phi_p + b\phi_p^2$	The effective conductivity of composites containing aligned spheroids of finite conductivity was modeled with the pair interaction The pair interaction was evaluated by solving a boundary value problem involving two aligned spheroids	Spherical and non-spherical particles. Spherical particles: $a = 2.25$, $b = 2.27$ for $\kappa = 10$; $a = 3$, $b = 4.51$ for $\kappa = \infty$

Source: Reprinted from K. Khanafer and K. Vafai. 2011, A critical synthesis of thermophysical characteristics of nanofluids, *International Journal of Heat and Mass Transfer* 54, 4410–4428, Copyright 2011, with permission from Elsevier.

TABLE 2.6
Summary of Theoretical Models for the Effective Thermal Conductivity of Nanofluids (Nano-Layer Effect)

Models	Expressions	Physical Model	Remarks
Yu and Choi [32]	$k_{\text{eff}} = \dfrac{k_{\text{pe}} + 2k_{\text{f}} + 2\phi_{\text{p}}(k_{\text{pe}} - k_{\text{f}})(1 + \beta)^3}{k_{\text{pe}} + 2k_{\text{f}} - \phi_{\text{p}}(k_{\text{pe}} - k_{\text{f}})(1 + \beta)^3} k_{\text{f}}$ $k_{\text{pe}} = \dfrac{2(1 - \gamma) + (1 + \beta)^3(1 + 2\gamma)\gamma}{-(1 - \gamma) + (1 + \beta)^3(1 + 2\gamma)} k_{\text{p}}$ $\beta = t/r_{\text{p}}$ and $\gamma = k_{\text{layer}}/k_{\text{p}}$	Modified Maxwell model [73]	Spherical particles Nano-layer
Yu and Choi [33]	$k_{\text{eff}} = \left(1 + \dfrac{n f_e A}{1 - f_e A}\right) k_{\text{f}}$ $A = \dfrac{1}{3} \displaystyle\sum_{j=a,b,c} \dfrac{(k_{\text{p}j} - k_{\text{f}})}{k_{\text{p}j} + (n - 1)k_{\text{f}}}$ $f_e = \dfrac{f \sqrt{(a^2 + t)(b^2 + t)(c^2 + t)}}{abc}$	Modified Hamilton–Crosser model [72]	Nonspherical particles. Nano-layer
Xue [35]	$9\left(1 - \dfrac{\phi_{\text{p}}}{\lambda}\right)\dfrac{k_{\text{eff}} - k_{\text{f}}}{2k_{\text{eff}} + k_{\text{f}}} + \dfrac{\phi_{\text{p}}}{\lambda}\dfrac{k_{\text{eff}} - k_{c,x}}{k_{\text{eff}} + B_{2,x}(k_{c,x} - k_{\text{eff}})}$ $+ \dfrac{\phi_{\text{p}}}{\lambda}\dfrac{k_{\text{eff}} - k_{c,y}}{4}\dfrac{}{2k_{\text{eff}} + (1 - B_{2,x})(k_{c,y} - k_{\text{eff}})} = 0$	Based on the Maxwell model and the average polarization theory and on the assumption that there is an interfacial shell between the nanoparticles and the base fluid	Spherical particles Nano-layer

continued

TABLE 2.6 (continued)

Summary of Theoretical Models for the Effective Thermal Conductivity of Nanofluids (Nano-Layer Effect)

Models	Expressions	Physical Model	Remarks
Xue and Xu [34]	$$\left(1 - \frac{\phi_p}{\kappa}\right)\frac{k_{eff} - k_f}{2k_{eff} + k_f}$$ $$+ \frac{\phi_p}{\kappa}\frac{(k_{eff} - k_{shell})(2k_{shell} + k_p) - \kappa(k_p - k_{shell})(2k_{shell} + k_{eff})}{(2k_{eff} + k_{shell})(2k_{shell} + k_p) + 2\kappa(k_p - k_{shell})(k_{shell} - k_{eff})} = 0$$	A modified Bruggeman model [74] including the effect of interfacial shells	Spherical particles Nano-layer
Xie et al. [36]	$$\frac{k_{eff} - k_f}{k_f} = 3\Theta\phi_T + \frac{3\Theta^2\phi_T^2}{1 - \Theta\phi_T}$$ $$\phi_T = \frac{4}{3}\pi(r_p + t)^3 N_p = \phi_p(1 + \beta)^3, \; \beta = \frac{t}{r_p}$$	Based on Fourier's law of heat conduction	Low particle loadings Nano-layer

Source: Reprinted from K. Khanafer and K. Vafai. 2011, A critical synthesis of thermophysical characteristics of nanofluids, *International Journal of Heat and Mass Transfer* 54, 4410–4428, Copyright 2011, with permission from Elsevier.

TABLE 2.7

Summary of Theoretical Models for the Effective Thermal Conductivity of Nanofluids (Brownian Effect)

Models	Expressions	Physical Model	Remarks
Wang et al. [19]	$$\frac{k_{\text{eff}}}{k_{\text{f}}} = \frac{(1-\phi_{\text{p}}) + 3\phi_{\text{p}}\int_0^\infty \frac{(k_{cl}(r)n(r)/k_{cl}(r)+2k_{\text{f}})\,dr}{(1-\phi_{\text{p}}) + 3\phi_{\text{p}}\int_0^\infty (k_{\text{f}}(r)n(r)/k_{cl}(r)+2k_{\text{f}})\,dr}}$$	Based on the effective medium approximation and the fractal theory for predicting the thermal conductivity of nanofluids	Accounts for the size effect and the surface adsorption of nanoparticles
Xuan et al. [79]	$$\frac{k_{\text{eff}}}{k_{\text{f}}} = \frac{k_{\text{p}} + 2k_{\text{f}} - 2\phi_{\text{p}}(k_{\text{f}} - k_{\text{p}})}{k_{\text{p}} + 2k_{\text{f}} + \phi_{\text{p}}(k_{\text{f}} - k_{\text{p}})} + \frac{\rho_{\text{p}}\phi_{\text{p}}C_{\text{p}}}{2k_{\text{c}}}\sqrt{\frac{k_BT}{3\pi r_{\text{c}}\mu}}$$	Based on Maxwell model The theory of Brownian motion and the diffusion-limited aggregation model are applied to simulate random motion and the aggregation process of the nanoparticles	Includes the effect of random motion, particle size, concentration, and temperature
Jang and Choi [10]	$$k_{\text{eff}} = k_{\text{f}}(1-\phi_{\text{p}}) + k_{\text{p}}\phi_{\text{p}} + 3C\frac{d_{\text{f}}}{d_{\text{p}}}k_{\text{f}}\,\text{Re}_{d_{\text{p}}}^2\,\text{Pr}\,\phi_{\text{p}}$$	A theoretical model was developed based on kinetics, Kapitza resistance, and convection A general expression for the thermal conductivity of nanofluids involving four modes of energy transport in nanofluids was derived	Considered four modes of energy transport: collision between fluid molecules, thermal diffusion of nanoparticles, collision between nanoparticles due to Brownian motion, and thermal interactions of dynamic nanoparticles with fluid molecules Collision of nanoparticles due to Brownian motion is neglected
Prasher et al. [80]	$$k_{\text{eff}} = (1 + A\,\text{Re}^m\,\text{Pr}^{0.333}\,\phi_{\text{p}}) \times \left[\frac{k_{\text{p}} + 2k_{\text{f}} + 2\phi_{\text{p}}(k_{\text{p}} - k)}{k_{\text{p}} + 2k_{\text{f}} - \phi_{\text{p}}(k_{\text{p}} - k_{\text{f}})}\right]$$	Based on Maxwell model and heat transfer in fluidized beds	Accounts for convection caused by the Brownian motion from multiple nanoparticles

continued

TABLE 2.7 (continued)
Summary of Theoretical Models for the Effective Thermal Conductivity of Nanofluids (Brownian Effect)

Models	Expressions	Physical Model	Remarks
Koo and Kleinstreuer [70,81]	$k_{\mathrm{eff}} = k_{\mathrm{static}} + k_{\mathrm{Brownian}}$ $$= \frac{k_p + 2k_f + 2\phi_p (k_p - k_f)}{k_p + 2k_f - \phi_p (k_p - k_f)} k_f$$ $$+ 5 \times 10^4 \beta \phi_p \rho_f c_{Pf} \sqrt{\frac{k_B T}{\rho_p D}} f(T, \phi_p)$$	Based on Maxwell model Curve fitting of the available experimental data to determine the effective conductivity due to Brownian motion	Considered surrounding liquid traveling with randomly moving nanoparticles
Chon et al. [82]	$$\frac{k_{\mathrm{eff}}}{k_f} = 1 + 64.7 \phi_p^{0.74} \left(\frac{d_f}{d_p} \right)^{0.369} \left(\frac{k_p}{k_f} \right)^{0.747}$$ $$\times \mathrm{Pr}^{0.9955}\, \mathrm{Re}^{1.2321}$$ $$\mathrm{Pr} = \frac{\mu_f}{\rho_f \alpha_f},$$ $$\mathrm{Re} = \frac{\rho_f V_{Br} d_p}{\mu_f} = \frac{\rho_f k_B T}{3\pi \mu_f^2 l_f}$$	Based on the curve fitting of the experimental data	Reynolds number is based on the Brownian motion velocity

Source: Reprinted from K. Khanafer and K. Vafai. 2011, A critical synthesis of thermophysical characteristics of nanofluids, *International Journal of Heat and Mass Transfer* 54, 4410–4428, Copyright 2011, with permission from Elsevier.

2.4.1 EXPERIMENTAL INVESTIGATIONS

Many studies have reported augmentation in the effective thermal conductivity of nanofluids at room temperature. Figures 2.6a and 2.6b illustrate the effective thermal conductivity measurements at ambient temperature for Al_2O_3–water and CuO–water nanofluids at various volume concentrations and nanoparticle diameters. Figure 2.6 shows that the effective thermal conductivity of nanofluids increases with an increase in the volume fraction of nanoparticles. In addition, the size of the particles is found to have a substantial effect on the thermal conductivity improvement. It should be noted that smaller particles exhibit larger surface area-to-volume ratio than the larger particles. As such, smaller particle diameters can possibly result in a

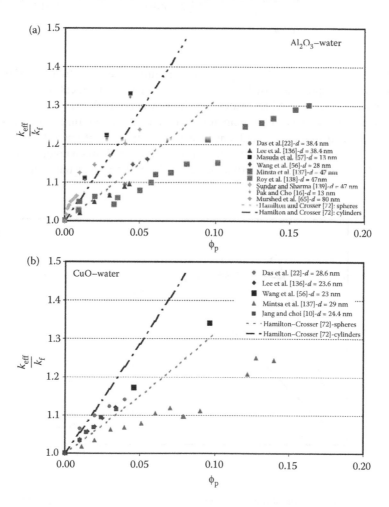

FIGURE 2.6 Effect of the volume fraction on the effective thermal conductivity measurements: (a) Al_2O_3–water; (b) CuO–water. (Reprinted from K. Khanafer and K. Vafai. 2011, A critical synthesis of thermophysical characteristics of nanofluids, *International Journal of Heat and Mass Transfer* 54, 4410–4428, Copyright 2011, with permission from Elsevier.)

larger augmentation in the effective thermal conductivity of nanofluids. It is interesting to note from Figures 2.6a and 2.6b that the Hamilton and Crosser model [72] may represent a good approximation for the effective thermal conductivity value for smaller volume fractions ($\phi_p \leq 4\%$).

A general correlation for the effective thermal conductivity of Al_2O_3–water and CuO–water nanofluids at ambient temperature accounting for various volume fractions and nanoparticles diameters was developed by Khanafer and Vafai [39] using various experimental data. This model was expressed as

$$\frac{k_{eff}}{k_f} = 1.0 + 1.0112\phi_p + 2.4375\phi_p \left(\frac{47}{d_p(nm)}\right) - 0.0248\phi_p \left(\frac{k_p}{0.613}\right); \quad (2.21)$$

$$R^2 = 96.5\%$$

where k_f is the thermal conductivity of water. Figure 2.7 demonstrates that the general correlation, represented by Equation 2.21, is in good agreement with the experimental measurements of Al_2O_3–water and CuO–water nanofluids.

Thermal conductivity measurements at different temperatures are important because the measurements at ambient temperature are not sufficient for estimating the heat transfer characteristics of nanofluids. Figure 2.8 shows a comparison of the relative effective thermal conductivity (ratio of the effective thermal conductivity of

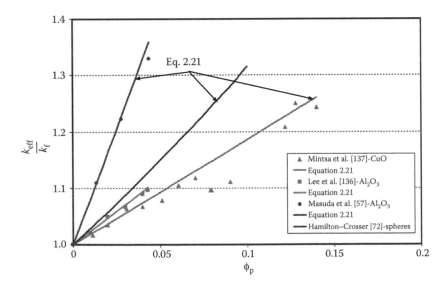

FIGURE 2.7 Comparison of the general correlation, Equation 2.21 developed by Khanafer and Vafai [39] with the experimental data (Al_2O_3–water, CuO–water) at room temperature . (Reprinted from K. Khanafer and K. Vafai. 2011, A critical synthesis of thermophysical characteristics of nanofluids, *International Journal of Heat and Mass Transfer* 54, 4410–4428, Copyright 2011, with permission from Elsevier.)

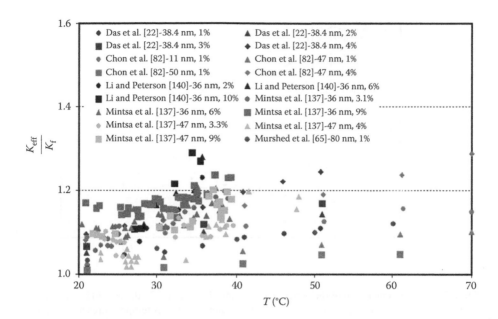

FIGURE 2.8 Comparison of the experimental data for the thermal conductivity enhancement of Al$_2$O$_3$–water nanofluid at different temperatures and volume fractions. (Reprinted from K. Khanafer and K. Vafai. 2011, A critical synthesis of thermophysical characteristics of nanofluids, *International Journal of Heat and Mass Transfer* 54, 4410–4428, Copyright 2011, with permission from Elsevier.)

the nanofluid to the thermal conductivity of the base fluid at the same temperature) results of Al$_2$O$_3$–water nanofluid obtained from various experimental results as a function of volume fraction and nanoparticle's diameter. Figure 2.8 illustrates that temperature has an important effect on the thermal conductivity augmentation.

A general correlation was developed by Khanafer and Vafai [39] for Al$_2$O$_3$–water nanofluid using the available experimental data at various temperatures, nanoparticle's diameter, and volume fraction. The developed correlation was given in terms of nanoparticle's diameter, volume fraction, dynamic viscosity of water, effective dynamic viscosity of the nanofluid, and temperature as follows:

$$\frac{k_{eff}}{k_f} = 0.9843 + 0.398\phi_p^{0.7383}\left(\frac{1}{d_p(nm)}\right)^{0.2246}\left(\frac{\mu_{eff}(T)}{\mu_f(T)}\right)^{0.0235}$$

$$- 3.9517\frac{\phi_p}{T} + 34.034\frac{\phi_p^2}{T^3} + 32.509\frac{\phi_p}{T^2} \tag{2.22}$$

$$0 \leq \phi_p \leq 10\%, 11\,nm \leq d \leq 150\,nm, 20\,°C \leq T \leq 70\,°C$$

where the dynamic viscosity (Pa.s) of water at different temperatures can be expressed as

$$\mu_f(T) = 2.414 \times 10^{-5} \times 10^{247.8/(T-140)} \tag{2.23}$$

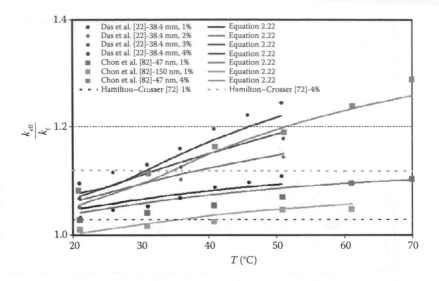

FIGURE 2.9 Comparison of the general correlation, Equation 2.22 developed by Khanafer and Vafai [39] with the experimental data (Al$_2$O$_3$–water) at various temperatures and volume fractions. (Reprinted from K. Khanafer and K. Vafai. 2011, A critical synthesis of thermophysical characteristics of nanofluids, *International Journal of Heat and Mass Transfer* 54, 4410–4428, Copyright 2011, with permission from Elsevier.)

where T in Kelvin. Figure 2.9 shows a very good agreement between the predicted relative effective thermal conductivity by Khanafer and Vafai [39] model and the experimental data.

Different models were developed in the literature for the effective thermal conductivity of a two-component mixture such as the Hamilton and Crosser model [72]. Although this model gave a good approximation for the effective thermal conductivity of the Al$_2$O$_3$–water and CuO–water nanofluids for small volume fractions at room temperature, it does not exhibit a good approximation of the effective thermal conductivity at various temperatures as depicted in Figure 2.9 because this model [72] as well as a number of other models in this area do not properly account for the variations of the effective thermal conductivity with temperature. Therefore, these analytical models cannot be used to estimate the effective thermal conductivity of nanofluids at various temperatures. Instead, Equation 2.22 developed by Khanafer and Vafai [39] may be used to give a better estimation of the effective thermal conductivity of Al$_2$O$_3$–water nanofluids at various temperatures.

2.5   NUCLEATE POOL BOILING AND CRITICAL HEAT FLUX OF NANOFLUIDS

Boiling heat transfer plays a significant role in a variety of technological and industrial applications such as heat exchangers, microchannel-cooling applications, cooling of high-power electronics and nuclear reactors. The use of nanofluids in enhancing boiling heat-transfer characteristics is of great interest [83,84,42]. Several experimental studies on the nucleate pool boiling and critical heat flux (CHF) characteristics of nanofluids have been conducted in the literature [18,85–96]. Conflicting results on the effect of nanoparticles on the nucleate boiling heat-transfer rate and CHF were reported. For example, Das et al. [21,22] conducted an experimental study on pool boiling characteristics of Al_2O_3–water nanofluids on smoother and roughened heating surfaces for various particle concentrations. Their results show that nanoparticles degraded the boiling performance with increasing particle concentration. You et al. [86] found that nucleate boiling heat-transfer coefficient remained unchanged with the addition of Al_2O_3 nanoparticles compared with water. Bang and Chang [87] experimentally studied boiling heat transfer characteristics of nanofluids on a smooth horizontal flat surface with nanoparticles suspended in water using different volume concentrations of Al_2O_3 nanoparticles. Their experimental results showed that nanofluids have poor heat-transfer performance compared with pure water in natural convection and nucleate boiling. Contrary to the above results, an experimental investigation into the pool boiling heat transfer of aqueous based γ-alumina nanofluids (primary particle size 10–50 nm) was carried out by Wen and Ding [18]. The results showed that alumina nanofluids can significantly enhance boiling heat transfer. The enhancement was shown to increase with increasing particle concentration up to approximatley 40% at a particle loading of 1.25% by weight. Ding et al. [27] showed that the boiling heat transfer was enhanced in the nucleate regime for both alumina and titania (TiO_2) nanofluids, and the enhancement is more sensitive to the concentration change for TiO_2 nanofluids.

Most CHF experimental studies using nanofluids have shown CHF enhancement under pool boiling conditions [86,87,91,92]. You et al. [86] investigated experimentally the effect of Al_2O_3 nanoparticles (tested concentrations of nanoparticles range from 0 to 0.05 g/L) on CHF of water in pool boiling. The measured pool boiling curves of nanofluids saturated at 60°C have demonstrated that the CHF increases dramatically (approximately 200%) compared with that of pure water. Kim et al. [91] conducted an experimental study on the CHF characteristics of nanofluids in pool boiling. Their results illustrated that the CHF of nanofluids containing TiO_2 or Al_2O_3 were enhanced up to 100% over that of pure water. Vassallo et al. [93] experimentally demonstrated a marked increase in the CHF (up to 60%) for both nano- and micro-solutions (silica–water) at the same concentration (0.5% volume fraction) compared with the base water. Bang and Chang [87] show that CHF performance using Al_2O_3–water nanofluids was enhanced to 32% and 13%, respectively, for both horizontal and vertical flat surfaces in the pool. They related the enhancement in CHF to the change of surface characteristics by the deposition of nanoparticles. Milanova and Kumar [97] conducted an experimental study to measure heat transfer characteristics of silica nanofluids at different acidity and base for various ionic concentrations in a pool boiling condition. They showed that nano-silioca suspensions increased the CHF by 200% times compared to when only pure water is utilized. In addition, they reported that nanofluids in a strong electrolyte exhibit a higher CHF than in buffer solutions because of the difference in surface areas. Figure 2.10 demonstrates a comparison of CHF enhancements between experimental results for various volume concentrations and nanoparticle material and diameter. Table 2.8 gives a summary of research studies on nucleate pool boiling heat transfer coefficient and CHF of nanofluids.

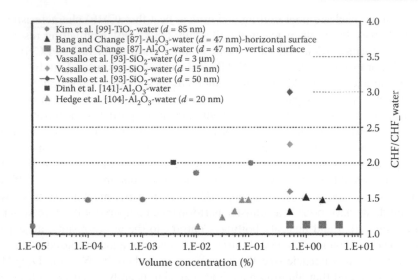

FIGURE 2.10 Comparison of CHF enhancements between experimental results for various volume concentrations, nanoparticles materials, and nanoparticles diameter.

2.5.1 NUCLEATE POOL BOILING HEAT TRANSFER AND CHF ENHANCEMENT MECHANISMS OF NANOFLUIDS

A number of investigations have been carried out to explore the augmentation mechanisms or deterioration of nucleate pool boiling heat-transfer coefficient using nanofluids. These mechanisms include development of nanoparticles coatings on the surface during pool boiling of nanofluids [87], decrease in active nucleation sites due to nanoparticle sedimentation on the boiling surface [103], and the wettability change of the surface [21,22]. The presented experimental results on nucleate pool boiling heat-transfer coefficient of nanofluids are in disagreement. Although the CHF enhancement results by nanofluids are consistent in the literature, the responsible mechanisms are not well established. For example, Golubovic et al. [101] concluded that the main reason behind the increase of CHF in pool boiling of nanofluids is a decrease in the static surface contact angle.

Many other studies consider the major reason for CHF augmentation is due to the surface coating effect [87,88,93,95,96,104,105]. For example, Bang and Chang [87] carried out an experimental study on boiling heat transfer characteristics of nanofluids with nanoparticles suspended in water using different concentrations of alumina nanoparticles (Al_2O_3). The CHF performance was improved for both horizontal (32%) and vertical (13%) flat surfaces and the authors associated this augmentation to a change of surface characteristics by the deposition of nanoparticles. If this reasoning is accepted, it might be easier to alter the boiling surface in pursuit of a greater number of nucleation sites per area rather than using nanofluids [106,107]. Anderson and Mudawar [106] demonstrated that the surfaces with microgrooves and square microstuds are highly effective in improving the nucleate boiling heat-transfer coefficient in Fluorinert electronic liquid (FC-72) resulting and increase in CHF values

TABLE 2.8

Summary of Research Studies on Nucleate Pool Boiling Heat Transfer Coefficient (BHT) and CHF of Nanofluids

Reference	Nanofluids	Remarks
Das et al. [21,22]	Al_2O_3–water	BHT degradation
Chopkar et al. [98]	ZrO_2–water	BHT enhancement at low-volume fraction of nanoparticles (<0.07%)
		BHT degradation (>0.07%)
You et al. [86]	Al_2O_3–water	No change in BHT coefficient
		CHF enhancement up to 200%
Bang and Chang [87]	Al_2O_3–water	BHT degradation
		CHF enhancement up to 32%
Wen and Ding [89]	γ-Al_2O_3–water	BHT enhancement up to 40%
Liu et al. [90]	Carbon nanotube, deionized water	Both BHT and CHF enhancement
		Decrease in pressure, increase in BHT and CHF enhancement
Ding et al. [27]	Al_2O_3–water TiO_2–water	BHT enhancement for both TiO_2 and Al_2O_3
Kim et al. [91]	TiO_2–water Al_2O_3–water	CHF enhancement up to 100%
Kim et al. [99]	TiO_2–water	CHF enhancement up to 200%
Vassallo et al. [93]	SiO_2–water	No change in BHT coefficient
		CHF enhancement up to 60%
Milanova and Kumar [97]	SiO_2–water (also in salt and strong electrolyte solution)	CHF enhancement: three times greater than pure water
Milanova and Kumar [100]	SiO_2–water	CHF enhancement: 50% with no nanoparticle deposition on wire
Golubovic et al. [101]	Al_2O_3–water, Bismuth oxide (Bi_2O_3)–water	CHF enhancement: up to 50% for Al_2O_3 and 33% for Bi_2O_3
Kwark et al. [102]	Al_2O_3–water, CuO–water, and diamond–water	BHT degradation
		CHF enhancement: increases with nanoparticles concentration until reaches an asymptotic value

by up to 2.5 times compared with a smooth surface. Honda et al. [108] and Wei et al. [109] illustrated that CHF values for the nano-roughened surface and micro-pin-finned surfaces were, respectively, 1.8 to 2.2 and 2.3 times those for a smooth silicon surface. Ujereh et al. [110] conducted experiments to evaluate the impact of coating silicon and copper substrates with nanotubes on pool boiling characteristics. Fully coating the substrate surface with carbon nanotubes was found to be highly effective at reducing the incipience superheat and significantly enhancing both the nucleate boiling heat-transfer coefficient and CHF.

More robust physical models are necessary to elucidate the influence of nanofluids on nucleate pool boiling and CHF. Detailed understanding of the thermophysical properties of nanofluids, coating of nanoparticles, and structure of the boiling surface can be helpful in resolving the controversies in the pool boiling heat-transfer coefficient of nanofluids as well as in illustrating the mechanisms that results in a substantial increase in CHF.

2.5.2 NUCLEATE POOL BOILING HEAT TRANSFER AND CHF CORRELATIONS

A number of studies in the literature have presented correlations in the absence of nanoparticles to explain the causes of CHF increase. Zuber's correlation [111], which was largely utilized to predict CHF for an infinite flat plate in the absence of nanoparticles is given by

$$q''_{CHF,Zuber} = 0.131 h_{fg} \rho_g^{1/2} \left[g\sigma \left(\rho_f - \rho_g \right) \right]^{1/4} \tag{2.24}$$

where σ is the surface tension, ρ_f and ρ_g are the liquid and vapor densities respectively, and h_{fg} is the latent heat. According to the above correlation, densities of liquid and vapor, surface tension, and heat of vaporization may affect CHF values. Later Lienhard and Dhir [112] modified Zuber's correlation to account for both size and geometrical effects. They provided hydrodynamic predictions of CHF from different finite bodies

$$q''_{CHF} = 0.149 h_{fg} \rho_g^{1/2} \left[g\sigma \left(\rho_f - \rho_g \right) \right]^{1/4} \tag{2.25}$$

This correlation shows that CHF is proportional to surface tension $\sigma^{1/4}$. This effect is rather weak. Kandlikar [113] extended Zuber's correlation to include the effect of contact angle (β) as follows:

$$\frac{q''_{CHF}}{h_{fg} \rho_g^{1/2} \left[g\sigma \left(\rho_f - \rho_g \right) \right]^{1/4}} = \left(\frac{1 + \cos\beta}{16} \right) \left[\frac{\pi}{4} (1 + \cos\beta) + \frac{2}{\pi} \right]^{1/2} \tag{2.26}$$

There are many studies reported in the literature associated with the effects of heating surface conditions on the pool boiling CHF. Ramilison et al. [114] studied the influence of surface conditions such as roughness and contact angle on CHF. Ramilison et al. [114] suggested the following correlation:

$$\frac{q''_{CHF}}{q''_{CHF,Zuber}} = A \left(\pi - \beta_r \right)^B \left(r \right)^{(C + D\beta_r)} \tag{2.27}$$

where β_r is the receding contact angle and r is the rms value of surface roughness. The above correlation shows that CHF is directly proportional to the surface roughness. Following a dimensional analysis, Kutateladze [115,116] proposed a correlation based on assumption that the critical condition is reached when the velocity in the vapor phase reaches a critical value. This correlation can be expressed as

$$\frac{q''_{CHF}}{h_{fg}\rho_g^{1/2}\left[g\sigma\left(\rho_f - \rho_g\right)\right]^{1/4}} = K \tag{2.28}$$

The value of K was found to be 0.16 from the experimental data.

Borishanskii [117] obtained an analytical expression for the constant K in Kutateladze correlation as

$$K = 0.13 + 4\left\{\frac{\rho_f\sigma^{3/2}}{\mu^2\left[g\left(\rho_f - \rho_g\right)\right]^{1/2}}\right\}^{-0.4} \tag{2.29}$$

Bubble crowding at heated surface was proposed by Rosenhow and Griffith [118]. They assumed that increased packing of the heating surface with bubbles at higher heat fluxes is responsible for stopping the flow of liquid to the heating surface leading to CHF. They proposed the following equation for CHF

$$\frac{q''_{CHF}}{h_{fg}\rho_g} = C_1\left(\frac{g}{g_s}\right)^{1/4}\left[\frac{\rho_f - \rho_g}{\rho_g}\right]^{0.6} \tag{2.30}$$

where $C_1 = 0.012$ m/s, g is the local gravitational acceleration, and g_s corresponds to the standard g value. Chang [119] considered the forces acting on the bubble and assumed that the CHF condition was achieved when the Weber number reached a critical value. The following correlation was developed by Chang [119] for vertical surfaces

$$q''_{CHF} = 0.098h_{fg}\rho_g^{1/2}\left[g\sigma\left(\rho_f - \rho_g\right)\right]^{1/4} \tag{2.31}$$

One can note from above correlations that the CHF depends on the physical properties such as surface tension, liquid and vapor densities, and viscosity as well as bubble dynamics and nucleation density site. Furthermore, structure of boiling surface and thermophysical properties of nanofluids may also affect nucleate boiling heat transfer and CHF.

2.6 MEDICAL APPLICATIONS OF NANOPARTICLES

The applications of nanoscience and nanotechnology in medicine, especially in diagnosis and treatment of diseases, have received considerable attention by many researchers and pharmaceutical companies [120]. One of the applications includes

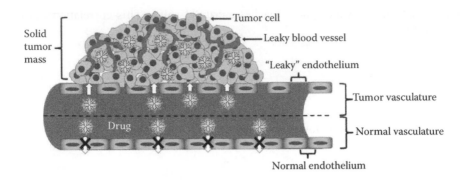

FIGURE 2.11 Diffusion of dendrimes-based drug-delivery systems across the tumor's leaky vasculature into the tumor tissue. (Reprinted with permission from S.H. Madina and M.E. El-Sayed. 2009, Dendrimers as carriers for delivery of chemotherapeutic agents, *Chem Rev.* 109, 3141–3157, Copyright 2009, with permission from ACS Publications.)

the use of nanoparticles (1–100 nm) in drug and gene delivery [121,122], detection of proteins [123], probing of DNA structure [124], tissue engineering [125], magnetic resonance imaging contrast enhancement [126], and tumor destruction via hyperthermia [127]. Nanoparticles have distinctive physicochemical properties such as ultra small size, large surface to mass ratio, high reactivity, and unique interactions with biological systems [128]. In drug-delivery applications, controlled-released drugs delivered to the site of action at a designed rate have numerous advantages over the conventional dosage forms. This interest stems from its importance in reducing dosing frequency, adverse side effects, and in achieving improved pharmacological activity as well as in maintaining constant and prolonged therapeutic effects [129,130]. Nanoparticles are engineered to bind to target cells and deliver high doses of therapeutic compounds which results in reducing damage to healthy cells in the body [122] (Figure 2.11). Nanoparticles used as drug delivery systems are made using a variety of materials such as polymers (polymeric nanoparticles, micelles, or dendrimers), lipids, viruses, and organometallic compound (nanotubes) [131]. Figure 2.12 shows a schematic diagram showing the composition of liposomes, dendrimers,

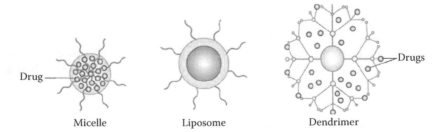

FIGURE 2.12 Schematic diagram of nanoparticles used as drug delivery systems. (Reprinted with kind permission from Springer: ElHazzat Jallal and E.H. El-Sayed Mohamed. 2010, Advances in targeted breast cancer, *Current Breast Cancer Reports* 2, 146–151.)

and polymeric micelles used for delivery of chemotherapeutic agents for treatment of breast cancer [132].

Another application of nanoparticles can be found in imaging. Recent advances in nanoparticle technology have led to the implementation of nanoparticles such as TiO_2, quantum dots, and gold nanoparticles in cellular imaging due to their distinctive properties compared with traditional fluorescent dyes and proteins. For instance, the smaller size and improved photostability of quantum dots allow for prolonged and enhanced visualization of biological components in fixed cells [133].

2.7 CONCLUSIONS

Thermophysical properties of nanofluids and their importance in biomedical applications and heat-transfer enhancement are discussed in this study. General correlations for the effective thermal conductivity and viscosity of nanofluids are developed in this study based on the experimental data in terms of various pertinent parameters. The experimental data reported by many authors for the effective thermal conductivity and dynamic viscosity of nanofluids are in disagreement. This study has illustrated that the results of the effective thermal conductivity and viscosity of nanofluids can be determined at room temperature using the classical equations at low-volume fractions. However, these models cannot predict the thermal conductivity at other temperatures. Moreover, for high heat-flux applications, the experimental results reported in the literature showed contradictory results in pool boiling heat-transfer characteristics while the CHF of nanofluids demonstrates a significant increase with the addition of nanoparticles. This review summarizes some of the potential applications of nanoparticles in biomedical applications related to cancer treatment, medical imaging, and drug-delivery systems.

NOMENCLATURE

c_{eff}	heat capacity of nanofluids
c_f	heat capacity of the base fluid
c_p	heat capacity of nanoparticles
C	proportional constant
d_f	diameter of the fluid molecule
d_p	nanoparticles diameter
h	inter-particle spacing
h_{fg}	latent heat
k	thermal conductivity
k_B	Stefan–Boltzmann constant
k_{eff}	thermal conductivity of nanofluids
k_H	Huggins Coefficient
k_{layer}	thermal conductivity of the nano-layer
m	mass
n	empirical shape factor
Pr	Prandtl number
Re	Reynolds number

r rms value of surface roughness
t thickness of the nano-layer
T temperature
V volume

GREEK SYMBOLS

β ratio of nano-layer thickness to radius of nanoparticle
β_{eff} thermal expansion coefficient of nanofluids
β_f thermal expansion coefficient of the base fluid
β_p thermal expansion coefficient of nanoparticle
β_r receding contact angle
ρ density
ρ_{eff} density of nanofluids
γ ratio of nano-layer thermal conductivity to nanoparticle thermal conductivity
κ ratio of nanoparticle thermal conductivity to fluid thermal conductivity
ψ particle sphericity
ϕ_p volume fraction of nanoparticles
$\phi_{p,max}$ maximum volume fraction of nanoparticles
μ_{eff} dynamic viscosity of nanofluids
μ_f dynamic viscosity of the base fluid
$\mu_{Brownian}$ dynamic viscosity due to Brownian motion
σ surface tension

SUBSCRIPTS

f   fluid
p   nanoparticle

REFERENCES

1. S.U.S. Choi. Nanofluids: From vision to reality through research, *Journal of Heat Transfer* 131, 2009, 1–9.
2. K.V. Wong and O. Leon. Applications of nanofluids: Current and future, *Advances in Mechanical Engineering.* 2010, 1–11.
3. V. Bianco, F. Chiacchio, O. Manca, and S. Nardini. Numerical investigation of nanofluids forced convection in circular tubes, *Applied Thermal Engineering* 29, 2009, 3632–3642.
4. M. Shafahi, V. Bianco, K. Vafai, and O. Manca. Thermal performance of flat-shaped heat pipes using nanofluids, *International Journal of Heat and Mass Transfer* 53, 2010, 1438–1445.
5. M. Shafahi, V. Bianco, K. Vafai, and O. Manca. An investigation of the thermal performance of cylindrical heat pipes using nanofluids, *International Journal of Heat and Mass Transfer* 53, 2010, 376–383.
6. K. Khanafer, K., Vafai, and M. Lightstone. Buoyancy-driven heat transfer enhancement in a two-dimensional enclosure utilizing nanofluids, *International Journal of Heat and Mass Transfer* 46, 2003, 3639–3653.
7. A.R.A. Khaled and K. Vafai, Heat transfer enhancement through control of thermal dispersion effects, *International Journal of Heat and Mass Transfer* 48, 2005, 2172.

8. J.A. Eastman, S.U.S. Choi, S. Li, L.J. Thompson, and S. Lee. Enhanced thermal conductivity through the development of nanofluids. In: 1996 Fall meeting of the Materials Research Society (MRS), Boston, USA.

9. J.A. Eastman, S.U.S. Choi, S. Li, W. Yu, and L.J. Thompson. Anomalously increased effective thermal conductivities of ethylene glycol-based nanofluids containing copper nanoparticles, *Applied Physics Letters* 78, 2001, 718–720.

10. S.P. Jang and S.U.S. Choi. Role of Brownian motion in the enhanced thermal conductivity of nanofluids, *Applied Physics Letters* 84, 2004, 4316–4318.

11. S. Lee and S.U.S. Choi. Application of metallic nanoparticle suspensions in advanced cooling systems. In: 1996 International Mechanical Engineering Congress and Exhibition, Atlanta, USA.

12. A. Ali, K. Vafai, and A.-R.A. Khaled. Comparative study between parallel and counter flow configurations between air and falling film desiccant in the presence of nanoparticle suspensions, *International Journal of Energy Research* 27, 2003, 725–745.

13. N. Putra, W. Roetzel, and S.K. Das. Natural convection of nanofluids. *Heat and Mass Transfer* 39, 2003, 775–784.

14. Y.M. Xuan and Q. Li. Heat transfer enhancement of nanofluids, *International Journal of Heat and Fluid Flow* 21, 2000, 58–64.

15. Y.M. Xuan and Q. Li. Investigation on convective heat transfer and flow features of nanofluids, *ASME Journal of Heat Transfer* 125, 2003, 151–155.

16. B.C. Pak and Y.I. Cho. Hydrodynamic and heat transfer study of dispersed fluids with submicron metallic oxide particles, *Experimental Heat Transfer* 11, 1999, 151–170.

17. Y. Yang, Z. Zhang, E. Grulke, W. Anderson, and G. Wu, G. Heat transfer properties of nanoparticle-in-fluid dispersions (nanofluids) in laminar flow, *International Journal of Heat and Mass Transfer* 48, 2005, 1107–1116.

18. D.S. Wen and Y.L. Ding. Experimental investigation into the pool boiling heat transfer of aqueous based alumina nanofluids, *Journal of Nanoparticle Research* 7, 2005, 265–274.

19. B.X. Wang, L.P. Zhou, and X.F. Peng. A fractal model for predicting the effective thermal conductivity of liquid with suspension of nanoparticles, *International Journal of Heat and Mass Transfer* 46, 2003, 2665–2672.

20. S.K. Das, N. Putra, and W. Roetzel. Pool boiling characteristics of nanofluids, *International Journal of Heat and Mass Transfer* 46, 2003a, 851–862.

21. S.K. Das, N. Putra, and W. Roetzel. Pool boiling of nanofluids on horizontal narrow tubes, *International Journal of Multiphase Flow* 29, 2003b, 1237–1247.

22. S.K. Das, N. Putra, P. Thiesen, and W. Roetzel, Temperature dependence of thermal conductivity enhancement for nanofluids, *Journal Heat Transfer* 125, 2003, 567–574.

23. J. Kim, Y.T. Kang, and C.K. Choi. Analysis of convective instability and heat transfer characteristics of nanofluids, *Physics of Fluids* 16, 2004, 2395–2401.

24. B. Ghasemi and S.M. Aminossadati. Natural convection heat transfer in an inclined enclosure filled with a water-CuO nanofluid, *Numerical Heat Transfer, Part A* 55, 2009, 807–823.

25. A.G.A. Nnanna, T. Fistrovich, K. Malinski, and S.U.S. Choi. Thermal transport phenomena in buoyancy-driven nanofluids, in *Proceedings of 2005 ASME International Mechanical Engineering Congress and RD&D Exposition*, November 15–17, 2004, Anaheim, CA, USA.

26. A.G.A. Nnanna and M. Routhu. Transport phenomena in buoyancy-driven nanofluids— Part II, *in Proceedings of 2005 ASME Summer Heat Transfer Conference*, July 17–22, 2005, San Francisco, California, USA.

27. Y. Ding, H. Chen, L. Wang et al., Heat transfer intensification using nanofluids, *Journal of Particle and Powder* 25, 2007, 23–36.

28. B.H. Chang, A.F. Mills, and E. Hernandez. Natural convection of micro-particle suspensions in thin enclosures, *International Journal of Heat and Mass Transfer* 51, 2008, 1332–1341.
29. P. Keblinski, S.R. Phillpot, S.U.S. Choi, and J.A. Eastman. Mechanisms of heat flow in suspensions of nano-sized particles (nanofluids), *International Journal of Heat and Mass Transfer* 45, 2002, 855–863.
30. J.A. Eastman, S.R. Phillpot, S.U.S. Choi, and P. Keblinski. Thermal transport in nanofluids, *Annual Review Of Materials Research* 34, 2004, 219–246.
31. W. Evans, J. Fish, and P. Keblinski. Role of Brownian motion hydrodynamics on nanofluid thermal conductivity, *Applied Physics Letters* 88(9), 2006, 93116.
32. W. Yu and S.U.S. Choi. The role of interfacial layers in the enhanced thermal of nanofluids: a renovated Maxwell model, *Journal of Nanoparticle Research* 5 (1–2), 2003, 167–171.
33. W. Yu and S.U.S. Choi, The role of interfacial layers in the enhanced thermal conductivity of nanofluids: A renovated Hamilton–Crosser model, *Journal of Nanoparticle Research* 6(4), 2004, 355–361.
34. Q. Xue and W.M. Xu. A model of thermal conductivity of nanofluids with interfacial shells, *Materials Chemistry and Physics* 90, 2005, 298–301.
35. Q. Xue. Model for effective thermal conductivity of nanofluids, *Physics Letters A* 307, 2003, 313–317.
36. H. Xie, M. Fujii, and X. Zhang. Effect of interfacial nanolayer on the effective thermal conductivity of nanoparticle-fluid mixture, *International Journal of Heat and Mass Transfer* 48, 2005, 2926–2932.
37. L. Xue, P. Keblinski, S.R. Phillpot, S.U.S. Choi, and J.A. Eastman. Effect of liquid layering at the liquid–solid interface on thermal transport, *Int. J. Heat Mass Transfer* 47(19–20), 2004, 4277–4284.
38. C.J. Ho, W.K. Liu, Y.S. Chang, and C.C. Lin. Natural convection heat transfer of alumina-water nanofluid in vertical square enclosures: An experimental study, *International Journal of Thermal Sciences* 49, 2010, 1345–1353.
39. K. Khanafer and K. Vafai. A critical synthesis of thermophysical characteristics of nanofluids, *International Journal of Heat and Mass Transfer* 54, 2011, 4410–4428.
40. S.P. Jang and S.U. Choi. Free convection in a rectangular cavity (Benard Convection) with nanofluids, *Proceedings of the 2004 ASME International Mechanical Engineering Congress and Exposition*, Anaheim, California, November 13–20.
41. L. Gosselin and A.K. da Silva. Combined heat transfer and power dissipation optimization of nanofluid flows, *Applied Physics Letters* 85, 2004, 4160.
42. J. Lee and I. Mudawar. Assessment of the effectiveness of nanofluids for single-phase and two-phase heat transfer in micro-channels, *International Journal of Heat and Mass Transfer* 50, 2007, 452–463.
43. S.Q. Zhou and R. Ni. Measurement of the specific heat capacity of water-based Al_2O_3 nanofluid, *Applied Physics Letters* 92, 2008, 093123.
44. K.S. Hwang, J.H. Lee, and S.P. Jang, Buoyancy-driven heat transfer of water-based Al2O3 nanofluids in a rectangular cavity, *International Journal of Heat and Mass Transfer* 50, 2007, 4003–4010.
45. C.J. Ho, M.W. Chen, and Z.W. Li. Numerical simulation of natural convection of nanofluid in a square enclosure: Effects due to uncertainties of viscosity and thermal conductivity, *International Journal of Heat and Mass Transfer* 51, 2008, 4506–4516.
46. A. Einstein. Eine neue bestimmung der molekuldimensionen, *Annalen der Physik, Leipzig*, 19, 1906, 289–306.
47. H.C. Brinkman. The viscosity of concentrated suspensions and solutions. *J. Chem Phys* 20, 1952, 571.

48. G. Batchelor. The effect of Brownian motion on the bulk stress in a suspension of spherical particles, *J. Fluid Mechanics* 83, 1977, 97–117.
49. T. Lundgren. Slow flow through stationary random beds and suspensions of spheres, *J. Fluid Mechanics* 51, 1972, 273–299.
50. A.L. Graham. On the viscosity of suspensions of solid spheres, *Appl. Sci. Res.* 37, 1981, 275–286.
51. R.A. Simha. Treatment of the viscosity of concentrated suspensions, *Journal of Applied Physics* 23, 1952, 1020–1024.
52. M. Mooney. The viscosity of a concentrated suspension of spherical particles, *Journal of Colloid Science* 6, 1951, 162–170.
53. V.H. Eilers. Die viskocitat von emulsionen hochviskoser stoffe als funktion der konzentration, *Kolloid-Zeitschrift* 97, 1941, 313–321.
54. N. Saito. Concentration dependence of the viscosity of high polymer solutions, *Journal of Physical Society of Japan* 5, 1950, 4–8.
55. N.A. Frankel and A. Acrivos. On the viscosity of a concentrate suspension of solid spheres, *Chemical Engineering Science* 22, 1967, 847–853.
56. X. Wang, X. Xu, and S.U.S. Choi. Thermal conductivity of nanoparticles–fluid mixture, *Journal of Thermophysics Heat Transfer* 13, 1999, 474–480.
57. H. Masuda, A. Ebata, K. Teramae, and N. Hishinuma. Alteration of thermal conductivity and viscosity of liquid by dispersing ultra-fine particles (dispersion of c-Al₂O₃, SiO₂ and TiO₂ ultra-fine particles), *Netsu Bussei* 4, 1993, 227–233.
58. S. Maiga, S.J. Palm, C.T. Nguyen, G. Roy and N. Galanis. Heat transfer enhancement by using nanofluids in forced convection flows, *International Journal of Heat and Fluid Flow* 26, 2005, 530–546.
59. S. Maiga, C.T. Nguyen, N. Galanis, G. Roy, T. Mar'e, and M. Coqueux. Heat transfer enhancement in turbulent tube flow using Al2O3 nanoparticle suspension. In: Lewis, R.W. (Eds.), *International Journal of Numerical Methods Heat and Fluid Flow* 16, 2006, 275–292.
60. P.K. Namburua, D.P. Kulkarni, D. Misra, and D.K. Das. Viscosity of copper oxide nanoparticles dispersed in ethylene glycol and water mixture, *Experimental Thermal Fluid Science* 32, 2007, 397–402.
61. C.T. Nguyen, F. Desgranges, G. Roy, N. Galanis, T. Mare, S. Boucher, and H.A. Mintsa. Temperature and particle-size dependent viscosity data for water-based nanofluids–hysteresis phenomenon, *International Journal of Heat and Fluid Flow* 28, 2007, 1492–1506.
62. W.J. Tseng and K.C. Lin, Rheology and colloidal structure of aqueous TiO2 nanoparticle suspensions, *Materials Science and Engineering* A355, 2003, 186–192.
63. D.P. Kulkarni, D.K. Das, and S.L. Patil. Effect of temperature on rheological properties of copper oxide nanoparticles dispersed in propylene glycol and water mixture, *Journal of Nanosciense and Nanotechnology* 7, 2007, 2318–2322.
64. D.P. Kulkarni, D.K. Das, and G. Chukwa. Temperature dependent rheological of copper oxide nanoparticles suspension (nanofluid), *Journal of Nanosciense and Nanotechnology* 6, 2006, 1150–1154.
65. S.M.S. Murshed, K.C. Leong, and C. Yang. Investigations of thermal conductivity and viscosity of nanofluids, *International Journal of Thermal Sciences* 47, 2008, 560–568.
66. J. Buongiorno. Convective transport in nanofluids ASME, *Journal of Heat Transfer* 128, 2006, 240–250.
67. S.J. Palm, G. Roy, and C.T. Nguyen, Heat transfer enhancement with the use of nanofluids in radial flow cooling systems considering temperature-dependent properties, *Applied Thermal Engineering* 26, 2006, 2209–2218.

68. W. Duangthongsuk and S. Wongwises. Measurement of temperature-dependent thermal conductivity and viscosity of TiO_2-water nanofluids, *Experimental Thermal and Fluid Science* 33, 2009, 706–714.

69. P. K. Namburu, D.K. Das, K.M. Tanguturi, and R.S. Vajjha. Numerical study of turbulent flow and heat transfer characteristics of nanofluids considering variable properties, *International Journal of Thermal Sciences* 48, 2009, 290–302.

70. J. Koo and C. Kleinstreuer, Laminar nanofluid flow in microheat-sinks, *International Journal of Heat and Mass Transfer* 48, 2005, 2652–2661.

71. K.B. Anoop, S. Kabelac, T. Sundararajan, and S.K. Das. Rheological and flow characteristics of nanofluids: Influence of electroviscous effects and particle agglomeration, *Journal of Applied Physics* 106, 2009, 034909.

72. R.L. Hamilton and O.K. Crosser. Thermal conductivity of heterogeneous two-component systems, *I&EC Fundam* 1, 1962, 182–191.

73. J.C.A. Maxwell *Treatise on Electricity and Magnetism*. Clarendon Press, Oxford, UK, 2nd edition, 1881.

74. D.A.G. Bruggeman. Berechnung verschiedener physikalischer konstanten von heterogenen substanzen, I. Dielektrizitatskonstanten und leitfahigkeiten der mischkorper aus isotropen substanzen, *Annalen der Physik, Leipzig*, 24, 1935, 636–679.

75. F.J. Wasp. *Solid-Liquid Slurry Pipeline Transportation*, 1977, Trans. Tech., Berlin, Germany.

76. D.J. Jeffrey. Conduction through a random suspension of spheres, *Proceedings of Royal Society (London)* A335, 1973, 355–367.

77. R.H. Davis. The effective thermal conductivity of a composite material with spherical inclusions, *International Journal of Thermophysics* 7, 1986, 609–620.

78. S. Lu and H. Lin. Effective conductivity of composites containing aligned spherical inclusions of finite conductivity, *Journal of Applied Physics* 79, 1996, 6761–6769.

79. Y. Xuan, Q. Li, and W. Hu. Aggregation structure and thermal conductivity of nanofluids, *AIChE Journal* 49(4). 2003, 1038–1043.

80. R. Prasher, P. Bhattacharya, and P.E. Phelan. Thermal conductivity of nanoscale colloidal solutions (nanofluids), *Physical Review Letters* 94(2), 2005, 025901.

81. J. Koo, and C. Kleinstreuer. A new thermal conductivity model for nanofluids, *Journal of Nanoparticle Research* 6(6). 2004, 577–588.

82. C.H. Chon, K.D. Kihm, S.P. Lee, and S.U.S. Choi, Empirical correlation finding the role of temperature and particle size for nanofluid (Al_2O_3) thermal conductivity enhancement, *Applied Physics Letters* 87, 2005, 153107.

83. R.A. Taylor. Phelan PE: Pool boiling of nanofluids. Comprehensive review of existing data and limited new data, *International Journal of Heat and Mass Transfer* 52, 2009, 5339–5347.

84. H.S. Ahn, H. Kim, H. Jo, S. Kang, W. Chang, and M.H. Kim. Experimental study of critical heat flux enhancement during forced convective flow boiling of nanofluid on a short heated surface, *International Journal of Multiphase Flow* 36(5), 2010, 375–384.

85. G. Prakash Narayan, K.B. Anoop, and Sarit K. Das. Mechanism of enhancement/deterioration of boiling heat transfer using stable nanoparticle suspensions over vertical tubes, *Journal of Applied Physics* 102, 2007, 074317.

86. S.M. You, J.H. Kim, and K.H. Kim. Effect of nanoparticles on critical heat flux of water in pool boiling heat transfer, *Applied Physics Letters* 83, 2003, 3374–3376.

87. I.C. Bang and S.H. Chang. Boiling heat transfer performance and phenomena of Al2O3-water nanofluids from a plain surface in a pool, *International Journal Heat and Mass Transfer* 48, (2005) 2407–2419.

88. J.P. Tu, N. Dinh, and T. Theofanous. An experimental study of nanofluid boiling heat transfer, in *Proceedings of 6th International Symposium on Heat Transfer*, Beijing, China, 2004.

89. D.S. Wen, Y.L. Ding, and R.A. Williams. Pool boiling heat transfer of aqueous based TiO_2 nanofluids, *Journal of Enhanced Heat Transfer* 13, 2006, 231–244.

90. Z.H. Liu, J.G. Xiong, and R. Bao. Boiling heat transfer characteristics of nanofluids in a flat heat pipe evaporator with micro-grooved heating surface, *Int. J. Multiphase Flow* 33, 2007, 1284–1295.

91. M.H. Kim, J.B. Kim, and H.D. Kim, Experimental studies on CHF characteristics of nano-fluids at pool boiling, *International Journal of Multiphase Flow* 33, 2007, 691–706.

92. S.J. Kim, T. McKrell, J. Buongiorno, and L.W. Hu, Experimental study of flow critical heat flux in alumina–water, zinc-oxide–water and diamond-water nanofluids, *Journal of Heat Transfer* 131(4), 2009, 043204-1-7.

93. P. Vassallo, R. Kumar, and S.D. Amico. Pool boiling heat transfer experiments in silica–water nanofluids, *International Journal of Heat and Mass Transfer* 47, 2004, 407–411.

94. C.H. Li, B.X. Wang, and X.F. Peng. Experimental investigations on boiling of nano-particle suspensions, in: *2003 Boiling Heat Transfer Conference*, 1993, Jamica, USA.

95. S.J. Kim, I.C. Bang, J. Buongiorno, and L.W. Hu. Effects of nanoparticle deposition on surface wettability influencing boiling heat transfer in nanofluids, *Applied Physics Letters* 89, 2006, 153107.

96. S.J. Kim, I.C. Bang, J. Buongiorno, and L. W. Hu. Surface wettability change during pool boiling of nanofluids and its effect on critical heat flux, *International Journal of Heat and Mass Transfer* 50, 2007, 4105–4116.

97. Milanova D, Kumar R. Role of ions in pool boiling heat transfer of pure and silica nano-fluids, *Applied Physics Letters* 87(23), 2005, 233107-1-3.

98. M. Chopkar, A.K. Das, I. Manna, and P.K. Das, Pool boiling heat transfer characteristics of ZrO_2-water nanofluids from a flat surface in a pool, *Heat and Mass Transfer* 44(8), 2007, 999–1004.

99. H.D. Kim, J.H. Kim, and M.H. Kim. Experimental study on CHF characteristics of water-TiO_2 nanofluids, *Nuclear Engineering Technology* 38(1), 2006, 61.

100. D. Milanova and R. Kumar. Heat transfer behaviour of silica nanoparticles in pool boiling experiment, *J Heat Transf* 130(4), 2008, 1–6.

101. M. Golubovic, H.D.M. Hettiarachchi, W.M. Worek, and W.J. Minkowycz. Nanofluids and critical heat flux, experimental and analytical study, *Applied Thermal Engineering* 29, 2009, 1281–1288.

102. S.M. Kwark, R. Kumar, G. Moreno, J. Yoo, and S.M. You. Pool boiling characteristics of low concentration nanofluids, *International Journal of Heat and Mass Transf* 53, 2010, 972–981.

103. Z-H. Liu, X-F. Yang, and J-G. Xiong. Boiling characteristics of carbon nanotube suspensions under sub-atmospheric pressures, *International Journal of Thermal Science* 49(7), 2010, 1156–1164.

104. R. Hegde, S.S. Rao, and R.P. Reddy. Critical heat flux enhancement in pool boiling using Alumina nanofluids, *Heat Transfer, Heat Transfer—Asian Research* 39, 2010, 323–331.

105. K. Sefiane. On the role of structural disjoining pressure and contact line pinning in critical heat flux enhancement during boiling of nanofluids, *Applied Physics Letters* 89, 2006, 044106.

106. T.M. Anderson and I. Mudawar. Microelectronic cooling by enhanced pool boiling of a dielectric fluorocarbon liquid, *ASME Journal of Heat Transfer* 111, 1989, 752–759.

107. I. Mudawar and T.M. Anderson, Optimization of enhanced surfaces for high flux chip cooling by pool boiling, *ASME Journal of Electronic Package* 115, 1993, 89–99.

108. H. Honda, H. Takamastu, and J.J. Wei. Enhanced boiling of FC-72 on silicon chips with micro-pin-fins and submicron-scale roughness, *ASME Journal of Heat Transfer* 124, 2002, 383–389.

109. J.J. Wei, H. Honda, and L.J. Guo. Experimental study of boiling phenomena and heat transfer performances of FC-72 over micro-pin-finned silicon chips, *Heat and Mass Transfer* 41, 2005, 744–755.

110. S. Ujereh, T. Fisher, and I. Mudawar. Effects of carbon nanotube arrays on nucleate pool boiling, *International Journal of Heat and Mass Transfer* 50, 2007, 4023–4038.
111. N. Zuber. On the stability of boiling heat transfer, *ASME Journal of Heat Transfer* 80, 1958, 711–720.
112. J.H. Lienhard and V.K. Dhir. Hydrodynamic prediction of peak pool boiling heat fluxes from finite bodies, *ASME Journal of Heat Transfer* 95, 1973, 477–482.
113. S.J. Kandlikar. A theoretical model to predict pool boiling CHF incorporating effects of contact angle and orientation, *Journal of Heat Transfer* 123, 2001, 1071–1079.
114. J.M. Ramilison, P. Sadasivan, and J. H. Lienhard. Surface factors influencing burnout on flat heaters, *Jouranl of Heat Transfer* 114, 1992, 287.
115. S.S. Kutateladze. On the transition to film boiling under natural convection, *Kotloturbostroenie* 3, 1948, 10–12.
116. S.S. Kutateladze. A hydrodynamic theory of changes in a boiling process under free convection,' *Izvestia Akademia Nauk, S.S.S.R., Otdelenie Tekhnicheski Nauk* 4, 1951, 529.
117. V.M. Borishanskii. *On the Problem of Generalizing Experimental Data on the Cessation of Bubble Boiling in Large Volume of Liquids*, 1955, Ts. K.I.T., 28, Moscow.
118. W. Rosenhow and P. Griffith. Correlation of maximum heat flux data for boiling of saturated liquids, *Chemical Engineering Progress Symposium Series* 52, 1956, 47–49.
119. Y.P. Chang. An Analysis of the Critical Conditions and Burnout in Boiling Heat Transfer, USAEC Rep. TID-14004, 1961, Washington, DC.
120. W.H. De Jong and P. Borm. Drug delivery and nanoparticles: Applications and hazards, *International Journal of Nanomedicine* 3, 2008, 133–149.
121. D.J. Irvine. Drug delivery: One nanoparticle, on kill, *Nature Materials* 10, 2011, 342–343.
122. D. Panatarotto, C.D. Prtidos, J. Hoebeke, F. Brown, E. Kramer, J.P. Briand, S. Muller, M. Prato, and A. Bianco. Immunization with peptide-functionalized carbon nanotubes enhances virus-specific neutralizing antibody responses, *Chemistry & Biology* 10, 2003, 961–966.
123. J.M. Nam, C.C. Thaxton, and C.A. Mirkin. Nanoparticles-based bio-bar codes for the ultrasensitive detection of proteins, *Science* 301, 2003, 1884–1886.
124. R. Mahtab, J.P. Rogers, and C.J. Murphy. Protein-sized quantum dot luminescence can distinguish between "straight", "bent", and "kinked" oligonucleotides, *Journal of American Chemical Society* 117, 1995, 9099–9100.
125. J. Ma, H. Wong, L.B. Kong, and K.W. Peng. Biomimetic processing of nanocrystallite bioactive apatite coating on titanium, *Nanotechnology* 14, 2003, 619–623.
126. R. Weissleder, G. Elizondo, J. Wittenburg, C.A. Rabito, H.H. Bengele, and L. Josephson. Ultrasmall superparamagnetic iron oxide: Characterization of a new class of contrast agents for MR imaging, *Radiology* 175, 1990, 489–493.
127. J. Yoshida and T. Kobayashi. Intracellular hyperthermia for cancer using magnetite cationic liposomes, *Journal of Magnetism and Magnetic Materials* 194, 1999, 176–184.
128. L. Zhang, F.X. Gu, J.M. Chan, A.Z. Wang, R.S. Langer, and O.C. Farokhzad. Nanoparticles in medicine: Therapeutic applications and developments, *Clinical Pharmacology and Therapeutics* 83, 2008, 761–9.
129. Y.W. Chien. *Novel Drug Delivery Systems*, 2nd edn. Marcel Dekker, New York, 1992.
130. K. Khanafer and K. Vafai. The role of porous media in biomedical engineering as related to magnetic resonance imaging and drug delivery, *Heat and Mass Transfer* 42, 2006, 939–953.
131. K. Cho, X. Wang, S. Nie, Z. Chen, and D.M. Shin. Therapeutic nanoparticles for drug delivery in cancer, *Clinical Cancer Research* 14, 2008, 1310–1316.
132. ElHazzat Jallal and E.H. El-Sayed Mohamed. Advances in targeted breast cancer, *Current Breast Cancer Reports* 2, 2010, 146–151.

133. K. Thurn, E. Brown, A. Wu, S. Vogt, B. Lai, J. Maser, T. Paunesku, and G. Woloschak. Nanoparticles for applications in cellular imaging, *Nanoscale Research Letters* 2, 2007, 430–441.
134. S.H. Madina and M.E. El-Sayed. Dendrimers as carriers for delivery of chemotherapeutic agents, *Chem Rev.* 109, 2009, 3141–3157.

3 Multiscale Simulation of Nanoparticle Transport in Deformable Tissue during an Infusion Process in Hyperthermia Treatments of Cancers

Ronghui Ma, Di Su, and Liang Zhu

CONTENTS

3.1 INTRODUCTION

Hyperthermia has been used in a variety of therapeutic procedures for patients with cancers in the past several decades [13]. It is suitable for patients diagnosed with previously unresectable tumors, or for patients who are considering an alternative to costly and risky surgical procedures. In hyperthermia treatment, thermal energy delivered to a tumor raises the tumor temperature above 43°C for durations of more than 60 min. It has been reported that such elevated temperatures may produce a heat-induced cytotoxic response and/or enhance the cytotoxic effects of radiation and drugs. Both the direct cell-killing effects of heat and the sensitization to other therapeutic agents by means of heat are phenomena strongly associated with the distribution of temperature elevations and durations of heating.

Nanoparticles have found important applications in novel hyperthermia treatments of cancers due to their ability to generate impressive levels of heat when excited by an external magnetic field or laser irradiation [56]. For example, magnetic nanoparticles dispersed in tumors can induce localized heating when agitated by an alternating magnetic field. The heat generation is mainly attributed to the Néel relaxation and/or Brownian motion of the particles. Iron oxide magnetite Fe_3O_4 and maghemite γ-Fe_2O_3 nanoparticles are the most commonly used to date due to their biocompatibility [18,29,35]. Smaller particles (10–40 nm) are preferred in magnetic-based hyperthermia applications due to their ability to produce impressive level of heating in relatively low magnetic fields [23]. In laser photothermal therapy, where heat generation in tumors is induced by near-infrared laser irradiation on the surface, the inclusion of gold nanoshells/nanorods in the tumor leads to maximized absorption of the laser energy to elevate local tumor temperatures [5,12,41]. Previous studies have demonstrated that the usage of gold nanoshells/nanorods can enhance laser energy absorption by several orders of magnitude compared with some traditional dyes such as indocynine green dye [52]. Despite the demonstrated potential of various types of nanoparticles in hyperthermia treatment, there exist a number of challenges to be addressed before their widespread applications in clinical studies. One leading issue is the limited knowledge and understanding of nanoparticle dispersion and anticipated temperature elevations in tumors. As nanoparticles serve as the heat-generating agents, the efficacy of the treatment depends largely on the spatial distribution of the nanoparticles in tumors. The lack of control of nanoparticle distribution may lead to under-dosage of heating in the tumor or overheating of the surrounding normal tissue.

Two techniques are currently used to deliver nanoparticles to a tumor. The first is systemic (venous) injection of dispersed nanoparticles in a biocompatible solution. The majority of the nanostructures eventually accumulate in tumors due to the small size of the nanostructures and the leaky nature of tumor vasculature. The efficiency of the systemic delivery can be improved by coating the nanostructures with some chemicals that target tumor cells [60]. However, systemic administration is not suitable for poorly perfused and large-sized tumors. The second approach, which is the focus of this study, is direct infusion by means of needles, also referred to as intratumoral infusion or convection-enhanced delivery. It is an important technique to deliver a variety of nanostructures in tumors by continuous injection of nanofluid under a positive pressure gradient [2,17,18,32]. It is so far the best method available for distributing large therapeutic agents in tumors in that it allows those agents to overcome some of the obstacles such as interstitial fluid pressure or brain–blood barrier, through enhanced convective transport [34]. This approach has been used to deliver ferrofluid to treat tumors in the liver [32] and breast [17,18]. In case of an irregular-shaped tumor, multiple-site injections can be exploited to cover the entire target region [50].

Quantitative characterization of the distribution of the nanoparticles infused in a tumor is very limited due to the opaque nature of the tissue and insufficient techniques to quantify the nanoparticle concentration distribution. Salloum et al. [48] studied the injection of a ferrofluid in semi-transparent agarose gels. This study demonstrated that the distribution volume of the ferrofluid is sensitive to both the injection rate and the gel/tissue properties. In addition, a higher infusion rate yields a more irregular-shaped distribution volume of the nanofluid in the gel. Heating patterns of the magnetic nanoparticles injected in gels/tissues were also quantified by placing the gels/tissues in an alternating magnetic field and measuring the specific absorption rate (SAR) at multiple locations [48,49]. These studies suggested that the SAR distributions can be used to characterize the nanoparticle concentration based on the hypothesis that the heating distribution is solely dependent on the spatial distribution of the nanoparticles, when other operational conditions, such as the type and concentration of the ferrofluid, the size and surface coating of the particles, the properties of the gel/tissue, and the strength of the magnetic field, are fixed. The nonuniform SAR distributions obtained in these studies illustrate a heterogeneous nanoparticle distribution in gels/tissues [48,49]. A Gaussian distribution of the SAR was proposed based on fitting the experimental measurements. Although a qualitative relation among the injection parameters, gel properties, particle distribution, and heat generation is established, the understanding of the nanofluid and nanoparticle transport in gels/tissue and its dependence on the major injection parameters remains poorly resolved.

The transport of nanoparticles in biological tissues during an infusion is a complex process that involves nanofluid flow through the deformable tissues, advection of particles in the extracellular space, particle binding to the cellular structure, and interactions among the particles. The shape and volume of the nanofluid distribution in a tumor during an infusion process is largely dependent on the velocity field, which is determined by both the hydraulic characteristics of the tissue and the elastic response of the tissue to the infusion pressure. As the particles travel in the interstitial space of the tissue, some may bind to the surface of the cells, causing limited nanoparticle penetration. The particle-binding rate is affected by the interstitial fluid velocity and tissue structure as well as the characteristics of the particles. Nanofluid flow and particle binding to the cellular structures are two interrelated mechanisms having substantial influence on the nanoparticle distribution during and after an infusion process.

In direct infusion, the infusion pressure can induce deformation in tissues which, in turn, affects the fluid flow as it changes the permeability of tumor tissues by altering the size and connectedness of the aqueous pathways [22,34]. The tissue deformation also affects the penetration of the nanoparticles in tissues due to the changes in the effective pore size of the extracellular structure [37]. McGuire et al. [34] developed a one-dimensional poroelastic model to describe the nonhomogeneous tissue deformation. Chen et al. [7] conducted a combined experimental and theoretical study of the poroelasticity in brain–tissue–equivalent phantom gels to characterize the influence of gel deformation on the infusion pressure, gel matrix dilation, and pore fraction. However, this study is limited to soft gels and low infusion rates in the range of 0.5–10 µL/min. Ivanchenko et al. [21] studied the mechanical response of a brain–tissue–equivalent phantom gel matrix to infusion through measurement of the deformation of the gels and analysis of the changes in the permeability and porosity of the gels close to the catheter tip. The results were obtained with a single infusion rate of 5 µL/min. In addition to their influence on fluid flow, the tissue deformation also contributes to the formation of a backflow of the infusate along the needle track, which has been considered an important issue for enhanced convection delivery of therapeutic agents in brain tumors [2]. Morrison et al. [36] and Raghavan et al. [42] derived a simplified analytical model to quantify the backflow distance as a function of infusion parameters and tissue properties. Although the analytical solution provides valuable insight into the process, it involves bold assumptions and does not give a prediction of the shape of the infusate distribution.

The convection and diffusion of large therapeutic agents such as antibodies and nucleotides by pressure-driven intratumoral infusion have been extensively studied in the past several decades [6,11,22,33,61]. However, the existing theory is not readily applicable to nanoparticles because the nano-sized particles are associated with strong surface interactions that could lead to particle binding on the cell surface. Su et al. [54] conducted a multi-scale study of nanoparticle transport in biological tissues during an injection process with the nanoparticle binding to the cellular structure being considered using a particle trajectory-tracking method. This model predicts nanoparticle penetration depths at the same order of magnitude as those indicated by the experimentally measured SAR distributions. This study also suggests that nanoparticle deposition on the extracellular structure is a leading factor for the nonuniform particle distribution in porous media. This model, however, is one-dimensional and the effects of tissue deformation on the particle transport are not considered. Neeves et al. [37] attempted to enhance penetration depths of infused polymer nanoparticles (54 nm) by dilation and degradation of the brain extracellular matrix. It was found that dilating the extracellular spacing by pre-infusion of a buffer solution offers an effective means to enlarge the nanoparticle distribution volume.

In light of the restrictive numerical and experimental studies of the direct infusion process, the understanding of nanoparticle transport in deformable tissues is very limited. The objective of this study is to investigate the interrelated mechanisms of nanofluid transport in tumors. Presented in this chapter is a multi-scale model that considers fluid flow and deformation of tumoral tissues during an infusion process, particle interaction with the cellular structure, and nanoparticle advection and deposition in tumors. The integration of the three components allows the study of the nanoparticle transport behavior during an infusion process. The influence of the deformation-induced backflow and change in porosity on particle distribution was quantified under a variety of process conditions. Parametric studies were also carried out to examine the effects of infusion rates, infusion volumes, needle sizes, and tissue properties on nanoparticle concentration distributions.

3.2 MATHEMATICAL MODEL

The behavior of nanoparticle transport in biological tissues stems from the complex chemicophysical processes occurring on largely disparate temporal and spatial scales. In this study, a multi-scale model that consists of three major components is developed to describe an intratumoral infusion of nanofluids: (a) a poroelastic model for fluid flow through a tumor and tumor deformation; (b) nanoparticle convection, diffusion, and deposition in a tumor; and (c) a particle trajectory-tracking model for particle interactions with the cell surface. The integrated model can be used to predict the distributions of nanoparticle concentration in tumors under a variety of infusion parameters.

3.2.1 FLUID FLOW AND TISSUE DEFORMATION DURING AN INFUSION PROCESS

Biological tissues are treated in the existing studies as poroelastic materials where the solid matrix is elastic and fluid is viscous [7,8,33,34]. We assumed that the tumor and normal tissue are homogenous before an injection. The nanofluid with a particle volumetric concentration less than 5% is considered as an incompressible dilute colloidal fluid where the presence of the particles does not significantly affect the transport properties of the fluid [62]. The effects of gravity, osmotic force, particle agglomeration, and fluid exchange between the interstitial fluid and blood or lymph vessels on fluid transport are not considered in this study [8,33,34]. Fluid flow and tissue deformation are considered in steady state [54]. Fluid flow in biological tissues has been described by various models, including Darcy's law, Brinkman equation, and Brinkman–Forchheimer–Darcy equation [25]. Owing to the slow fluid motion and lack of solid boundaries in a tumor, it is suggested that Darcy's equation is suitable for an intratumoral infusion process [25,34], which are

$$\nabla \cdot (\varepsilon \mathbf{v}) = 0 \tag{3.1}$$

$$\nabla p_f = -\frac{\varepsilon \mu}{K} \mathbf{v}, \tag{3.2}$$

where ε is porosity, $\mathbf{v}$ is interstitial fluid velocity vector, p_f is fluid pressure, μ is the viscosity of the fluid, and K is permeability of the tissue.

During an infusion, a backflow forms as the hydraulic pressure opens an annular space surrounding the needle outer surface. The bulk fluid flow in the annular space is governed by the conservations of mass and momentum, which are

$$\nabla \cdot \mathbf{v} = 0$$
$$\mathbf{v} \cdot \nabla \mathbf{v} = -\frac{1}{\rho} \nabla p_f + \frac{\mu}{\rho} \nabla^2 \mathbf{v}. \tag{3.3}$$

The same variables $\mathbf{v}$ and p_f are used to represent fluid velocity and pressure, respectively, in the free space and porous structure. In the poroelastic model, the solid phase of the medium is assumed to be Hookian material, incompressible, isotropic, and fully saturated with fluid. The deformation during the infusion is infinitesimal. The steady-state constitutive equation for tissue deformation yields:

$$G\nabla^2\mathbf{u} + (\lambda + G)\nabla(\nabla \cdot \mathbf{u}) = \nabla(\varepsilon p_f) \qquad (3.4)$$

where $\mathbf{u}$ is the displacement vector, and G and λ are the Lamé constants. Given Young's modulus and Poisson ratio, G and λ can be calculated by using the following two relationships $\lambda = E\upsilon/((1 + \upsilon)(1 - 2\upsilon))$ and $G = E/(2(1 + \lambda))$.

It should be noted that the porosity and permeability in Equations 3.1 and 3.2 are spatially varying in the presence of tissue deformation, both are functions of dilatation e, where $e = \nabla \cdot \mathbf{u}$. With the assumption that the volume of the solid phase of the porous medium does not change for small deformation, the porosity ε is calculated by the expression [7,8]

$$\varepsilon = \frac{\varepsilon_0 + e}{1 + e}, \qquad (3.5)$$

where ε_0 is the tissue porosity in the absence of deformation. Various empirical relationships have been proposed to quantify the permeability K as a function of dilatation e [15,19,28]. In this study, a commonly used relation proposed by Lai and Mow [28] is employed, which is

$$K = K_0 \exp(Me), \qquad (3.6)$$

where K_0 is the permeability in the absence of deformation, and M is a material constant that governs the variation of permeability with the dilatation. Typically, M is related to tissue properties, the densities of the cells, and the structure of the extracellular matrix [8]. Through curve-fitting of experimental data, previous studies suggest that M for tumor tissues is in the range from 0 to 5 [8,28,53]. As result of the largely diverse nature of biological tissues, there are no sufficient experimental data available for estimating the values of M for various types of tissues/tumors. Chen and Sarntinoranont [8] obtained reasonable results using an M value of two in their Finite Element model for the infusion process. Thereby, we choose $M = 2$ in this study. The effect of M on nanoparticle transport will be investigated using unified theoretical and experimental approach in future study.

3.2.2 NANOPARTICLE CONVECTION AND DIFFUSION IN TUMORS

The transport of large therapeutic agents in tissue is typically described by convection and diffusion equations with extra terms to consider mechanisms such as internalization of drugs by cells, collection of macromolecules by the circulation, reactions on the cell surfaces, etc. In this study, the short injection duration (less than 1 h) renders the extravascular exchange of nanoparticles through a capillary wall insignificant. The nanoparticle internalization does not alter the macroscale particle concentration profile, and, thereby can be neglected. The binding of nanoparticles on the cellular structure, in contrast, reduces the particle concentration in the fluid phase, and substantially affects the nanoparticle distribution in the tissue [37,54]. The attachment of nanoparticles to the cellular structure can be treated as a concentration-dependent reaction that consumes particles in the fluid phase [58], and represented by a volumetric deposition rate in the convection and diffusion equation. A deposition rate coefficient, a quantity analogous to the rate constant of a chemical reaction, is used to quantify the dependence of the volumetric deposition rate on local particle concentration, particle size and properties, local velocity, porous structure, etc. With the assumption that the particle binding to the cell surfaces is irreversible, the equation for nanoparticle transport in a porous tissue is

$$\frac{\partial C}{\partial t} + \nabla \cdot (\mathbf{v}C) = \nabla \cdot \left[D_e \nabla C \right] - k_f C, \tag{3.7}$$

where C is the molar concentration of the particles in the fluid, and k_f is the deposition rate coefficient. The term $k_f C$ represents the volumetric deposition rate of the particles on the cellular structure that reduces the particle concentration in the liquid phase. D_e is the effective diffusion coefficient which is calculated based on the diffusion coefficient of the particles in an unbounded liquid phase and on the tortuosity of the tissue [14]:

$$D_e = D_0 \frac{L}{F\tau(\varepsilon)} \quad \text{and} \quad D_0 = \frac{k_B T}{3\pi\mu d_p}, \tag{3.8}$$

where D_0 is the diffusion coefficient in an unbounded liquid phase, k_B is Boltzmann's constant, T is the absolute temperature, d_p is the particle diameter, L is the factor responsible for hydrodynamic reduction of the diffusion coefficient in the pore, $\tau(\varepsilon)$ is the tortuosity which is the function of porosity ε, and F is a shape factor valued from one to four [44,51]. The volumetric concentration of the nanoparticles bounded to the solid structure S can be calculated by the equation

$$\frac{\partial S}{\partial t} = k_f C. \tag{3.9}$$

3.2.3 PARTICLE TRAJECTORY-TRACKING MODEL FOR CALCULATION OF k_f

The deposition rate coefficient k_f is dependent on many factors related to the characteristics of fluid flow, particles, and porous structure. k_f can be determined by experimental measurement and theoretical calculation in the studies of colloidal particle transport in sand or soil bank [58,59]. When experimental measurement is not available or feasible, a particle trajectory tracking method is widely used to determine the deposition rate coefficient [40]. In this method, the suspended particles are treated as individual entities. Given the local velocity field of the interstitial fluid, the size and surface charges of the particles, and the geometry of the unit structural cell of the porous medium, the trajectories of a number of particles introduced into the unit structural cell can be determined based on the forces acting on these particles. A binding is counted when a particle is in contact with the solid surface [38]. The ratio of the particles captured by the solid surface to those brought into a unit cell gives the collector efficiency η_s, which measures the probability of the particle interception by the solid structure in classical colloidal theory. Once the collector efficiency is determined, the deposition rate coefficient k_f for a porous medium consisting of spherical bodies with a diameter of d_c can be calculated by the equation [58]:

$$k_f = \frac{3(1-\varepsilon)}{2\varepsilon d_c}\eta_s \, |\mathbf{v}|, \qquad (3.10)$$

where $|\mathbf{v}|$ is the magnitude of the local fluid velocity.

Happel's sphere-in-cell model [16] is widely used as the unit structural cell for granular porous media [38,43,59]. As schematically shown in Figure 3.1, it consists of a solid spherical body representing a cell and a uniform layer of fluid that envelops the sphere. The thickness of the fluid layer is calculated as $\gamma = a_c[(1-\varepsilon)^{-1/3} - 1]$, where a_c is the radius of the solid sphere [38]. Selection of a unit structural cell for tissue is a challenge due to the existence of the extracellular matrix. Although tissues are typically conceptualized as porous structures consisting of nearly spherical cells in the studies of drug delivery [7], Netti et al. [39] reported that collagen can significantly hinder the diffusion of large molecules. It is unclear to what extent the extracellular matrix affects the convective transport of the particles due to limited study on this topic. Furthermore, it is a formidable task to include the collagen fibers in a unit cell at the current stage. Considering that the focus of the study is the effect of particle deposition on particle transport, we focus on an ideal situation where the presence of the extracellular matrix can be neglected. This idealization allows the usage of Happel's model as the unit structural cell for tissue.

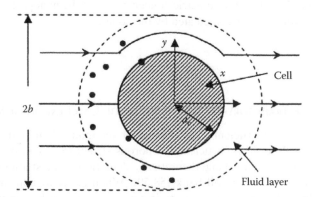

FIGURE 3.1 Happel's sphere-in-cell for the particle trajectory tracking analysis, $b = a_c(1-\varepsilon)^{-1/3}$, where ε is porous porosity. Black dots refer to nanoparticles. (Adapted from KE. Nelson and TR. Ginn, *Langmuir*, 21(6), 2173–2184, 2005.)

The details of the particle tracking model can be found in the literature [54]. In brief, we assume that the particles are small (nanosized particles), spherical, chemically inert, solid, and very dilute in a liquid that flows with a low Re number in the laminar regime. Hydrodynamic interactions among the particles are neglected. Typically, the particles are properly coated to prevent agglomeration.

The forces that act on a particle near a solid surface immersed in the moving fluid include van der Waals attractive force, electrostatic double layer force, hydrodynamic drag force, lift force, buoyancy force, and Brownian motion. The van der Waals force and electrostatic double layer force act along the normal direction of a surface and only become significant at close separation distance between the particles and a surface. The lift force pushes particles away from the surface toward the direction of increasing velocity [24,31]. A particle partially or fully immersed in a fluid also experiences an upward buoyancy force. For a nanoparticle, the buoyancy force is insignificant when compared with other forces due to its small size. Virtual mass effect and Basset force are unsteady forces due to acceleration of a body with respect to the fluid. For the laminar flow conditions used in this study, they are not considered in the model [24]. Magnus force, which arises as the fluid stream against a rotating and yawing body interacts with its boundary layer, is considered negligible when compared with the drag force because of the small particle size [24].

For submicron-sized colloids whose relaxation time is small, the inertial force can be neglected, which implies that that the particles relax to the fluid velocity instantaneously. The colloidal particle trajectory is then governed by the Stochastic Langevin equation [30,38,61]:

$$dr_j = \left(\frac{D}{k_B T} \sum_i F_i + v \right) \Delta t + (\Delta r)_j^B, \tag{3.11}$$

where dr_j is the displacement vector of jth particle, D is the particle diffusivity, F_i represents the forces acting on the particle, v is the fluid velocity, and $(\Delta r)_j^B$ is the random Brownian displacement. Δt is the time interval used in the integration.

The velocity expressions in a Happel's sphere-in-cell shown in Figure 3.1 are obtained directly from the stream function of Stokes flow in Happel's model. The radial and angular velocities, respectively, are [16]

$$v_{f,r} = -V_\infty \cos\theta \left(\frac{K_1}{r^{*3}} + \frac{K_2}{r^*} + K_3 + K_4 r^{*2} \right),$$

$$v_{f,\theta} = \frac{1}{2} V_\infty \sin\theta \left(-\frac{K_1}{r^{*3}} + \frac{K_2}{r^*} + 2K_3 + 4K_4 r^{*2} \right), \tag{3.12}$$

where $p = (1 - \varepsilon)^{1/3}$, $r^* = 2r/d_c$, $w = 2 - 3p + 3p^5 - 2p^6$, $K_1 = 1/w$, $K_2 = -(3 + 2p^5)/w$, $K_3 = (2 + 3p^5)/w$, and $K_4 = -p^5/w$. V_∞ is the free stream velocity approaching the sphere. The local interstitial velocity is employed as V_∞ in this study.

When a freely moving particle travels near a rigid surface, the viscous resistance exerted by the wall and the rotation of the particle can substantially modify both the velocity and mobility of the particle. This is referred to as hydrodynamic retardation. The correction factors for the velocity of the particles used in this study are those given in reference [54].

3.2.3.1 Viscous Lift Force

Particles traveling across a velocity gradient caused by the presence of a wall can experience a lift force that directs a particle away from the wall. Saffman force, which is defined as the lift force due to shear, is insignificant for nanoparticles having a very small Stokes number [47]; rather, the pressure difference across the particle and particle rotation can cause an appreciable lift force on a noninertial particle [31]. Cox and Hsu [10] derived the following expression to calculate the lift velocity for noninertial spherical particles in a laminar parabolic flow field near a single vertical plane:

$$\mathbf{v}_{\text{lift}} = \frac{55}{144} \frac{\rho d_p v_{f,\text{max}}^2}{2\mu} \left(\frac{d_p}{2h_{\text{max}}} \right)^2 \left(1 - \frac{h}{h_{\text{max}}} \right) \left(1 - \frac{73}{22} \frac{h}{h_{\text{max}}} \right), \tag{3.13}$$

where h is the distance between the particle center and the wall, h_{max} is the distance at which the velocity profile reaches its maximum $v_{f,\text{max}}$. More details about the calculation of the lift force can be found in reference [1]. Given the lift velocity $\mathbf{v}_{\text{lift}}$, the lift force can be obtained accordingly:

$$\mathbf{F}_{\text{lift}} = 3\pi\mu d_p \mathbf{v}_{\text{lift}}. \tag{3.14}$$

3.2.3.2 van der Waals Force and Electrostatic Double Layer Force

The potential for particle–surface interactions within the interaction range (few tens of nanometers for submicron colloidal particles) is calculated according to the Derjaguin, Landau, Verwey, and Overbeek (DLVO) theory [46]. The DLVO potential for interaction forces can be derived through the differentiation of the potential interaction energies [20]:

$$\mathbf{F}_{\text{potential}} = -\frac{\partial}{\partial h}(A_{\text{elec}} + A_{vdW}), \tag{3.15}$$

where A_{elec} is the potential due to electrostatic interaction, and A_{vdW} is the potential due to the van der Waals force between particles and a surface. As the size of a nanoparticle is much smaller than that of a cell, their interactions can be approximated as those between a particle and a flat wall.

The van der Waals interaction energy between a sphere and a wall at a distance of h is expressed as [20]:

$$A_{vdW} = -A_H d_p/(12h) \tag{3.16}$$

where A_H is the Hamaker constant, which can be calculated by an empirical formulation provided by reference [46]. According to the Gouy–Chapmann model of a diffuse double layer and the electrostatic Poisson–Boltzmann equation, A_{elec} between a spherical particle and a flat surface with the zeta-potentials of ψ_1 and ψ_2, respectively, is given by the following equation:

$$A_{elec} = 64\pi\varepsilon_r\varepsilon_0\left(\frac{k_BT}{zE}\right)^2\frac{d_p}{2}\times\left(\tanh\left(\frac{zE\psi_1}{4k_BT}\right)\tanh\left(\frac{zE\psi_2}{4k_BT}\right)\right)e^{-\kappa h}, \quad (3.17)$$

where ε_0 is the vacuum permittivity, ε_r is the relative dielectric constant of the water, I is the electron charge, and z is the valence of the electrolyte. The effect of the aqueous environment is reflected by the Debye–Hückel parameter κ.

3.2.3.3 Brownian Motion

Brownian motion is formulated through the Brownian displacement $(\Delta\mathbf{r})^B$, a random value taken from a Gaussian white-noise distribution with a zero mean $W(t)$ and a specific intensity that relates to the mean square displacement (MSD). The Stokes–Einstein equation is used to calculate the MSD [20], which is

$$MSD^2(t) = \left\langle(\Delta\mathbf{r})^2\right\rangle = \frac{2k_BT}{f}|\Delta t| = 2D|\Delta t|, \quad f = 3\pi\mu d_p. \quad (3.18)$$

A freely moving particle may experience rotations near a solid surface. In the case of spheroid and ellipsoidal particles, the determination of the rotational velocity is essential because the interaction energy is dependent on the particle orientation. However, spherical particles are symmetric and hence are less likely to be affected by the particle rotation [61]. Therefore, the particle rotational velocities are neglected in the trajectory-tracking analysis. The effect of particle rotation on the motion of a particle near a surface is considered through the lift force and corrections of the fluid velocity and particle mobility.

The particle trajectories are determined by integrating Equation 3.11 using the predictor–corrector method [27]. At the beginning of a particle trajectory analysis, a large number of particles are distributed randomly over the curved segment extending from $y = 0$ to $y = b$ as shown in Figure 3.1 [45]. The vertical position of a particle is determined by

$$y_0 = \xi_i b, \quad (3.19)$$

where ξ_i is a sequence of uniformly distributed random numbers in the range of zero to one, and b is the radius of the fluid shell shown in Figure 3.1. Once its vertical position is determined, the x coordinate of an entering particle can be determined as

$$x_0 = -(b^2 - y_0^2)^{1/2}. \quad (3.20)$$

A particle deposition is counted if the calculated trajectory of a particle reaches the solid surface. Through the calculation the trajectories of a number of particles, the collector efficiency can be determined for various combinations of operational parameters such as particle size, surface properties, and local fluid velocity. Note that the number of particles used in the simulation should be sufficiently large to ensure that the collector efficiency is independent of the particle number and that the result is statistically meaningful.

The time step Δt should be sufficiently small such that the deterministic forces remain constant during each time interval. Also, the assumption of negligible particle inertia requires that the time step should be much greater than the particle relaxation time $\tau_p = m_p f^{-1}$. Thus, the requirement of the time step may be written as $\tau_p \ll \Delta t \ll \tau_u$, where τ_u is the time increment at which the deterministic velocity is considered constant. This study used a Δt of 10^{-5} s. Integration is completed when all the particles introduced into the unit structural cell either deposit on the solid surface or exit the unit cell.

3.2.4 MODEL INTEGRATION

The three components of the model are integrated to simulate nanofluid transport during an infusion process by conducting the following tasks in sequence: (a) Equations 3.1 through 3.6 were solved iteratively to determine the fluid velocity field, tissue deformation, and the distribution of the porosity in the tumor and surrounding tissue. This provided information on the range of the velocity and porosity for a given set of infusion conditions; (b) the trajectory-tracking model is used to calculate the deposition rate coefficient for various velocities and tissue porosities in the range prescribed in step (a). Curving fitting is used to derive the relationship of the deposition rate coefficient to the velocity and porosity; (c) Equation 3.7 for nanoparticle convection, diffusion, and deposition in the tumor and tissue can be solved with the velocity field obtained in the previous calculation. The dependence of particle deposition on velocity and porosity is accounted for by using the function obtained in step (b). As Equation 3.7 is transient, the injection duration is determined by injection amount and injection rate. The geometries and the boundary conditions used in the simulations are given in Sections 3.4 and 3.5. A commercially available multiphysics software COMSOL® was used in this study. Major simulation parameters and tissue and particle properties are given in Table 3.1.

3.3 PREDICTION OF DEPOSITION RATE COEFFICIENT

In this study, the dependence of the deposition rate coefficient k_f on particle surface charge and porosity at various velocities was studied using the particle trajectory tracking model. Shown in Figure 3.2a are the variations of the deposition rate coefficient k_f with the fluid velocity for nanoparticles of 20 nm diameter with different surface charges: −20, −40, and −60 mV. In this simulation, the cell diameter is 20 μm and the tissue porosity is 0.2. The local fluid velocity affects the particle deposition through two mechanisms having opposite effects on the particle deposition. A higher fluid velocity increases the rate of the particles approaching the cell surface. In contrast, an elevated fluid velocity reduces the probability of a particle being captured by the cell surface because it yields a stronger particle rotation and larger lift force that direct a particle away from the surface. The results shown in Figure 3.2a demonstrate that the lift force dominates the particle deposition

TABLE 3.1

Major Parameters and Properties in the Simulation

Properties and Parameters	Values
Infusion amount	0.1 cc
Needle size	22, 26, and 32-gauge, Hamilton needle
Infusion rate	5 ~ 15 μL/min
Nanofluid concentration	0.75 M ferrofluid (3% by volume)
Magnetic nanoparticle	Fe_3O_4
Nanoparticle density	5240 kg/m³
Nanoparticle diameter	20 nm
Nanoparticle diffusivity	$10^{-11} \sim 10^{-12}$ m²/s [54]
Nanoparticle surface zeta potential	−20 mV ~ −60 mV
Tumor diameter	10 mm
Young's module of tissue (E)	60 KPa [9]
Young's module of tumor (E)	0.2–0.5 MPa [4,8]
Tumor permeability	5×10^{-16} m² [8,39]
Tissue permeability	1×10^{-14} m² [57]
Tissue/tumor Poisson ratio	0.35 [8,34]
Tissue porosity	0.4 [34]
Tumor porosity	0.2 [39]
Cell diameter	20 μm
Cell surface zeta potential	−20 mV [54]

process, and increased velocity leads to reduced deposition rate for all the surface charges. In addition to the velocity field of the fluid, the electrostatic force also plays an important role in the particle–cell interactions due to the repulsive electrostatic force that keeps particles away from the cell surface. Figure 3.2a shows that increasing the surface charge of the particles can significantly reduce the deposition rate coefficient. Figure 3.2b shows the dependence of the deposition rate coefficient on the local velocity for three tissue porosities: 0.2, 0.3, and 0.4. There is a clear indication that porosity is an important parameter than affects the rate of particle deposition on the cell surface. A higher porosity leads to lower deposition coefficients.

3.4 ONE-DIMENSIONAL NANOPARTICLE DISTRIBUTIONS IN NONDEFORMABLE TUMORS

The multiscale model was used to calculate one-dimensional particle distributions in a nondeformable spherical tumor of 10 mm diameter for various infusion rates, particle surface charges, and tissue porosities. The nanofluid is assumed to be delivered from a point source at the center of a tumor. The boundary conditions at the center of the tumor are constant infusion rate and nanoparticle concentration. At the tumor outer boundary $\partial C/\partial r = 0$ are applied due to the large tumor region. Figure 3.3 shows the particle distributions along the radial direction for various infusion rates. It is observed that the total particle concentration decreases exponentially in the radial direction

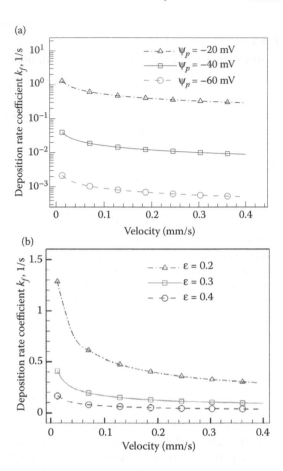

FIGURE 3.2   (a) Variations of deposition rate coefficient k_f with velocity for various surface charges of the particles ($\psi_c = -20$ mV, $d_p = 20$ nm, $a_c = 10$ μm, $\varepsilon = 0.2$). (b) Variations of the deposition rate coefficient k_f with velocity for various porosities ($\psi_c = -20$ mV, $\psi_p = -20$ mV, $d_p = 20$ nm, $a_c = 10$ μm).

from the center of the tumor. The particle concentration distributions for various surface charges and tissue porosities at the infusion rate of 5 μL/min are shown in Figures 3.4 and 3.5, respectively. It is evident that nanoparticle surface potential plays a vital role in the infiltration of the nanoparticles in tumors. Consistent with the relationship between deposition rate coefficient and tissue porosity demonstrated in Figure 3.2b, the nanoparticle distribution is sensitive to the change in tissue porosity.

3.5   NANOFLUID INFUSION IN DEFORMABLE TUMORS

We studied the nanofluid infusion process in deformable tumors in an axisymmetric domain depicted in Figure 3.6, which schematically shows a needle inserted into a spherical tumor embedded in 20-mm-thick normal tissue. The boundary conditions used in this study are as follows: a finite pressure P_{inf} and constant concentration of

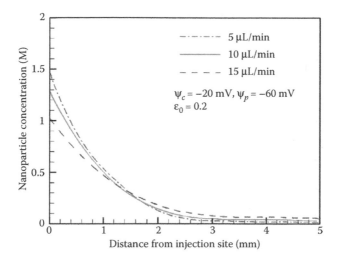

FIGURE 3.3 Distributions of nanoparticle concentration for various injection rates.

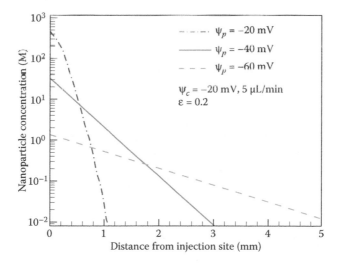

FIGURE 3.4 Distributions of nanoparticle concentration for various particle surface charges.

the nanofluid C_0 are applied at the needle tip. Considering the large domain of the tumor and surrounding tissue, the conditions at the outer boundary, $r = R$, are set as $P = 0$ and $\partial C / \partial \mathbf{n} = 0$, where $\mathbf{n}$ is the unit normal vector at the outer boundary. The value of P_{inf} is adjusted in the simulation so that it yields the desired infusion rate. The no-slip condition is applied on the surfaces of the tissue and needle exposed to the backflow. For tissue deformation, the constant hydraulic pressure P_{inf} is applied on the needle tip, which yields

$$(2G + \lambda)\frac{\partial u_z}{\partial z} = -P_{inf}. \qquad (3.21)$$

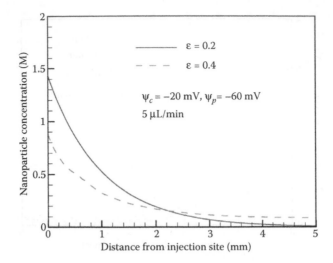

FIGURE 3.5   Distributions of nanoparticle concentration for various tumor porosities.

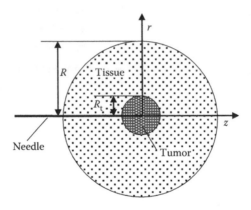

FIGURE 3.6   Nanofluid infusion in a tumor embedded in 20-mm-normal tissue.

At $r = R$, we consider a case of deep-seated tumor and apply a fixed boundary condition $\mathbf{u} = 0$. The interface between the tissue and needle is considered a deformable interface that allows the tissue to recede from the needle surface. The original porosities of the tumor and normal tissue are 0.2 and 0.4, respectively. The surface charge of the particle is −60 mV. Different permeability and elastic properties are assigned to the tumor and normal tissue. The nanofluid concentration, tumor size, tumor and tissue properties, and other major parameters used in the model are given in Table 3.1.

A baseline case with the infusion rate of 5 μL/min, the needle size of 26 gauge, and the Young's modulus E of 0.5 MPa was simulated first. The predicted nanoparticle distribution in a tumor was shown in Figure 3.7a. It can be seen that the particle

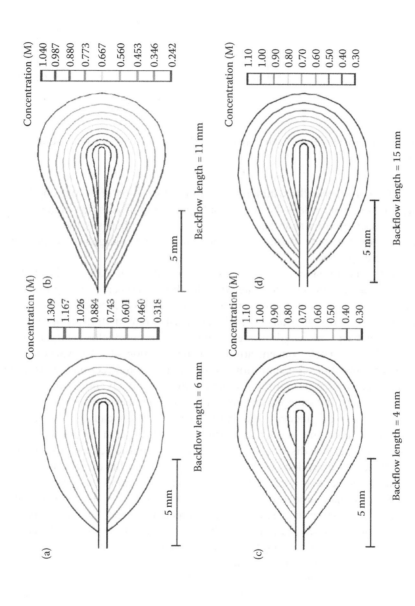

FIGURE 3.7 Distributions of nanoparticle concentration for (a) the baseline case with infusion rate of 5 μL/min, $E = 0.5$ MPa, and 26-gauge needle, (b) infusion rate of 10 μL/min, (c) 32-gauge needle, and (d) $E = 0.2$ MPa.

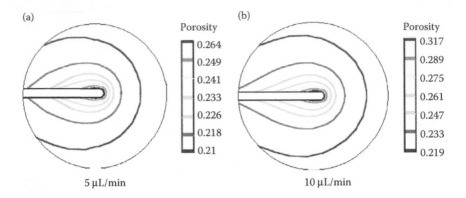

FIGURE 3.8 Distributions of tissue porosity for the infusion rates of (a) 5 μL/min and (b) 10 μL/min ($E = 0.5$ MPa, $\varepsilon_0 = 0.2$, $M = 2$, 26-gauge needle).

distribution is not spherically symmetric due to the formation of 6-mm-long back-flow. Depicted in Figure 3.8a is the corresponding porosity distribution in the tumor. It shows increased porosity near the needle tip. The similar shapes of the concentration and porosity distributions suggest the influence of the enlarged tissue porosity on the particle dispersion.

The effect of the infusion rate on the tissue deformation and particle distribution is studied by changing the infusion rate while holding other parameters prescribed in the baseline case constant. Shown in Figures 3.7b and 3.8b, respectively, are the particle and porosity distributions for a higher infusion rate of 10 μL/min. In addition to a longer backflow length, the elevated infusion rate yields a larger tissue porosity and lower particle concentration near the injection site. The variations of porosity and nanoparticle concentration along the injection direction for different infusion rates are shown in Figures 3.9a and 3.9b, respectively. For both infusion rates, tissue porosity and nanoparticle concentration decrease monotonously in the radial direction from the injection site. A higher infusion rate causes a deeper particle penetration depth and lower nanoparticle concentration at the needle tip. In comparison to the particle distribution in the absence of tissue deformation (Figure 3.3), the particle concentration decays less sharply in the radial direction.

A 32-gauge needle was used in the simulation to examine the effects of the needle size on the nanofluid infusion. It should be noted that the diameter of a 32-gauge needle is smaller than a 26 one. The predicted distribution of the nanoparticle concentration is shown in Figure 3.7c. It demonstrates that reducing the needle size yields a shorter backflow length (4 mm) despite the elevated infusion pressure required for delivering the same amount of the infusate. This result is consistent with the study of Morrison et al. [36]. Figures 3.10a and 3.10b compare the variations of the porosity and particle concentration along the injection direction, respectively, for two different needle sizes, 26 and 32 gauge. There is clear indication that the smaller needle size is associated with a more enlarged tissue pore size, and a lower particle concentration at the needle tip. Also, a deeper penetration depth occurs when the smaller needle size is used.

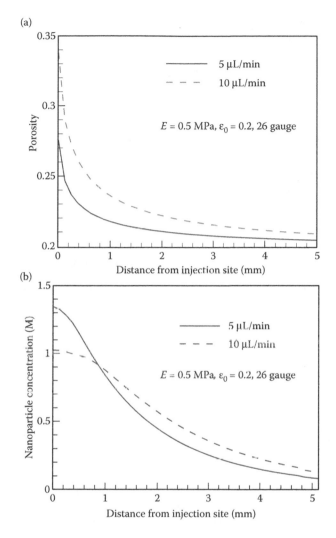

FIGURE 3.9 Variations of porosity (a) and particle distribution (b) along the injection direction for various infusion rates ($E = 0.5$ MPa, $\varepsilon_0 = 0.2$, 26-gauge needle).

Figure 3.7d displays the nanoparticle distribution for tumors with a different value of Young's modulus, $E = 0.2$ MPa. In comparison to Figure 3.7a where $E = 0.5$ MPa, a flexible tissue makes it easier to deform, leading to elongated backflow length, more enlarged porosity near the needle tip, and reduced particle concentration near the needle tip (Figure 3.11).

We also compared the nanoparticle concentration distributions in a tumor with various amounts of nanofluid infused. The pattern of nanoparticle distribution in a tumor with 0.2-cc nanofluid is similar to that shown in Figure 3.7a and therefore, is not displayed. The variations of nanoparticle distributions along the injection direction are shown in Figure 3.12.

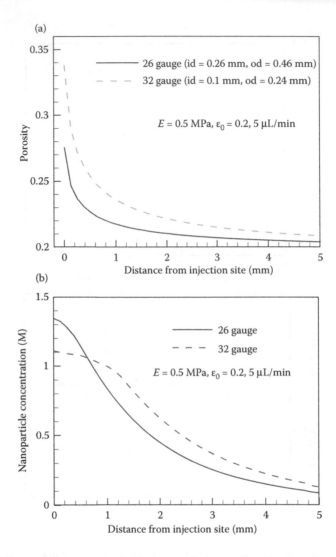

FIGURE 3.10   Variations of porosity (a) and concentration (b) along the injection direction for various needle sizes.

3.6   DISCUSSIONS

In hyperthermia treatments of cancers, tumor cells are killed by maintaining sufficiently high-temperature elevations for a period of time. Typically, 43°C (6°C above 37°C) is the minimal temperature threshold to induce cytotoxic responses if the heating time is longer than 1 or 2 hours. Salloum et al. [50] proposed an optimization algorithm for irregular large-sized tumors with magnetic nanoparticles being delivered at multiple sites. The developed algorithm allows adjustments of injection parameters such as injection sites, injection rates, and injection amounts for elevating at least 90% of a tumor above certain threshold temperature (43°C), whereas less

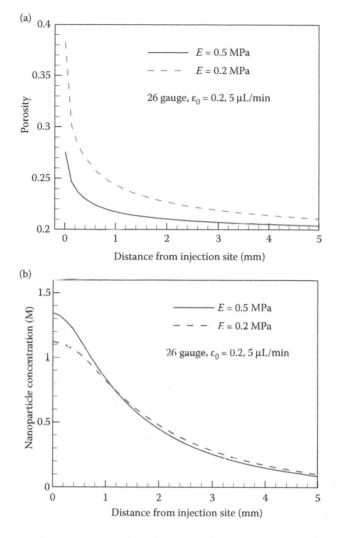

FIGURE 3.11 Variations of (a) porosity and (b) particle concentration along the infusion direction for various Young's modulus.

than 10% of the normal tissue temperatures exceed this threshold. The success of this optimization algorithm requires a controlled particle delivery at each injection site. Understanding the particle transport mechanism and the effects of injection parameters are critical for designing the multi-site injection strategy.

Particle binding to the cellular structure has been identified as an important mechanism that causes the nonuniform particle distribution in tissues and limits the penetration depth of the particles after an infusion [54]. The interstitial fluid velocity, tumor porosity, and particle surface properties are considered as leading factors that affect the rate of particle deposition on the cellular structure. Local interstitial velocities affect the particle binding through two competing mechanisms. A high

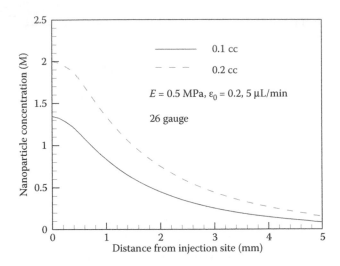

FIGURE 3.12 Variations of particle concentration distribution along the injection direction for various injection amounts.

fluid velocity enhances the rate at which the particles are brought to the cell surface; however, it also yields a greater lift force that keeps particle away from the cell surface. Our simulation results show that the lift force dominates the deposition process for the operational conditions used in this study. Thereby, increasing the local fluid velocity leads to a decreased deposition rate coefficient. In addition to the lift force, the repulsive electrostatic force due to surface coating also plays an important role in the particle–cell interactions. Proper coating of the particles provides an effective means of avoiding nanoparticle accumulation at the injection site. Furthermore, larger tissue porosities are associated with larger pore sizes and larger distances between the cells. The deposition rate coefficient decreases substantially with an increase in the tumor porosity, especially for low interstitial velocities.

Tissue deformation can significantly affect the transport of nanoparticles during an infusion process in that it affects the velocity field, forms a backflow, and alters the effective pore size of the tumor. In particular, the formation of the backflow causes the leakage of the nanoluifd into the surrounding normal tissue and reduces the particle concentration near the injection site. The length of the backflow is dependent on the infusion parameters and tumor properties. Our study also shows that tissue deformation enlarge the effective pore size of the tumor near the injection site and mitigate the particle accumulation near the injection site.

Infusion rates, needle sizes, and the elastic properties of a tumor are vital parameters that affect both particle deposition and tissue deformation. A higher infusion rate yields three consequences: a longer backflow length and the spreading of the particles along the needle track, a lower deposition rate coefficient, and a larger tissue porosity near the needle tip. All these contribute to a reduced particle deposition at the needle tip, leading to deeper particle penetrations and lower particle concentrations at the injection site. The relationship of the particle penetration depth to the infusion rates observed in this study is consistent with previous micro-computed tomographic imagining results and SAR measurements in gels and tissues [3,48,49]. For a given infusion rate, it is the needle size that defines the infusion pressure and the velocity of the infusate at the needle tip. In fact, large-gauge needles having a smaller diameter result in a higher infusion pressure at the needle tip. The elevated infusion pressure causes faster velocity and more enlarged pore size near the needle tip, both facilitating particle infiltration into tissue. Also, a smaller needle size can cause a shorter backflow length which is more favorable in confining the therapeutic agents inside of a tumor. However, large-gauge needles also have adverse effect on the infusion. For instance, the reduced needle size only affects an area in the close vicinity of the needle tip and the elevated infusion pressure may cause tumor breakage.

The response of tissue to infusion pressure is largely dependent on the tissue elastic properties such as Young's modulus. A stiffer tissue yields a shorter backflow length and less substantial change in tissue porosity near the needle tip. Thereby, more nanoparticles are accumulated near the needle tip. Tissue elastic properties should be considered when selecting infusion parameters. Increasing the injection amount can enhance the nanoparticle concentration in the tumor. However, most nanoparticles will be confined in the tumor if the leakage due to backflow can be minimized.

As nanoparticle transport in biological tissues is such a complex matter, a number of simplifications were made when developing this multiscale model. Firstly, one major simplification is that the hindrance of the extracellular matrix to particle transport is neglected when predicting the deposition rate coefficient. Consequently, the prediction made by this model may overestimate the particle penetration depth. In contrast, this model does not consider the effect of limited binding sites on the cell surfaces [37]. Neglecting the availability of the binding sites may cause overestimation of the particle deposition on the cell surface. Secondly, the physicochemical theory underlying the calculation of particle interactions with the cell surface insofar shares the general properties of all colloidal particles. However, the interaction of particles with the living cells is of course more complex than that of nonliving, inert, smooth, and spherical bodies. Thirdly, this study assumes that the entire surface of a cell is exposed to the particle deposition. In fact, the complex cellular structure renders the flow path tortuous and the random packing of the cells causes some volume fraction inaccessible to the nanoparticles. Lastly, it should also be noted that in the currently model, dilute nanofluid is employed and particle agglomeration is not considered. It is unclear to what extent the particle–particle interactions affect the particle transport, especially when highly concentrated nanofluids are employed. These limitations should be addressed in the future through experimental and theoretical study.

In addition, the poroelastic model developed in this study is limited to spherical tumors of homogenous properties and isotropic deformation. However, nanofluid transport in tumors and the resulting nanofluid distributions are dependent on the heterogeneous and anisotropic tissue properties, irregular tumor shape, and cracks and necrotic tissues which are common features in large tumors. Also, elevated infusion pressures may break tumors and cause the formation of cracks during an infusion. These mechanisms should be investigated in future studies.

Despite the assumptions used in this study, the simulation predicts the penetration depth at the same order of magnitude as those obtained through micro-computed tomographic imaging [3] and indicated by SAR measurements [48,49], suggesting that the modeling framework presented in this study captures the main features of the complex process of nanoparticle transport in tissues. A quantitative relationship among fluid flow, tissue deformation, nanoparticle deposition, and nanoparticle transport is established and the effects of major infusion parameters on nanoparticle distribution are characterized.

3.7 CONCLUSION

In nanoparticle hyperthermia, controlling the nanoparticle distribution delivered in tumors is vital for achieving an optimum distribution of temperature elevations to maximize the damage of the tumor cells while minimizing the heating in the surrounding healthy tissues. A multi-scale model was developed in this study to investigate the spatial concentration distribution of nanoparticles in a tumor after an intratumoral infusion. Simulations were performed using this model to quantify the effects tumor deformation and particle binding to the cellular structure on the behavior of particle transport. The simulation results show that the rate of particle binding to the cell surface can be reduced by increasing interstitial fluid velocity, tissue porosity, and surface charge of the particles. Infusion-induced tissue deformation causes backflow and change in tissue porosity near the needle tip, which substantially affects the distribution and penetration depth of the nanoparticles. A longer backflow length occurs when higher infusion rates and larger needle sizes are used on tumors with lower Young's modulus. In contrast, high infusion rates, small needle sizes, and low values of Young's modulus cause reduced nanoparticle deposition near the injection site. The nanoparticle concentration in a tumor can be increased if more nanofluid is infused.

This study will be continued by progressively including further considerations such as heterogeneous tumor structure, the presence and formation of cracks in a tumor, and correction factors for the deposition rate coefficient to address the limitations of the current model. We anticipate that continuous improvement in the model will advance the design of treatment protocols for large-sized and irregular-shaped tumors using nanoparticle hyperthermia.

ACKNOWLEDGMENTS

This research is supported in part by an NSF research grant CBET-0828728, an NSF MRI grant CBET-0821236, and a research grant from the University of Maryland Baltimore County (UMBC) Research Seed Funding Initiative.

NOMENCLATURE

A	Potential
b	Radius of the fluid shell shown in Figure 3.1
F	Shape factor in Equation 3.8
C	Molar concentration of the particles in the fluid
a_c	Cell radius
d_p	Particle diameter
e	Dilatation

D	Diffusion coefficient
E	Young's modulus
F	Forces acting on the particle
G	Lamé constants
h	Distance between the particle centre and the wall
I	Electron charge
K	Permeability of the tissue
k_f	Deposition rate coefficient
k_B	Boltzmann's constant
m	Mass
M	Material constant in Equation 3.6
p_f	Fluid pressure
r	Displacement vector of a particle
S	Volumetric concentration of the nanoparticles bounded to the solid structure
T	Absolute temperature
t	Time
u	Displacement vector of solid structure
v	Interstitial fluid velocity vector
V_∞	Free stream velocity
x, y	Particle position

GREEK SYMBOLS

γ	Thickness of the fluid layer
ε	Porosity
ε_0	Vacuum permittivity/initial porosity
ε_r	Relative dielectric constant of the water
ξ	Uniformly distributed random number
η_s	Collector efficiency
z	Valence of the electrolyte
κ	Debye–Hückel parameter
λ	Lamé constant
μ	Viscosity of the fluid
ρ	Density
τ	Tortuosity
υ	Poisson ratio

REFERENCES

1. P. Adomeit and U. Renz, Correlation for the particle deposition rate accounting for lift forces and hydrodynamic mobility reduction, *Canadian Journal of Chemical Engineering*, 78, 32–39, 2000.
2. E. Allard, C. Passirani, and J.P. Benoit, Convection-enhanced delivery of nanocarriers for the treatment of brain tumors, *Biomaterials*, 30(12), 2302–2318, 2009.
3. A. Attaluri, R. Ma, Y. Qiu, W. Li, and L. Zhu, Nanoparticle distribution and temperature elevations in prostatic tumors in mice during magnetic nanoparticle hyperthermia, *International Journal of Hyperthermia*, 27(5), 491–502, 2011.

4. PJ. Basser, Interstitial pressure, volume, and flow during infusion into brain tissue, *Microvascular Research*, 44, 143–165, 1992.

5. RJ. Bernardi, AR. Lowery, PA. Thompson, SM., Blaney, and JL. West, Immunonanoshells for targeted photothermal ablation in medulloblastoma and glioma: An *in vitro* evaluation using human cell lines, *Journal of Neuro-oncology*, 86, 165–172, 2008.

6. RH. Bobo, D.W. Laske, A. Akbasak, P.F. Morrison, R.L. Dedrick, and E.H. Oldfield, Convection-enhanced delivery of macromolecules in the brain, *Proceedings of the National Academy of Sciences of the United States of America*, 91(6), 2076–2080, 1994.

7. ZJ. Chen, W.C. Broaddus, R.R. Viswanathan, R. Raghavan, and G.T. Gillies, Intraparenchymal drug delivery via positive-pressure infusion: Experimental and modeling studies of poroelasticity in brain phantom gels, *IEEE Transactions on Biomedical Engineering*, 49(2), 85–96, 2002.

8. XM. Chen and M. Sarntinoranont, Biphasic finite element model of solute transport for direct infusion into nervous tissue, *Annuals of Biomedical Engineering*, 35(12), 2145–2158, 2007.

9. APC. Choi and YP. Zheng, Estimation of Young's modulus and poisson's ratio of soft tissue from indentation using two different-sized indentors: Finite element analysis of the finite deformation effect, *Medical and Biolgical Engineering and Computing*, 43, 258–264, 2005.

10. RG. Cox and SK. Hsu, The lateral migration of solid particles in a laminar flow near a plane, *International Journal of Multiphase Flow*, 3(3), 201–222, 1977.

11. LE. Dillehay, Decreasing resistance during fast infusion of a subcutaneous tumor, *Anticancer Research*, 17(1A), 461–466, 1997.

12. IH. El-Sayed, X. Huang, and MA. El-Sayed, Selective laser photo-thermal therapy of epithelial carcinoma using anti-EGFR antibody conjugated gold nanoparticles, *Cancer Letter*, 239, 129–135, 2006.

13. RK. Gilchrist, R. Medal, WD. Shorey, RC. Hanselman, JC. Parrott, and CB. Taylor, Selective inductive heating of lymph nodes, *Annals of Surgery*, 146(4), 596–606, 1957.

14. TT. Goodman, J. Chen, K. Matveev, and SH. Pun, Spatio-temporal modeling of nanoparticle delivery to multicellular tumor spheroids, *Biotechnology and Bioengineering*, 101(2), 388–399, 2008.

15. WY. Gu, H. Hao, CY. Huang, and HS. Cheung, New insight into deformation-dependent hydraulic permeability of gels and cartilage, and dynamic behavior of agarose gels in confined compression, *Journal of Biomechanics*, 36(4), 593–598, 2003.

16. J. Happel, Viscous flow in multiparticle systems: Slow motion of fluids relative to beds of spherical particles, *AIChE J*, 4, 197–201, 1958.

17. I. Hilger, W. Andra, R. Hergt, R. Hiergeist, H. Schubert, and WA. Kaiser, Electromagnetic heating of breast tumors in interventional radiology: *in vitro* and *in vivo* studies in human cadavers and mice, *Radiology*, 218(2), 570–575, 2001.

18. I. Hilger, R. Hergt, and WA. Kaiser, Towards breast cancer treatment by magnetic heating, *Journal of Magnetism and Magnetic Materials*, 293(1), 314–319, 2005.

19. MH. Holmes and V.C. Mow, The nonlinear characteristics of soft gels and hydrated connective tissues in ultrafiltration, *Journal of Biomechanics*, 23(11), 1145–1156, 1990.

20. JN. Israelachvili, *Intermolecular and Surface Forces*, Academic Press, London, 1991.

21. O. Ivanchenko, N. Sindhwani, and A. Linninger, Experimental Techniques for Studying Poroelasticity in Brain Phantom Gels Under High Flow Microinfusion, *Journal of Biomechanical Engineering—Transactions of the ASME*, 132(5), 2010.

22. RK. Jain, Delivery of molecular and cellular medicine to solid tumors, *Advanced Drug Delivery Reviews*, 26(2–3), 71–90, 1997.

23. A. Jordan, R. Scholz, K. Maier-Hauff, FKH. van Landeghem, N. Waldoefner, U. Teichgraeber, J. Pinkernelle et al., The effect of thermotherapy using magnetic nanoparticles on rat malignant glioma, *Journal of Neuro-Oncology*, 78(1), 7–14, 2006.

24. R. Johnson, *The Handbook of Fluid Dynamics*, CRC Press, Boca Raton, pp. 18–29, 18–32, 1998.
25. A-RA. Khaled and K. Vafai, The role of porous media in modeling flow and heat transfer in biological tissues, *International Journal of Heat and Mass Transfer*, 46(26), 4989–5003, 2003.
26. B. Khlebtsov, V. Zharov, A. Melnikov, V. Tuchin, and N. Khlebtsov, Optical amplification of photothermal therapy with gold nanoparticles and nanoclusters, *Nanotechnology*, 17(20), 5167–5179, 2006.
27. PE. Kloeden and E. Platen, Numerical solution of stochastic differential equations, Springer, Berlin, 110–154, 1992.
28. WM. Lai and VC. Mow, Drug-induced compression of articular cartilage during a permeation experiment, *Biorheology*, 17, 111–123, 1980.
29. FJ. Lazaro, AR. Abadia, MS. Romero, L. Gutierrez, J. Lazaro, and MP. Morales, Magnetic characterisation of rat muscle tissues after subcutaneous iron dextran injection, *Biochimica et Biophysica (BBA)—Molecular Basis of Disease,* 1740(3), 434–445, 2005.
30. R. Maniero and P. Canu, A model of fine particles deposition on smooth surfaces: I—theoretical basis and model development, *Chemical Engineering Science*, 61(23), 7626–7635, 2006.
31. JP. Matas, JF. Morris, and E. Guazzelli, Lateral forces on a sphere, *Oil Gas Science and Technology—Rev Inst Fr Pet*, 59(1), 59–70, 2004.
32. H. Matsuki, T. Yanada, T. Sato, K. Murakami, and S. Minakawa, Temperature sensitive amorphous magnetic flakes for intratissue hyperthermia, *Material Science and Engineering*, A181(182), 1366–1368, 1994.
33. S. McGuire and F. Yuan, Quantitative analysis of intratumoral infusion of color molecules, *American Journal of Physiology. Heart and Circulatory Physiology,* 281(2), H715–H721, 2001.
34. S. McGuire, D. Zaharoff, and F. Yuan, Nonlinear dependence of hydraulic conductivity on tissue deformation during intratumoral infusion, *Annals of Biomedical Engineering,* 34(7), 1173–1181, 2006.
35. P. Moroz, SK. Jones, and BN. Gray, Magnetically mediated hyperthermia: Current status and future directions, *International Journal of Hyperthermia*, 18(4), 267–284, 2002.
36. PF. Morrison, MY. Chen, RS. Chadwick, RR. Lonser, and EH. Oldfield, focal delivery during direct infusion to brain: Role of flow rate, catheter diameter, and tissue mechanics, *American Journal of Physiology—Regulatory Integrative and Comparative Physiology*, 277(4), R1218–R1229, 1999.
37. KB. Neeves, AJ. Sawyer, CP. Foley, WM. Saltzman, and WL. Olbricht, dilation and degradation of the brain extracellular matrix enhances penetration of infused polymer nanoparticles, *Brain Research*, 1180, 121–132, 2007.
38. KE. Nelson and TR. Ginn, Colloid filtration theory and the Happel sphere-in-cell model revisited with direct numerical simulation of colloids, *Langmuir*, 21(6), 2173–2184, 2005.
39. PA. Netti, LT. Baxter, and Y. Boucher, Macro-and microscopic fluid transport in living tissues: Application to solid tumors, *Bioengineering, Food and Natural Products*, 43(3), 818–834, 1997.
40. CR. O'Melia and W. Stumn, Theory of water filtration, *Journal of American Water Works Association,* 59, 1393–1412, 1967.
41. DP. O'Neal, LR. Hirsch, NJ. Halas, JD. Payne, and JL. West, Photo-thermal tumor ablation in mice using near infrared-absorbing nanoparticles, *Cancer Letters*, 209, 171–176, 2004.
42. R. Raghavan, S. Mikaelian, M. Brady, and ZJ. Chen, Fluid infusions from catheters into elastic tissue: I. Azimuthally symmetric backflow in homogeneous media, *Physics in Medicine and Biology*, 55, 281–304, 2010.

43. R. Rajagopalan, and C. Tien, Trajectory analysis of deep-bed filtration with sphere-in-cell porous media model, *AIChE Journal*, 22(3), 523–533, 1976.
44. S, Ramanujan, A. Pluen, TD. McKee, EB. Brown, Y. Boucher, and RK. Jain, Diffusion and convection in collagen gels: Implications for transport in the tumor interstitium, *Biophysical Journal*, 83(3), 1650–1660, 2002.
45. BV. Ramarao, C. Tien, and S. Mohan, Calculations of single fiber efficiencies for interception and impaction with superposed Brownian motion, *Journal of Aerosol Science*, 25(2), 295–313, 1994.
46. W. Russel, AD. Saville, and W. Schowalter, *Colloidal Dispersions*, Cambridge University Press, UK, 1989.
47. PG. Saffman, The lift on a small sphere in a slow shear flow, *Journal of Fluid Mechanics Digital Archive*, 22, 385–400, 1965.
48. M. Salloum, RH. Ma, D. Weeks, and L. Zhu, Controlling nanoparticle delivery in magnetic nanoparticle hyperthermia for cancer treatment: Experimental study in agarose gel, *International Journal of Hyperthermia*, 24(4), 337–345, 2008a.
49. M. Salloum, RH. Ma, and L. Zhu, An in-vivo experimental study of temperature elevations in animal tissue during magnetic nanoparticles hyperthermia, *International Journal of Hyperthermia*, 24(7), 589–601, 2008b.
50. M. Salloum, RH. Ma, and L. Zhu, Enhancement in treatment planning for magnetic nanoparticle hyperthermia: Optimization of the heat absorption pattern, *International Journal of Hyperthermia*, 25(4), 309–321, 2009.
51. CN. Satterfield, *Mass Transport in Heterogeneous Catalysis*, MIT Press, Cambridge, 1970.
52. S.E. Skrabalak, J. Chen, L. Au, and X. Lu, Gold nanocages for biomedical applications, *Advanced Materials (Deerfield)*, 19, 3177–3184, 2007.
53. I. Sobey and B. Wirth, Effect of non-linear permeability in a spherically symmetric model of hydrocephalus, *Mathematical Medicine and Biology*, 23(4), 339–361, 2006.
54. D. Su, RH. Ma, M. Salloum, and L. Zhu, Multi-scale study of nanoparticle transport and deposition in tissues during an injection process, *Medical and Biological Engineering and Computing,* 48, 853–863, 2010.
55. D. Su, RH. Ma, and L. Zhu, Numerical study of nanofluid infusion in deformable tissue during hyperthermia treatment of cancers, *Medical and Biological Engineering and Computing,* 49(11), 1233–1240, 2011.
56. K. Subramani, H. Hosseinkhani, A. Khraisat, M. Hosseinkhani, and Y. Pathak, Targeting Nanoparticles as Drug Delivery Systems for Cancer Treatment, *Current Nanoscience*, 5(2), 135–140, 2009.
57. MA. Swartz and ME Fleury, Interstitial flow and its effects in soft tissues, *Annuals of Reviews in Biomedical Engineering*, (9), 229–256, 2007.
58. C. Tien and BV. Ramarao, *Granular Filtration of Aerosols and Hydrosols*, 2nd edn. Elsevier, Oxford, 2007.
59. N. Tufenkji and M. Elimelech, Correlation equation for predicting single-collector efficiency in physicochemical filtration in saturated porous media, *Environmental Science adn Technoogy,* 38(2), 529–536, 2004.
60. M. Wang and M. Thanou, Targeting nanoparticles to cancer, *Pharmacological Research*, 62, 90–99, 2010.
61. Y. Wang, H. Wang, CY. Li, F. Yuan, Effects of rate, volume, and dose of intratumoral infusion on virus dissemination in local gene delivery, *Molecular Cancer Therapeutics*, 5(2), 362–366, 2006.
62. P. Warszynski, Coupling of hydrodynamic and electric interactions in adsorption of colloidal particles, *Advances in Colloid and Interface Science*, 84(1–3), 47–142, 2000.

4 Superparamagnetic Iron Oxide Nanoparticle Heating
A Basic Tutorial

Michael L. Etheridge, Navid Manuchehrabadi,
Rhonda R. Franklin, and John C. Bischof

CONTENTS

4.1 INTRODUCTION

Nanoparticles are being used in a rapidly increasing variety of biomedical applications, including detection, imaging, and treatment of disease. These particles have controllable dimensions in the nanometer range, matching the scale of biological entities and facilitating intimate interactions with cells and molecular constituents. They exhibit remarkable physical properties that can be finely tuned by adjusting their composition, size, and shape. One of the special features of iron oxide nanoparticles is their ability to serve as colloidal mediators for heat generation in externally applied, alternating magnetic fields. This application has been termed magnetic fluid hyperthermia (MFH) and has attracted growing research interest for treatment of malignant tumors due to its potential for highly specific energy delivery through a minimally invasive (or potentially noninvasive) platform. In this method, magnetic particles delivered to tissue induce localized heating when exposed to an alternating magnetic field, leading to thermal damage concentrated to the tumor [1]. Although cancer therapy has been the overwhelming focus in development thus far, magnetically heated nanoparticles have also demonstrated promise for application in thermoresponsive drug delivery [2,3], activation of ion channels and neurons [4], remote-controlled microfluidic valves [5], and heat-initiated shape-memory alloys [6]. As a result of the well-demonstrated bio-compatibility of iron oxide nanoparticles, magnetite (Fe_3O_4) and maghemite (Fe_2O_3) are the most popular materials for *in vivo* investigations [7].

In any hyperthermia application, the distribution of temperature elevation due to the specific absorption rate (SAR) is an important factor in determining the therapeutic outcome. SAR for MFH can be estimated experimentally and theoretically, while the exact mechanisms by which heating is derived (eddy current, hysteresis, and relaxation processes) can vary. Experimental, *in vitro* SAR values previously reported for different magnetic colloids show strong sample/protocol dependence (Table 4.1). In many other studies, the conditions of the tests (properties of system/ tissue, nanoparticles, or magnetic field) are not published or are unclear. This wide variability throughout the literature demonstrates the need for standard methods of measuring and predicting SAR for magnetic nanoparticles. The high prevalence of iron oxide in experimental study should also be noted.

One of the most important aspects in using MFH for cancer treatment in clinical approaches is controlling the heat distribution and temperature rise. Numerous articles predict MFH outcomes based on presumed SARs in phantoms and tissues assuming idealized nanoparticle distributions [16–19]. While this is a useful academic exercise, it will not translate into clinical use unless there is an accurate method to (1) measure the concentration of the nanoparticles within the tissue; and (2) translate this concentration to an effective SAR. The capability to measure nanoparticle concentration in phantoms and tissues has been demonstrated through computed tomography (CT) imaging [20,21] and so the conversion of concentration to SAR for superparamagnetic nanoparticles is the focus of this tutorial.

In order to accurately model SAR, it is useful to consider certain factors. First, most samples of iron oxide nanoparticles are polydisperse (having broad size distributions). As will be discussed in subsequent sections, the power generated is a strong function of size and is, therefore, not uniform for all the nanoparticles in a sample. Second, the effective temperature rise observed is the macroscopic temperature change of the suspending medium. The connection between nanoscale heat response and macroscopic heat response is an intense area of research and must be understood before accurate quantitative models can be developed [22,23]. Finally, the thermophysical properties (viscosity, density, specific heat, thermal conductivity) of tissues and *in vitro* model systems are complicated and highly dependent on

TABLE 4.1

Values of *in vitro* SAR$_{Fe}$, as Reported for a Number of Magnetic Colloids having Different Nanoparticle Properties [8–15]

Core Material	Magnetic Core Diameter (nm)	Coating	Suspending Medium	H_a (kA/m)	f (kHz)	SAR$_{Fe}$ (W/g Fe)	Reference
Iron oxide	3.1	Dextran	Water	0.5	200–1000	0.15–0.8	[8]
		Dextran				10–235	
MnZnFeO	7.6	Dextran	Water	0.2–13.2	520	0.05–0.5	
Maghemite	3–15	Surfactant	Water	6.84	1100	60	[9]
		Dextran				140–370	
Magnetite	10	Uncoated	Kerosene	6.5	300	60	[10]
	10	Uncoated	Ether			40	
	8	Uncoated	Water			29	
	6	Dextran	Water			< 0.1	
Maghemite	100–150	Uncoated	Water	7.2	880	42	[11]
Magnetite	100–150	Uncoated				45	
Ferrite	6–12	Dextran				12 ~ 140	
	6–12	Dextran				90	
	10–12	Dextran				210	
Magnetite	8	Surfactant	Water	6.5	400	84	[12]
	3–10	Starch				56	
	3–10	Starch				31	
	3–10	Starch				54	
Magnetite	7.5	Dextran	Water	32.5	80	15.6	[13]
	13	Dextran				39.4	
	46	Dextran				75.6	
	81	Dextran				63.7	
Magnetite	10	Surfactant	Water	14	175	42–54	[35]
		Collagen				27	
Iron oxide	11.2*	Carboxyl	Water	5.66	900	67	[15]
	12.6*	Starch				77	
	10.5*	Dextran				89	
	11.8*	Carboxyl				90	

Note: The nanoparticles with diameters <20 nm are likely heated through relaxation mechanisms, while the larger particles are likely subject to hysteresis losses. Some studies only described the core material as "iron oxide" or "ferrite" without providing the specific makeup. The diameter values marked with (*) were reported as crystallite size and the core structure may be composed of multiple crystallites.

the tissue microenvironment and must be factored in to accurate modeling. Many of the properties affecting magnetic nanoparticle heating are included in Table 4.2 and their interactions will be discussed in detail in the following sections. This analysis provides values for predicting SAR, which are then compared with experimental results.

TABLE 4.2

Summary of Important Parameters for SAR Calculation in Magnetic Nanoparticle Heating Applications

Nanoparticle	Medium	Applied Magnetic Field
Saturation magnetization	Nanoparticle concentration	Field strength
Magnetic anisotropy	Viscosity	Applied frequency
Magnetic remanence/ coercivity		
Size	Specific heat	Uniformity
Polydispersity	Conductivity	
Coating	Homogeneity	
Aggregation		

4.2 THEORETICAL BACKGROUND AND UNDERLYING PHYSICS

Although magnetic fields demonstrate minimal interaction and attenuation in tissues compared with other forms of electromagnetic radiation, alternating magnetic fields will still induce some eddy current losses and this limits usable fields in biological systems. Applications utilizing magnetic nanoparticles attempt to minimize this field–tissue interaction while maximizing interactions between the field and the energy absorbing nanoparticle deposits. Energy conversion in the particles occurs through hysteresis losses in multidomain particles or through relaxation losses (Brownian and Néelian) in superparamagnetic, single-domain particles. Domain and superparamagnetic behavior is determined by the magnetic material and particle size, with the latter mode generally appearing below about 20 nm for iron oxide [10]. These concepts are discussed in detail in subsequent sections, including conceptual illustrations in Figure 4.3.

4.2.1 Effects of Alternating Magnetic Fields in Human Application

Electromagnetic radiation, such as radiofrequency, microwaves, and lasers, produce very strong interactions with tissues, even in the absence of nanoparticles. This allows direct application in various forms of thermal therapies, but also leads to high attenuation in surface layers, which limits treatment of deep-seated tumors. However, alternating magnetic fields with frequencies up to 10 MHz demonstrate negligible attenuation in tissue equivalents [24,25], offering a platform for uniformly penetrating deep tissue areas with external fields.

The components of human tissue are largely diamagnetic (very low magnetic susceptibility) and so magnetic effects are generally negligible. However, application of an electromagnetic induction field will induce eddy currents in any conducting medium, including biological tissue [25,26]. These eddy currents increase radially and will be subject to resistive losses, so maximum losses are expected in regions with the greatest cross-sectional area (such as the torso). Assuming a uniform field and treating the torso as a cylinder, the volumetric power generation

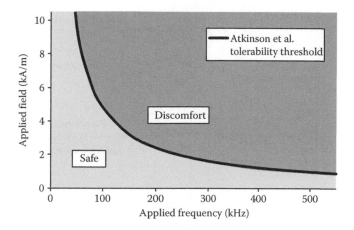

FIGURE 4.1 Clinical limits of alternating magnetic field parameters for the torso proposed by Atkinson et al. [27].

(P) can be estimated by integrating the time averaged current density over the area, giving [27]:

$$P = \sigma\,(\pi\mu_0\,f\,H_a)^2\,r^2 \tag{4.1}$$

where σ is the bulk tissue conductivity, μ_0 is the permeability of free space, f is the applied frequency, H_a is the applied field strength, and r is the effective torso outer radius. The eddy current losses demonstrate three critical quadratic dependencies with frequency, field strength, and radius. Thus, losses will increase significantly with increases in field strength or frequency and will be most prominent in larger cross-sections of tissue.

Atkinson et al. performed a series of clinical studies to determine the range of tolerable parameters for alternating magnetic field-based treatments [27] and very similar limits were verified by Wust et al. in later clinical studies [28,29]. The results indicated that field tolerance could be roughly estimated by limiting the product of frequency and field strength ($f H_a$) to <4.85 × 10⁸ A/m-s. The approximate range of usable field parameters is illustrated in Figure 4.1. As will be discussed in the next section, SAR for magnetic nanoparticles depends strongly on the applied field and significant heating is not expected for field strengths <1 kA/m. This limit therefore constrains the applied frequency to <500 kHz and all subsequent discussions will focus on physical phenomena occurring in this lower frequency range. It should also be noted that this threshold was developed with respect to the torso and higher values should be tolerable for fields constrained to regions of the body with smaller cross sections (such as the cranium or extremities).

4.2.2 Physical Mechanisms of Heat Generation

Detailed descriptions of the fundamental principles of magnetics and magnetic materials can be found in textbooks by Cullity [30] and O'Handley [31], and Gubin

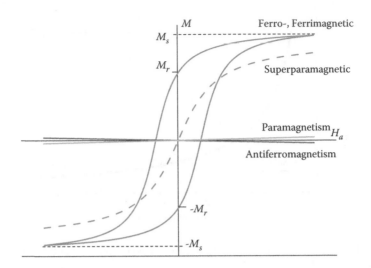

FIGURE 4.2 Characteristic response of material magnetization to an applied field for various types of magnetism. Diamagnetism is not shown due to its very weak magnetic response.

has recently published a textbook focusing specifically on magnetism in nanoparticles [32]. The different types of magnetism can be best described by a material's response to an applied field, as demonstrated in Figure 4.2. Magnetism arises at the atomic level from unpaired electron spins, which behave like atomic dipole moments. Ferromagnetism is the strongest form of magnetism and is due to strong exchange interactions between atomic moments in metals (most commonly, Fe, Ni, and Co). Ferrimagnetism is similar to ferromagnetism, but results from exchange interactions in ionic solids, such as metallic oxides (iron oxide). Both ferromagnetic and ferrimagnetic materials demonstrate strong enough internal interactions to maintain magnetization in the absence of an applied field, but when a strong external field is applied, the atomic moments will align to the applied field direction. Diamagnetism, paramagnetism, and antiferromagnetism generally demonstrate fairly weak interactions. Superparamagnetism is a unique form of magnetic behavior that arises in nanoscale particles and will be described in more detail below.

Like many physicochemical properties, a material's magnetic behavior can change as its characteristic dimensions approach the nanoscale and this affects the loss mechanisms in an alternating magnetic field [8,33,34]. Heat generation in magnetic materials under alternating magnetic fields can be generally ascribed to one of three different mechanisms, depending on size scales [35], as illustrated in Figure 4.3:

(i)   Eddy current generation in bulk magnetic materials
(ii) Hysteresis loss in bulk and multidomain magnetic materials
(iii) Relaxation loss in superparamagnetic, single-domain nanoparticles

4.2.2.1 Eddy Current Generation

As discussed in the previous section, eddy current induction takes place whenever a conductor is exposed to an alternating magnetic field and this can result in resistive

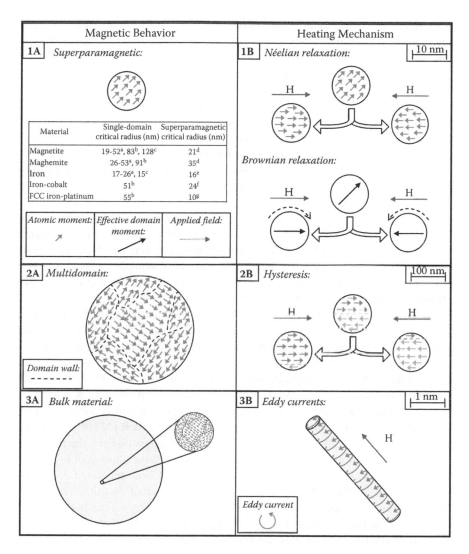

FIGURE 4.3 Size-dependent magnetic behavior and mechanisms of heat generation. Values for single-domain critical radii are taken from: [32][a], [36][b], and [34][c]. Values for super-paramagnetic critical radii are calculated from properties in: [37][d], [31][e], [38][f], and [39][g].

heating. Bulk magnetic material and eddy currents induced by an alternating, external magnetic field are illustrated in Panel 3A and 3B of Figure 4.3. For magnetic materials, significant eddy current heating is observed only for bulk magnetic materials (i.e., >1 mm for magnetite) [19]. For individual iron oxide nanoparticles (10–100 nm), eddy current effects can be neglected.

4.2.2.2 Hysteresis Loss

Typical magnetic materials demonstrate unique domains of magnetism (areas of parallel magnetic moments), separated by narrow zones of transition termed domain walls. Simplified domains are illustrated in Panel 2A of Figure 4.3. Domains form to minimize the overall magnetostatic energy of the material, but as dimensions approach the nanoscale, the energy reduction provided by multiple domains is overcome by the energy cost of maintaining the domain walls and it becomes energetically favorable to form a single magnetic domain. A number of methods for estimating the critical radius for single-domain behavior have been proposed [32,34], and the results can vary notably depending on the approach. Some estimated values from literature have been included in the table in Panel 1A of Figure 4.3, with typical diameters on the order of tens of nanometers.

Hysteresis loss can occur in multidomain particles (and bulk materials). Ferromagnetic and ferrimagnetic materials, when placed in an alternating magnetic field, will produce heat due to hysteresis losses. When exposed to the external field, the magnetic moments tend to align in the direction of the applied field. This is the phenomenon of magnetization. Essentially, domain walls can move in the presence of an applied magnetic field such that many single domains combine and create larger domains (domain growth). In other words, those domains whose magnetic moments are aligned with the external field expand at the expense of the other surrounding domains. This domain wall displacement continues until the point of magnetic saturation (M_s), at which point the domain walls are maximally displaced. This domain wall response to an alternating field is shown in Panel 2B of Figure 4.3.

Figure 4.2 presents the relationship between applied magnetic field strength (H_a) and magnetization (M). If the relationship between the two is plotted for increasing levels of field strength, the magnetization will increase up to a point, then saturate. If the magnetic field is now reduced, the plotted relationship will follow a different part of the curve back to zero field strength, at which point it will be offset from the original curve by an amount called the remanent magnetization (M_r) or remanence. If this relationship is plotted for all strengths of applied magnetic field, the result is a hysteresis loop. The width of the middle section describes the amount of hysteresis, related to the coercivity of the material. The amount of heat generated is directly related to the area of the hysteresis loop [31]. Although much of the initial research in MFH has focused on superparamagnetic relaxation heating, hysteresis has also demonstrated notable heating capability [40,41]. Hergt et al. [42] utilized experimental data for various ferromagnetic particles ranging in size from 30 to 100 nm to produce phenomenological expressions which closely predicted losses based on the applied field parameters and particle size distributions. The experimental values and theoretical predictions offered heating rates comparable to those of superparamagnetic nanoparticles and this is likely going to be an area of significant future development.

4.2.2.3 Relaxation Loss

In addition to single-domain behavior, magnetic nanoparticles can exhibit another type of unique behavior, superparamagnetism, in which thermal motion causes the magnetic moments to randomly flip directions, eliminating any remnant magnetization in the absence of an applied field. Thus, a normally ferromagnetic or ferrimagnetic material will only exhibit magnetism under an applied field. This behavior arises because below a critical volume, the anisotropic energy barrier $(K_u V_m$, where K_u is the material anisotropy constant and V_m is the magnetic volume) of the magnetic crystal is reduced to the point where it can be overcome by the energy of random thermal motion ($k_B T$). The definition of superparamagnetism is somewhat ambiguous, in that it relies on the choice of a measurement time (τ_m) for which the behavior is observed and is generally taken to be 100 s. The approximate critical radius (r_c) for superparamagnetic behavior can be determined by assuming a spherical geometry ($V_m = 4\pi r_c^3/3$) and modifying the equation describing the probability of thermal relaxation [31]:

$$\frac{\tau_m}{\tau_0} = \exp\left(\frac{K_u V_m}{k_B T}\right) \tag{4.2a}$$

$$r_c = \left[\frac{3}{4\pi} \ln\left(\frac{\tau_m}{\tau_0}\right) \frac{k_B T}{K_u}\right]^{\frac{1}{3}} \tag{4.2b}$$

where τ_0 is the attempt time (generally 10^{-9} s), V_m is the volume of magnetic material, k_B is Boltzmann's constant, and T is the absolute temperature. The approximate critical radii for several common magnetic nanoparticle materials are included in the table in Panel 1A of Figure 4.3. Although remnant magnetization and hysteresis behavior are eliminated in superparamagnetic particles, significant losses can still occur through moment relaxation mechanisms.

The physical mechanisms of relaxation leading to losses in superparamagnetic iron oxide nanoparticles are reviewed by Rosensweig [37]. These losses fall into two modes: Brownian and Néelian, which are conceptually demonstrated in Panel 1B of Figure 4.3. The Brownian mode represents the rotational friction component in a given suspending medium. As the whole particle oscillates towards the field, the suspending medium opposes this rotational motion resulting in heat generation. The Néelian mode represents the rotation of the individual magnetic moments towards the alternating field. Upon application of an alternating magnetic field, the magnetic moment rotates away from the crystal axis towards the field to minimize potential energy. The remaining energy is released as heat into the system. The theoretical contribution of each relaxation mechanism is described in more detail below.

4.2.3 Relaxation Time Constants

Brownian relaxation is due to orientation fluctuations of the particle itself in the carrier fluid, assuming the magnetic moment is locked onto the crystal anisotropy axis. The time taken for a magnetic nanoparticle to align with the external magnetic field is given by the Brownian relaxation time constant (τ_B):

$$\tau_B = \frac{3\eta V_H}{k_B T} \tag{4.3}$$

where η is the fluid viscosity and V_H is the hydrodynamic volume of the particle (including coatings).

Néelian relaxation refers to the internal thermal rotation of the particle's magnetic moment within the crystal, which occurs when the anisotropy energy barrier is overcome. The typical time between orientation changes is given by the Néelian relaxation time. Néelian rotation occurs even if the particle movement is blocked. The relaxation time of such a process is

$$\tau_N = \frac{\sqrt{\pi}}{2}\tau_0\frac{e^\Gamma}{\sqrt{\Gamma}}, \ \Gamma = \frac{K_uV_M}{k_BT} \tag{4.4}$$

where Γ is the ratio of anisotropy energy to thermal energy.

As these two relaxation processes are occurring in parallel, the overall behavior can be described by an effective relaxation time (τ), given by

$$\tau = \left(\frac{1}{\tau_B} + \frac{1}{\tau_N}\right)^{-1} = \frac{\tau_B\tau_N}{\tau_B + \tau_N} \tag{4.5}$$

Note that this relationship is very similar to that of current flow through electronic resistors in parallel and that the shorter time constant will dominate the overall behavior.

Effective design and application of MFH requires a clear understanding of Néelian and Brownian contributions to distinguish the portion of heat produced by each mechanism and achieve the highest possible SAR. Néelian relaxation time demonstrates a complicated dependence on magnetic anisotropy and volume, while Brownian relaxation time varies linearly with particle volume and fluid viscosity. Relative relaxation mechanisms for magnetite particles in water (low viscosity, $\eta = 0.0009$ kg/m-s) and glycerol (high viscosity, $\eta = 0.3$ kg/ms) are represented in Figure 4.4. The effective relaxation time is represented as a solid line. The crossover

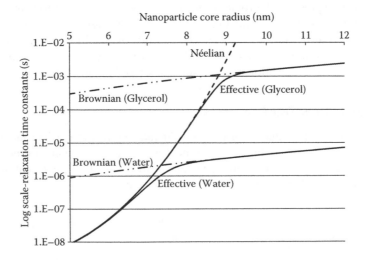

FIGURE 4.4 Characteristic relaxation time constants of magnetite suspended in water and glycerol as a function of particle radius, assuming a 2-nm nonmagnetic coating.

between Néelian and Brownian regimes (which also corresponds to the maximum value of SAR) occurs roughly at a magnetite core radius of 7.4 nm in water and 8.8 nm for glycerol. Relaxation times on the order of 1.5×10^{-5} s $< \tau < 1.5 \times 10^{-6}$ generally result in the most significant heat dissipation.

It is clear that the Néelian relaxation time depends on the anisotropy constant, which is material and crystal dependent. To achieve the maximum SAR in the range of preferred excitation frequencies, there is then an ideal core size for the Néelian contribution. For magnetite ($K_u = 23$ kJ/m³), the ideal radius varies within a range of about $6 < r < 8$ nm in water. Maghemite has a lower anisotropy constant ($K_u = \sim 5$ kJ/m³), which leads to an ideal radius ranging from about $10 < r < 12$ nm in water. Materials with higher anisotropies, such as cobalt–ferrite ($K_u = 200$ kJ/m³) or barium–ferrite ($K_u = 300$ kJ/m³) will have peak Néelian contributions with core radii in the range of $3 < r < 5$ nm [37]. Frequency and viscosity will have the most impact on shifting the peak within these ranges, but all the properties listed in Table 4.2 will contribute at some level.

4.2.4 POWER DISSIPATION FOR SUPERPARAMAGNETIC NANOPARTICLE HEATING

The analytical relationships by which suspensions of superparamagnetic nanoparticles generate heat due to an alternating magnetic field have been described in detail by Rosensweig [37] and will be summarized below. The differential form of the first law of thermodynamics suggests that the internal energy (u) for a system of constant density and unit volume is equal to the sum of the heat added (q) and the work done on the system (w):

$$du = dq + dw \qquad (4.6)$$

Assuming that no heat is added to the magnetic nanoparticles (adiabatic process, $dq = 0$) and the only work done is that of the magnetic field ($dw = H \cdot dB$) then

$$du = H \cdot dB \qquad (4.7)$$

where H is the magnetic field intensity and B is the induction. H and B are collinear and so can be expressed as scalar values (H and B). The general equation for magnetic induction is

$$B = \mu_0 (H + M) \qquad (4.8)$$

where μ_0 is the permeability of free space and M is the material magnetization. Substituting into the previous equations and applying integration by parts leads to

$$\Delta u = -\mu_0 \oint M \, dH \qquad (4.9)$$

If the particles are noninteracting, the response of the magnetization of a ferrofluid to an alternating magnetic field can be described in terms of its complex magnetic susceptibility. This suggests that magnetic work is converted to internal energy when magnetization lags the applied field. If the applied magnetic field is sinusoidal, it can be expressed by Euler's form

$$H(t) = H_a \cos(2\pi f t) \tag{4.10}$$

and the magnetization then becomes

$$M(t) = H_a (\chi_m' \cos(2\pi f t) + \chi_m'' \sin(2\pi f t)) \tag{4.11}$$

where χ_m' is the in-phase component and χ_m'' is the out-of-phase component of magnetic susceptibility. Substituting back into Equation 4.9 and integrating yields

$$\Delta u = \mu_0 \pi H_a^2 \chi_m'' \tag{4.12}$$

where only the out-of-phase component of susceptibility survives (thus it is also termed the "loss component"). This is the change in internal energy per cycle, so the total volumetric power generation (P) can be found by multiplying this value by the applied frequency:

$$P = f \Delta u = \left(\mu_0 \pi f H_0^2 \right) \chi_m'' \tag{4.13}$$

Thus, the rate of heating is dependent on the loss component of susceptibility and the incident power density (which is the term in parentheses). This expression is equivalent to SAR in watts per cubic meter of fluid (or tissue) (that is, SAR = P in Equation 4.13). This value can be easily converted into more standard units of cubic centimeters or grams tissue. In addition, absorption for magnetic nanoparticles is often expressed in terms of watts per mass iron, which is obtained by dividing this expression by the product of the bulk density of iron in the nanoparticles (ρ_{Fe}) and the nanoparticle volume fraction (φ). This value is often termed SAR_{Fe} or specific loss power (SLP). Both SAR and SAR_{Fe} will be used throughout the remainder of the chapter and it is important to keep the distinction straight.

The ferrofluid susceptibility is dependent on both nanoparticle and field properties, so it is helpful to express this quantity through more fundamental parameters. Frequency dependence of the susceptibility is given by

$$\chi_m'' = \frac{2\pi f \tau}{1 + \left(2\pi f \tau \right)^2} \chi_{m,0} \tag{4.14}$$

where $\chi_{m,0}$ is the static susceptibility, which can conservatively be taken as the chord susceptibility corresponding to the Langevin equation:

$$\chi_{m,0} = \chi_{m,i} \frac{3}{\xi} \left[\cot h(\xi) - \frac{1}{\xi} \right] \tag{4.15a}$$

$$\xi = \frac{\mu_0\, M_s\, H_a\, V_m}{k_B\, T} \tag{4.15b}$$

For nanoparticles with bulk magnetic saturation (M_s) and a ferrofluid volume fraction (φ), the initial susceptibility ($\chi_{m,i}$) is determined from differentiation of the Langevin equation:

$$\chi_{m,i} = \frac{\mu_0\, \phi\, M_s^2\, V_m}{3 k_B\, T} \tag{4.16}$$

Finally, the SARs for monodisperse particles (that is, no variation in size) are commonly expressed as

$$SAR = \left(\mu_0\, \pi f\, H_a^2 \right)\left(\chi_{m,0}\, \frac{2\pi f \tau}{1 + \left(2\pi f \tau\right)^2} \right) \sim \text{Units}\left[\frac{W}{m^3} \right] \tag{4.17}$$

$$SAR_{Fe} = \frac{SAR}{\rho_{Fe}\, \phi} \sim \text{Units}\left[\frac{W}{gFe} \right] \tag{4.18}$$

The nanoparticle volume fraction appears in the numerator of Equation 4.16 and so the SAR will be directly proportional to nanoparticle concentration. However, it appears again in the denominator of Equation 4.18 and so SAR_{Fe} is independent of nanoparticle concentration (assuming noninteracting particles).

4.2.5 EFFECTS OF PROPERTIES ON SUPERPARAMAGNETIC HEATING

As discussed, the heating effects of superparamagnetic nanoparticles subjected to alternating magnetic fields are due to several types of loss processes and the relative contributions of each will depend strongly on particle size and other properties.

Figure 4.5a shows how SAR values vary as a function of the particle radius and excitation frequency for monodisperse, magnetite and maghemite particles. Both materials demonstrate a clear peak heating efficiency due to the size-dependent relaxation behavior, which becomes more pronounced as the frequency increases.

In addition, the strong size dependence indicates that polydispersity will be an important consideration. Most magnetic nanoparticle populations demonstrate significant polydispersity and often follow log-normal distributions with (r_0, σ_{ln}). The effective polydisperse SAR can be solved for by numerically integrating across the probability distribution function, as shown below in Equation 4.19. The general effects of polydispersity are illustrated for magnetite in Figure 4.5b. It is clear that polydispersity can significantly flatten the peak, but this broadening will also reduce the sensitivity to small shifts in the mean size.

$$SAR = \int_0^\infty SAR(r)\, g(r)\, dr \tag{4.19a}$$

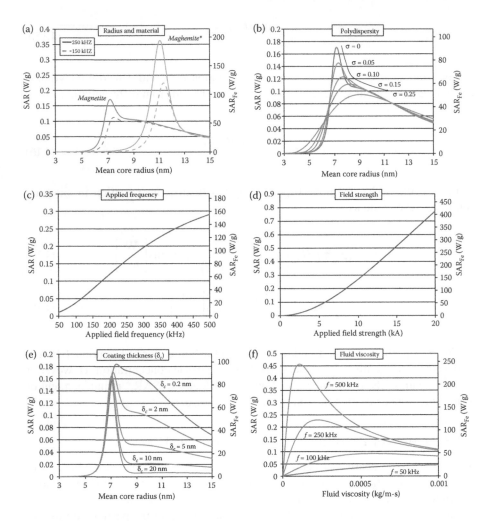

FIGURE 4.5 Calculated dependence of SAR and SAR_{Fe} suspended in water for: (a) radius and material (at 5 kA/m), (b) size polydispersity (at 5 kA/m, 250 kHz), (c) applied frequency (at 7 nm radius, 5 kA/m), (d) field strength (at 7 nm radius, 150 kHz), (e) coating thickness (at 5 kA/m, 250 kHz), and (f) fluid viscosity (8 nm radius, 5 kA/m). Volume fraction for all cases was 0.0005 and surface coating thickness was 2 nm. Maghemite* SAR_{Fe} values were scaled to iron content of magnetite for comparison on a single plot, but calculated values should be 15% higher, due to lower relative iron content.

$$g(R) = \frac{1}{\sigma_{ln} R \sqrt{2\pi}} \exp\left[-\frac{\ln\left(R/r_0\right)^2}{2\sigma_{ln}^2} \right] \qquad (4.19b)$$

Differences in the heating performance for various ferrofluids are due largely to the material-specific anisotropy and saturation magnetization. Values for these properties and the size-dependent SAR performance are listed for several magnetic materials in Table 4.3. As has already been discussed, anisotropy has significant effects on Néelian relaxation behavior and so will be the major determinant in the peak radius for heating. Saturation magnetization mainly affects the magnitude of the peak, with higher magnetizations generally producing higher losses. This also demonstrates the importance of well-controlled synthesis techniques for producing nanoparticle populations with both uniform size distributions and magnetocrystalline properties. Kappiyoor et al. [43] recently provided a complete comparison of predicted *in vivo* heating performance among several magnetic nanoparticle systems.

Although the power conversion efficiency is largely a function of nanoparticle properties, the applied field is a major determinant of the total power generated by a magnetic nanoparticle system. Referring back to Equation 4.13, the volumetric power input depends directly on the applied frequency and square of the applied field strength. However, the applied field also affects the magnetic susceptibility through the complex relations in Equations 4.14 and 4.15.

Figure 4.5c and 4.5d illustrates the overall field effects on SAR performance. The power generation is roughly linearly dependant on both frequency and field strength, within the usable field parameters. Frequency effects will approach a plateau for frequencies on the order of MHz's, but again, other unintended heating effects will dominate in these ranges, limiting clinical use for MFH below this level.

The nanoparticle coating supports colloidal stability, but also contributes to the power absorption through Brownian relaxation. As was shown from the theoretical point of view, the core exhibits a physical oscillation depending on the excitation frequency and core magnetization. Therefore, some optimization parameters related to the viscosity and structure of the coating material are expected.

Figure 4.5e demonstrates this conclusion. With increases in coating size, the SAR value reduces dramatically due to the increase in Brownian relaxation time as a consequence of the increase in hydrodynamic volume.

From the relaxation time constant graph and model, a strong dependence of the Brownian contribution on viscosity of the carrier fluid is observed (Figure 4.4). In media with high viscosity (glycerol), the Brownian agitation can be slowed to the point where the particles cannot align themselves with the switching external magnetic field, which in turn will reduce the overall SAR. However, varying the magnetic field frequency will change the graph's peak.

Figure 4.5f shows high Brownian dominated peaks at very low viscosities, but these viscosity values are typically lower than those relevant in biological systems. The effect of solvent viscosity can be well demonstrated by comparing heating performance in water and glycerol. Varying the viscosity allows the Brownian contribution to be approximated, with the SAR value diminishing if the viscosity is increased significantly. Comparisons of the model predictions and experimental data can be found in subsequent sections.

4.3 EXPERIMENTALLY CHARACTERIZING SAR

As displayed in Table 4.1, experimentally measured SAR values for magnetically heated nanoparticles can vary significantly depending on the system and it is important to have robust methods to ensure consistent data. The following section will

TABLE 4.3

Saturation Magnetization (M_s), Anisotropy (K_u), Specific Heat (c_p), Density (ρ), Iron Mass Fraction, Optimal Radius for Heating, and Peak SARs for some Potential Magnetic Nanoparticle Materials

Material	M_s (kA/m)	K_u (kJ/m^3)	C_p (J/kg-K)	ρ (kg/m^3)	Iron Mass Fraction	Optimal Radius (nm)	Peak SAR (W/g)	Peak SAR$_{Fe}$ (W/g Fe)
Magnetite[a]	446	23	670	5180	0.72	7.5	0.11	59
Maghemite[a]	414	4.7	746	4600	0.70	11.3	0.22	137
Iron[b]	1707	48	450	7870	1.00	5.9	0.70	178
Iron-cobalt[c]	1815	15	172	8031	0.49	8.3	1.13	575
FCC iron-platinum[d]	1140	206	327	15,200	0.22	7.2	0.35	210

Note: Property values taken from [37][a], [31][b], [38][c], and [39][d]. Optimal radius and peak SAR calculated for monodisperse particles at 150 kHz, 5 kA/m, and 0.0005 volume fraction. SAR for total nanoparticle mass can be found by multiplying SAR$_{Fe}$ by the iron mass fraction.

describe experimental procedures adapted from Jordan et al. [8] and convey results which show close agreement with the theory discussed in previous sections.

4.3.1 MAGNETIC NANOPARTICLES

Heating was experimentally characterized for magnetite nanoparticles from Ferrotec GmbH (Unterensingen, Germany). The iron oxide nanoparticles were in an aqueous suspension with an anionic surfactant coating to prevent aggregation. Manufacturer's specifications suggested an average size of 10 nm with log-normal standard-deviation of approximately $\sigma_{\ln} = 0.25$. The surfactant coating used will likely preclude the nanoparticles from unmodified *in vivo* application, but they do provide an excellent platform for characterizing superparamagnetic heating behavior.

4.3.2 HEATING STUDIES AND EXPERIMENTAL SETUP

The methods used to characterize the heating of the ferrofluids followed those demonstrated by Jordan et al. [8]. Samples of nanoparticle solutions were prepared in concentrations ranging from 1 to 9 mg Fe/mL by dilution with distilled water, glycerol, or agarose gel. One milliliter samples contained in 1.5 mL microcentrifuge tubes were placed in an insulated 4-turn copper coil (360 KHz, 14 kA/m) powered by an inductive heating system (1 KW Hotshot, Ameritherm Inc, Scottsville, NY), shown in Figure 4.6. The coil was water-cooled to prevent external heating effects. A majority of developmental work is performed in fields created by inductive coils due to the ease of application and high field uniformity within the coil. The strength of the uniform magnetic field within an inductive coil can be approximated by [31]

$$H_a = \frac{NI}{L} \tag{4.20}$$

where N is the number of turns in the coil, I is the coil current, and L is the coil height but more accurate methods of characterizing field inhomogeneity in non-ideal coils should be applied.

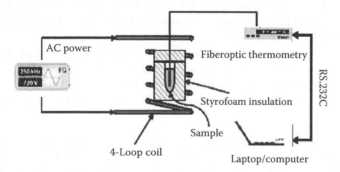

FIGURE 4.6 Experimental setup for alternating magnetic field heating of 1 mL magnetite nanoparticle sample solutions.

The temperature was measured every 5 s using a fluoroptic thermometry system (Luxtron 3100, Luxtron Inc, Santa Clara, CA) positioned in the middle of the sample. Each sample was heated for 15 min and the temperature response was studied in three different carrier liquids: water, glycerol, and 1% agarose gel. Heating at all concentrations was repeated at least five times in all suspending media.

4.3.3 MEASURING SAR

If the heavily insulated samples are treated as lumped systems, SAR can be estimated from the initial slope of the temperature response curve by the rate of temperature rise method [8,13,44,45]. The lumped assumption can be validated through basic scaling arguments.* The sample SAR was then estimated as

$$\text{SAR} = c_p \frac{\Delta T}{\Delta t} \tag{4.21}$$

where c_p is the mass-average specific heat and the initial temperature–time slope ($\Delta T/\Delta t$) was determined from a linear curve fit for the first minute of heating. This method assumes that there is little or no diffusive loss during this period, which is supported by simple scaling arguments.* Representative heating curves are illustrated in Figure 4.7, demonstrating the initial linear trend and eventual plateau as the system begins to equilibrate.

4.3.4 EXPERIMENTAL RESULTS

The heating response of magnetite nanoparticles in water, glycerol, and 1% agarose gel were studied. Samples were allowed to equilibrate to room temperature (24°C) before heating. The temperature changes in glycerol were smaller than those in water, but the differences were not significant for all the concentrations studied. Temperature changes of <3°C were observed in the control samples (pure water, pure glycerol, or pure 1% agarose gel) under the same conditions. A maximum temperature rise of about 70°C was observed in glycerol for a concentration of 9 mg Fe/mL after 15 min of heating using a 14 kA/m alternating field amplitude operating at 360 kHz. The temperature rise under the same conditions was still almost 40°C with the concentration at 5 mg Fe/mL.

Heating response has a strong dependence on the suspending medium viscosity. An increase in viscosity produces a longer Brownian time constant. As a consequence, particle rotation may be too slow to take effect and this can effectively decrease or eliminate Brownian contributions. While the differences in temperatures

---

* Scaling arguments support the use of the lumped, temperature rise method. Assuming convective and conductive transfer coefficients on the order of $h \sim 1$ W/(m²-°C) and $k = 0.5$ W/(m-°C), the Biot number for the characteristic dimensions of the sample will be around 0.01, suggesting that the convective losses dominate. Even at the largest temperature differences observed in the first 60 s (about 30°C), the expected convective losses will be a small fraction of the SAR from iron oxide heating and so are negligible.

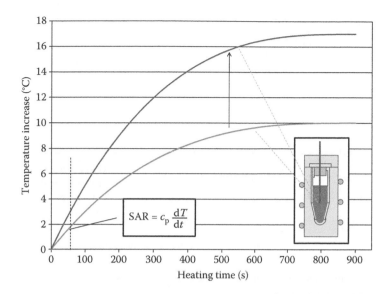

FIGURE 4.7 Representative temperature–time curves for estimating SAR from alternating magnetic field heating experiments. Insulation around the sample provides negligible losses during initial heating.

between water and glycerol were nominal, the temperatures were notably lower in agarose. In addition, referring back to Equation 4.21, the measured SAR will also depend on the specific heat of each sample, so although the temperature changes were very similar for water and glycerol, water has a higher heat capacity and so more energy is required for equivalent temperature changes.

Higher concentrations of ferrofluids are expected to produce larger SAR values, following a first-order relationship. Measured and predicted SAR values for water and glycerol are included in Figure 4.8. Values for agarose were not included because a reasonable estimation of the viscosity in the gel matrix was not available and so SAR could not be predicted for this case. Comparison of the experimental data with theoretical SAR shows good agreement in water and reasonable agreement in glycerol for the full range of concentrations investigated. Optimal linear fit of the data points suggested SAR_{Fe} for water and glycerol of 88 W/g Fe ($R^2 = 0.97$) and 60 W/g Fe ($R^2 = 0.98$), respectively. Theoretical SAR_{Fe} was predicted at 84 and 72 W/g Fe for the same respective mediums and settings.

4.4 CONCLUSION

Many investigators have reported experimental SAR values for superparamagnetic nanoparticle heating hyperthermia. The methods are well established, data are plentiful, and predictive models are increasing in the literature. However, comparison of SAR values across studies is still difficult because of the lack of a global protocol for experimental measurement and theoretical model formation. Heating efficacy can range over several orders of magnitude due to variability in size, polydispersity,

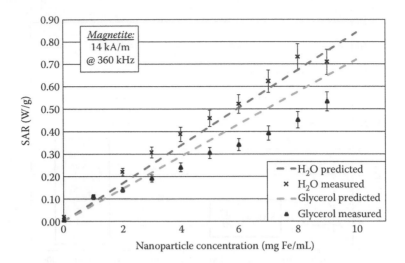

FIGURE 4.8   Measured and predicted specific absorption rates for various concentrations of magnetite dispersed in water and glycerol [46].

aggregation, coatings, anisotropy, concentration, and tissue/suspension medium, as well as magnetic field properties (refer back to Table 4.2). Based on all these factors, it is essential to develop an explicit, comprehensive approach for the use of nanoparticles in hyperthermia. A parametric study has been performed for magnetite nanoparticles in which parameters were varied independently of Brownian and Néelian loss mechanisms, through both theoretical and experimental investigation. Quantitative SAR data have been shown to agree well with theoretical predictions based on relevant parameters (including nanoparticle, tissue/system, and field characteristics). This demonstrates the validity of both the experimental and analytical methodologies for future study and optimization of MFH applications. Most significantly, SAR was shown to vary linearly with nanoparticle concentration, facilitating predictive modeling of expected treatment temperatures, based on CT-imaged concentration data [20,21]. However, this analysis was for an idealized system and there is still significant opportunity for developments in the field, with the most notable including characterization of biological interactions and improvements in clinical translation.

Although these approaches have shown good agreement *in vitro*, significant investigation is still required before achieving accurate *in vivo* models. Many forms of coating and surface functionalization provide adequate colloidal stability, but magnetic nanoparticles have demonstrated a strong tendency to form aggregates in biological systems [47]. The theoretical discussions in this chapter reflected systems of noninteracting particles, and descriptions of systems of close-packed, interacting particles become significantly more complicated. In addition, any binding between particles or biological entities is going to affect Brownian motion and this needs to be factored in as well. Further development is thus needed to allow numerical models to better predict the *in vivo* environment. Some authors have also demonstrated enhanced therapeutic effects of intracellularly heated iron oxide [48,49], while scaling models suggest MFH is only capable of bulk heating and not localized temperature increases [22,23]. It will, therefore, be important to clearly illuminate the modes of biological damage that are in play.

While superparamagnetic nanoparticle heating has already demonstrated clear clinical success [50], much development is still required to take full advantage of this exciting platform. Further experimental work is needed to optimize the applied field response. The theoretical discussion outlined the complicated interactions between the applied field and the tissue and nanoparticles. One of the major limitations encountered in clinical application has been the limits to the applied field due to patient discomfort (and in some cases, injury) arising from the unwanted eddy current heating of tissue in the body. In addition to optimizing the nanoparticles for peak response, there is also significant opportunity for fully exploiting the field strength and frequency combinations, as well as improving field delivery. Varying frequency through the range of usable values may provide an optimal response for a polydisperse nanoparticle population. In addition, designing field applicators to deliver more focused fields to specific areas of the body [51] or investigating high energy waveforms (square waves [52] or pulsed waves [53]) could provide higher incident power densities and lead to increased heating. Advancements in imaging techniques may also provide better estimates of *in vivo* nanoparticle concentrations (and thus better predictive modeling), as well as the potential for real time, noninvasive monitoring of treatment temperatures through magnetic resonance (MR) [54], ultrasound [55], CT [56], and other techniques [57]. However, it should be noted that MR currently demonstrates some compatibility issues, due to artifacts created by the high local moments of the magnetic nanoparticle deposits.

NOMENCLATURE

B	Magnetic induction (T)
c_p	Specific heat (J/g–K)
f	Applied frequency (Hz)
H or H_a	Applied magnetic field (A/m)
k_B	Boltzmann constant (1.38×10^{-23} J/K)
K_u	Magnetic anisotropy constant (J/m^3)
M	Magnetization (A/m)
M_r	Remnant magnetization (A/m)
M_s	Saturation magnetism (A/m)
P	Volumetric power generation (W/m^3)
r, r_0	Mean nanoparticle radius (m)
SAR	Specific absorption rate (W/cm^3 or W/g)
SAR$_{Fe}$	Specific absorption rate—iron (W/g$_{Fe}$)
SLP	Specific loss power (W/g$_{Fe}$)
T	Absolute temperature (K)
t	Time (s)
U	Internal energy (J)
V_m	Magnetic volume (m^3)

| V_H | Hydrodynamic volume (m^3) |
| W | Magnetic work (J) |

GREEK SYMBOLS

χ_m	Magnetic susceptibility (–)
χ'_m	In-phase susceptibility (–)
χ''_m	Out-of-phase susceptibility (–)
$\chi_{m,0}$	Static susceptibility (–)
$\chi_{m,i}$	Initial susceptibility (–)
δ_c	Surface coating thickness (M)
φ	Nanoparticle volume fraction (–)
η	Fluid viscosity (Pa-s)
μ_0	Permeability of free space ($4\pi \times 10^{-7}$ N/A^2)
ρ	Density (g/m^3)
σ	Bulk conductivity (s/m)
σ_{ln}	Log-normal standard deviation (–)
τ	Effective relaxation time (s)
τ_0	Attempt time (s)
τ_B	Brownian relaxation time (s)
τ_N	Néelian relaxation time (s)

REFERENCES

1. Gilchrist R. K., Medal R., Shorey W. D., Hanselman R. C., Parrott J. C., and Taylor C. B., 1957, Selective inductive heating of lymph nodes, *Annals of Surgery*, **146**(4), 596.
2. Deng Y. H., Yang W. L., Wang C. C., and Fu S. K., 2003, A novel approach for preparation of thermoresponsive polymer magnetic microspheres with core–shell structure, *Advanced Materials*, **15**(20), 1729–1732.
3. Zhang J. and Misra R. D. K., 2007, Magnetic drug-targeting carrier encapsulated with thermosensitive smart polymer: Core-shell nanoparticle carrier and drug release response, *Acta Biomaterialia*, **3**(6), 838–850.
4. Huang H., Delikanli S., Zeng H., Ferkey D. M., and Pralle A., 2010, Remote control of ion channels and neurons through magnetic-field heating of nanoparticles, *Nature Nanotechnology*, 5, 602–606.
5. Satarkar N. S., Zhang W., Eitel R. E., and Hilt J. Z., 2009, Magnetic hydrogel nanocomposites as remote controlled microfluidic valves, *Lab on a Chip*, **9**(12), 1773–1779.
6. Mohr R., Kratz K., Weigel T., Lucka-Gabor M., Moneke M., and Lendlein A., 2006, Initiation of shape-memory effect by inductive heating of magnetic nanoparticles in thermoplastic polymers, *Proceedings of the National Academy of Sciences of the United States of America*, **103**(10), 3540.
7. Moroz P., Jones S. K., and Gray B. N., 2002, Magnetically mediated hyperthermia: current status and future directions, *International Journal of Hyperthermia*, **18**(4), 267–284.
8. Jordan A., Wust P., Fähling H., John W., Hinz A., and Felix R., 1993, Inductive heating of ferrimagnetic particles and magnetic fluids: physical evaluation of their potential for hyperthermia, *International Journal of Hyperthermia*, **25**(7), 499–511.

9. Chan D. C. F., Kirpotin D. B., and Bunn P. A., 1993, Synthesis and evaluation of colloidal magnetic iron oxides for the site-specific radiofrequency-induced hyperthermia of cancer, *Journal of Magnetism and Magnetic Materials*, **122**(1–3), 374–378.

10. Hergt R., Andra W., d' Ambly C. G., Hilger I., Kaiser W. A., Richter U., and Schmidt H. G., 1998, Physical limits of hyperthermia using magnetite fine particles, *IEEE Transactions on Magnetics*, **34**(5), 3745–3754.

11. Brusentsov N. A., Gogosov V. V., Brusentsova T. N., Sergeev A. V., Jurchenko N. Y., Kuznetsov A. A., Kuznetsov O. A., and Shumakov L. I., 2001, Evaluation of ferromagnetic fluids and suspensions for the site-specific radiofrequency-induced hyperthermia of MX11 sarcoma cells *in vitro*, *Journal of Magnetism and Magnetic Materials*, **225**(1–2), 113–117.

12. Hilger I., Frühauf K., Andrä W., Hiergeist R., Hergt R., and Kaiser W. A., 2002, Heating potential of iron oxides for therapeutic purposes in interventional radiology, *Academic Radiology*, **9**(2), 198–202.

13. Ma M., Wu Y., Zhou J., Sun Y., Zhang Y., and Gu N., 2004, Size dependence of specific power absorption of Fe_3O_4 particles in AC magnetic field, *Journal of Magnetism and Magnetic Materials*, **268**(1–2), 33–39.

14. Kalambur V. S., Han B., Hammer B. E., Shield T. W., and Bischof J. C., 2005, *In vitro* characterization of movement, heating and visualization of magnetic nanoparticles for biomedical applications, *Nanotechnology*, **16**, 1221.

15. Kallumadil M., Tada M., Nakagawa T., Abe M., Southern P., and Pankhurst Q. A., 2009, Suitability of commercial colloids for magnetic hyperthermia, *Journal of Magnetism and Magnetic Materials*, **321**(10), 1509–1513.

16. Salloum M., Ma R. H., Weeks D., and Zhu L., 2008, Controlling nanoparticle delivery in magnetic nanoparticle hyperthermia for cancer treatment: Experimental study in agarose gel, *International Journal of Hyperthermia*, **24**(4), 337–345.

17. Xu R., Zhang Y., Ma M., Xia J., Liu J., Guo Q., and Gu N., 2007, Measurement of specific absorption rate and thermal simulation for arterial embolization hyperthermia in the maghemite-gelled model, *IEEE Transactions on Magnetics*, **43**(3), 1078–1085.

18. Johannsen M., Thiesen B., Jordan A., Taymoorian K., Gneveckow U., Waldöfner N., Scholz R., Koch M., Lein M., Jung K., and Loening S.A., 2005, Magnetic fluid hyperthermia (MFH) reduces prostate cancer growth in the orthotopic Dunning R3327 rat model, *The Prostate*, **64**(3), 283–292.

19. Chen Z. J., Broaddus W. C., Viswanathan R. R., Raghavan R., and Gillies G. T., 2002, Intraparenchymal drug delivery via positive-pressure infusion: Experimental and modeling studies of poroelasticity in brain phantom gels, *IEEE Transactions on Biomedical Engineering*, **49**(2), 85–96.

20. Johannsen M., Gneveckow U., Taymoorian K., Hee Cho C., Thiesen B., Scholz R., Waldöfner N., Loening S. A., Wust P., and Jordan A., 2007, Thermal therapy of prostate cancer using magnetic nanoparticles, *Actas Urológicas Espanolas*, **31**, 660–667.

21. Kalambur V. S., Han B., Hammer B. E., Shield T. W., and Bischof J. C., 2005, *In vitro* characterization of movement, heating and visualization of magnetic nanoparticles for biomedical applications, *Nanotechnology*, **16**, 1221.

22. Keblinski P., Cahill D. G., Bodapati A., Sullivan C. R., and Taton T. A., 2006, Limits of localized heating by electromagnetically excited nanoparticles, *Journal of Applied Physics*, **100**, 054305.

23. Qin Z., Etheridge M., and Bischof J., In submission., Nanoparticle heating: Nanoscale to bulk effects of electromagnetically heated iron oxide and gold for biomedical applications, Proceedings of *SPIE*, **7901–11**, 79010C.

24. Young J. H., Wang M. T., and Brezovich I. A., 2007, Frequency/depth-penetration considerations in hyperthermia by magnetically induced currents, *Electronics Letters*, **16**(10), 358–359.

25. Foster K. R., 2000, Thermal and nonthermal mechanisms of interaction of radio-frequency energy with biological systems, *IEEE Transactions on Plasma Science*, **28**(1), 15–23.
26. Zhu L., Xu L. X., and Chencinski N., 1998, Quantification of the 3-D electromagnetic power absorption rate in tissue during transurethral prostatic microwave thermotherapy using heat transfer model, *IEEE Transactions on Biomedical Engineering*, **45**(9), 1163–1172.
27. Atkinson W. J., Brezovich I. A., and Chakraborty D. P., 2007, Usable frequencies in hyperthermia with thermal seeds, *IEEE Transactions on Biomedical Engineering*, **31** (1), 70–75.
28. Wust P., Nadobny J., Fähling H., Riess H., Koch K., John W., and Felix R., 1990, The influencing factors and interfering effects in the control of the power distributions with the BSD-20000 hyperthermia ring system. 1. The clinical observables and phantom measurements, *Radiotherapy and Oncology: Journal of the German Radiological Society*, **166**(12), 822.
29. Wust P., Nadobny J., Fähling H., Riess H., Koch K., John W., and Felix R., 1991, Determinant factors and disturbances in controlling power distribution patterns by the hyperthermia-ring system BSD-2000. 2. Measuring techniques and analysis, *Radiotherapy and Oncology: Journal of the German Radiological Society,* **167**(3), 172.
30. Cullity B. D., and Graham C. D., 2009, *Introduction to Magnetic Materials*, Wiley-IEEE, Hoboken, NJ.
31. O'Handley R. C., 2000, *Modern Magnetic Materials: Principles and Applications*, Wiley, New York.
32. Gubin S. P., 2009, *Magnetic Nanoparticles*, Wiley-VCH, Weinheim.
33. Hergt R., Andra W., d' Ambly C. G., Hilger I., Kaiser W. A., Richter U., and Schmidt H. G., 2002, Physical limits of hyperthermia using magnetite fine particles, *IEEE Transactions on Magnetics*, **34**(5), 3745–3754.
34. Lu A. H., Salabas E. L., and Schüth F., 2007, Magnetic nanoparticles: synthesis, protection, functionalization, and application, *Angewandte Chemie International Edition*, **46**(8), 1222–1244.
35. Kalambur V. S., Longmire E. K., and Bischof J. C., 2007, Cellular level loading and heating of superparamagnetic iron oxide nanoparticles, *Langmuir*, **23**(24), 12329–12336.
36. Krishnan K. M., 2010, Biomedical nanomagnetics: A spin through possibilities in imaging, diagnostics, and therapy, *IEEE Transactions on Magnetics*, **46**(7), p. 2523.
37. Rosensweig R. E., 2002, Heating magnetic fluid with alternating magnetic field, *Journal of Magnetism and Magnetic Materials*, **252**, 370–374.
38. Kline T. L., Xu Y.-H., Jing Y., and Wang J.-P., 2009, Biocompatible high-moment FeCo-Au magnetic nanoparticles for magnetic hyperthermia treatment optimization, *Journal of Magnetism and Magnetic Materials*, **321**(10), 1525–1528.
39. Maenosono S., and Saita S., 2006, Theoretical assessment of FePt nanoparticles as heating elements for magnetic hyperthermia, *IEEE Transactions on Magnetics*, **42**(6), 1638–1642.
40. Hergt R., Hiergeist R., Hilger I., Kaiser W. A., Lapatnikov Y., Margel S., and Richter U., 2004, Maghemite nanoparticles with very high AC-losses for application in RF-magnetic hyperthermia, *Journal of Magnetism and Magnetic Materials*, **270**(3), 345–357.
41. Dennis C. L., Jackson A. J., Borchers J. A., Hoopes P. J., Strawbridge R., Foreman A. R., Lierop J., Grüttner C., and Ivkov R., 2009, Nearly complete regression of tumors via collective behavior of magnetic nanoparticles in hyperthermia, *Nanotechnology*, **20**, 395103.
42. Hergt R., Dutz S., and Röder M., 2008, Effects of size distribution on hysteresis losses of magnetic nanoparticles for hyperthermia, *Journal of Physics: Condensed Matter*, **20**(38), 385214.
43. Kappiyoor R., Liangruksa M., Ganguly R., and Puri I. K., 2010, The effects of magnetic nanoparticle properties on magnetic fluid hyperthermia, *Journal of Applied Physics*, **108**(9), 094702.

44. Chou C. K., 1990, Use of heating rate and specific absorption rate in the hyperthermia clinic, *International Journal of Hyperthermia*, **6**(2), 367–370.
45. Babincová M., Leszczynska D., Sourivong P., Cicmanec P., and Babinec P., 2001, Superparamagnetic gel as a novel material for electromagnetically induced hyperthermia, *Journal of Magnetism and Magnetic Materials*, **225**(1–2), 109–112.
46. Manuchehrabadi N., 2009, *Experimental and Theoretical Characterization of Specific Absorption Rate (SAR) of Iron Oxidenanoparticles for Biomedical Heating Applications*, MS, University of Hannover.
47. Jordan A., Scholz R., Wust P., Fähling H., Krause J., Wlodarczyk W., Sander B., Vogl T., and Felix R., 1997, Effects of magnetic fluid hyperthermia (MFH) on C3H mammary carcinoma *in vivo*, *International Journal of Hyperthermia*, **13**(6), 587–605.
48. Jordan A., Scholz R., Wust P., Schirra H., Schiestel T., Schmidt H., and Felix R., 1999, Endocytosis of dextran and silan-coated magnetite nanoparticles and the effect of intracellular hyperthermia on human mammary carcinoma cells in vitro, *Journal of Magnetism and Magnetic Materials*, **194**(1–3), 185–196.
49. Ogden J. A., Tate J. A., Strawbridge R. R., Ivkov R., and Hoopes P. J., 2009, Comparison of iron oxide nanoparticle and waterbath hyperthermia cytotoxicity, *Proceedings of SPIE*, 7181, 71810K.
50. Maier-Hauff K., Ulrich F., Nestler D., Niehoff H., Wust P., Thiesen B., Orawa H., Budach V., and Jordan A., 2010, Efficacy and safety of intratumoral thermotherapy using magnetic iron oxide nanoparticles combined with external beam radiotherapy on patients with recurrent glioblastoma multiforme, *Journal of Neurooncology*, **81**(1), 53–60.
51. Wust P., Gneveckow U., Johannsen M., Böhmer D., Henkel T., Kahmann F., Sehouli J., Felix R., Ricke J., and Jordan A., 2006, Magnetic nanoparticles for interstitial thermotherapy—Feasibility, tolerance and achieved temperatures, *International Journal of Hyperthermia*, **22**(8), 673–685.
52. Morgan S. M., and Victora R. H., 2010, Use of square waves incident on magnetic nanoparticles to induce magnetic hyperthermia for therapeutic cancer treatment, *Applied Physics Letters*, **97**, 093705.
53. Foner S., and Kolm H. H., 2009, Coils for the production of high-intensity pulsed magnetic fields, *Review of Scientific Instruments*, **28**(10), 799–807.
54. Poorter J. D., Wagter C. D., Deene Y. D., Thomsen C., Staahlberg F., and Achten E., 1995, Noninvasive MRI thermometry with the proton resonance frequency (PRF) method: *in vivo* results in human muscle, *Magnetic Resonance in Medicine*, **33**(1), 74–81.
55. Amini A. N., Ebbini E. S., and Georgiou T. T., 2005, Noninvasive estimation of tissue temperature via high-resolution spectral analysis techniques, *IEEE Transactions on Biomedical Engineering*, **52**(2), 221–228.
56. Fallone B. G., Moran P. R., and Podgorsak E. B., 1982, Noninvasive thermometry with a clinical x-ray CT scanner, *Medical Physics*, **9**, 715.
57. Weaver J. B., Rauwerdink A. M., and Hansen E. W., 2009, Magnetic nanoparticle temperature estimation, *Medical Physics*, **36**(5), 1822.

5 Light-Induced Energy Conversion in Liquid Nanoparticle Suspensions

Patrick E. Phelan, Robert Taylor, Ronald J. Adrian, Ravi S. Prasher, and Todd P. Otanicar

CONTENTS

5.1 INTRODUCTION

Liquid nanoparticle suspensions, popularly called "nanofluids" in the heat-transfer literature, have been widely investigated for their interesting thermal conduction and convection properties (see, e.g., [1–6]). The radiative properties of nanofluids, however, have been much less studied [7]. It is readily apparent to anyone who has observed a prepared nanofluid, though that nanofluids strongly scatter and absorb visible light. This observation motivated the concept that nanofluids can be used as direct volumetric absorbers for solar radiation [8–16], leading potentially to more efficient solar energy conversion that may be termed *volumetric photothermal*

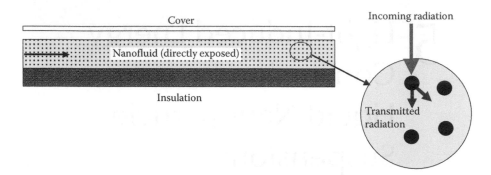

FIGURE 5.1 Schematic diagram of a nanofluid-based volumetric direct-absorption solar receiver.

energy conversion. The basic idea is schematically presented in Figure 5.1. A flowing nanofluid is exposed to incident light, which can be concentrated or not, through a transparent cover. Absorption of light within the nanofluid, largely by the nanoparticles, leads to a temperature rise in the nanofluid that can then be exploited as thermal energy. Continuing research demonstrates that not only can sensible nanofluid temperature increases occur, but also boiling can take place around the nanoparticles. This gives rise to the possibility that direct-absorption nanofluids, like those under investigation here, can be utilized for concentrating-solar, direct-steam generation, a promising direction for concentrating solar power technology [11,15,17–20].

Here recent results from the authors' own work is reviewed, with emphasis on boiling induced by laser heating of the nanofluid. A brief discussion of the radiative properties of nanofluids is also presented, followed by suggestions for future directions.

5.2 NANOFLUIDS AND SOLAR THERMAL ENERGY

Photographs of representative nanofluids are presented in Figure 5.2. From these pictures it is clear that even nanoparticle volume fraction φ as low as 0.1% yield very dark, strongly scattering/absorbing nanofluids. Such fluids can then serve to absorb sunlight directly, as compared with the conventional approach in which a solid receiver surface absorbs sunlight and then has to transfer that energy to a fluid. For liquids, this can be accomplished by the introduction of a black dye, a concept dating back to the 1970s (see, e.g., [21]), or by creating a colloidal suspension (see, e.g., [22–24]). Alternatively, particles can be suspended in gases for high-temperature solar applications (see, e.g., [25–29]). From these approaches, it seems that nanofluids can be utilized for solar thermal energy conversion, as presented schematically in Figure 5.3, which shows how concentrated sunlight is absorbed directly in the nanofluid. This approach can then serve as the heat source to drive a heat engine or other thermal process. Another intriguing possibility is direct steam generation in water-based nanofluids, which is discussed in the following section.

FIGURE 5.2 Representative nanofluid samples tested in micro-collector (CNT, carbon nanotubes).

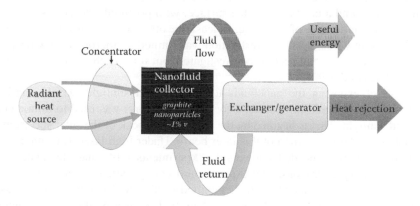

FIGURE 5.3 Proposed nanofluid-based photothermal energy conversion.

Does the use of direct-absorption nanofluids leads to higher solar collector efficiencies? The answer is yes. For flat-plate (nonconcentrating) collectors, calculated efficiencies were as much as 10% higher, on an absolute basis, than comparable conventional flat-plate collectors [8].

Experimental data on microscale solar collectors yielded a maximum collector efficiency of 57% for a nanofluid consisting of $\varphi = 0.25\%$ 20-nm Ag nanoparticles in water, compared with a maximum collector efficiency of 52% for pure water flowing over a black surface. In both cases, the fluid thickness was 150 μm [9,14].

Calculations of the efficiency of a proposed nanofluid-based receiver, appropriate for a "power tower" concentrating solar power type of powerplant (see, e.g., [30]), revealed a maximum receiver efficiency of ~75%, compared with a measured receiver efficiency of ~65% at the same temperature (300°C) [12,15,31].

It appears that there is plenty of motivation to pursue nanofluids as the basis for direct-absorption solar energy conversion. This gives rise to a number of fundamental research questions that need to be addressed for successful technology development. In the following sections, ongoing work related to two such questions will be discussed: (i) how is boiling generated in nanofluids that are exposed to concentrated light? and (ii) how can the radiative properties of nanofluids be controlled to maximize light absorption and minimize losses, that is, to optimize collector/receiver efficiency?

5.3 LIGHT-INDUCED BOILING IN NANOFLUIDS

A number of investigators have examined pool boiling in nanofluids, in which heat is added via an immersed wire or other type of heater (see, e.g., [32–35]). Although conflicting results were reported in many studies, most researchers concluded that nanoparticle deposition on the heated surface played an important role. Exposing a nanofluid to concentrated light, however, enables heat—in the form of radiative energy absorbed by the nanoparticles—to be introduced into the nanofluid without a conventional immersed heater. Rather, the nanoparticles themselves, or nanoparticle aggregates, serve as the heaters (Figure 5.4). This allows an investigation of boiling phenomena around a nanoscale heated surface, with potential effects arising from, say, the very small radius of curvature. At least one other study measured size effects on boiling from wires, but that investigation was limited to micron-sized wires [36]. Such phenomena have not been fully investigated, but some of the authors' results to date are discussed below.

Instead of exposing the nanofluids to concentrated sunlight, which entails a number of experimental difficulties, a focused laser beam was used to mimic concentrated sunlight. This is justified by the fact that the peak of the solar spectrum is near 530 nm—the wavelength of this laser beam. Under high laser irradiation, only small sample sizes are needed, allowing the experiments to be conducted under controlled laboratory conditions. The experimental configuration is shown in Figure 5.5. Light energy is provided by a diode pumped solid state (Coherent—DPSS) laser that gives about 120 mW at a 532 nm wavelength. The laser produces a continuous column of light having a Gaussian intensity profile ($TEM_{0,0}$) in the radial direction.

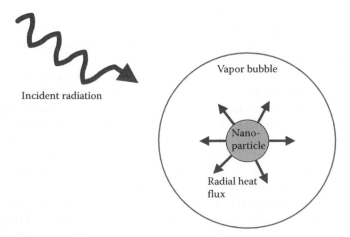

Incident radiation

Vapor bubble

Nano-particle

Radial heat flux

FIGURE 5.4 Cartoon of boiling induced around a radiatively heated nanoparticle.

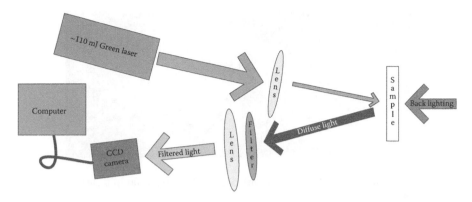

FIGURE 5.5　Experimental configuration for laser-induced boiling in nanofluids.

It is focused by a 40-mm positive lens into a thin cuvette containing a 100-μm-thick fluid sample between quartz walls 1.2- and 1.3-mm thick, respectively. In the region of the sample, the focusing produces a constant diameter, 0.4-mm-beam waist which is subsequently absorbed in the nanofluid sample. In the absence of convection, the intensity of light passing through a stationary, homogeneously absorbing fluid should attenuate exponentially, according to Beer's law. Under conditions leading to phase change in these experiments, the absorption is strong enough to absorb over 90% of the laser energy by the time it leaves the test cell, and the majority of the absorption occurs in a portion of the nanofluid layer closest to the laser.

The fluid behavior was observed by back-lighting the fluid layer with white light, and microscopically imaging the laser side of the cell with a Questar QM1 long-range microscope (~3× magnification) using a Thorlabs USB 2.0, 1280 × 1024 CCD. The laser light scattered toward the microscope was removed with a low-pass optical filter. Light intensity entering and exiting the cell was measured by a Coherent (FieldMaxII TOP) power meter with 0.1% accuracy.

5.3.1　Base Fluid Measurements

The control fluid in this study is de-ionized water with 0.1% by volume-added surfactant (Polysorbate 80 or sodium dodecyl sulfate [SDS/NaDS]). Surfactant must be used in order to create a semi-stable nanofluid. Each sample was also degassed in a vacuum chamber to remove air. Two control tests were run (a) water/surfactant in a clear cuvette and (b) water/surfactant in a cuvette with a black backing (made with three coats of "satin black" Krylon spray paint on a microscope slide) attached to the exterior rear surface of the cuvette. Images of the control fluids are shown in Figure 5.6.

For a clear cuvette, over 99% of the light passed through the sample, absorbing very little laser light. For a black backing, the majority of light transmitted to the backing was absorbed. Heat was then conducted through the quartz of the back wall and subsequently into the fluid. Another control fluid which absorbed light well, a black dye from Pylam products, was also tested (not shown in Figure 5.6). This dye

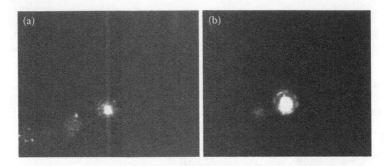

FIGURE 5.6 Water exposed to ~770 W/cm², 532-nm laser light: (a) in a clear cuvette and (b) in a cuvette with a black backing.

is composed of large-sized particles (20–800 µm as tested by a Nicomp dynamic light scattering system).

5.3.2 NANOFLUID MEASUREMENTS

The tested nanofluids were copper, graphite, silver and multiwalled carbon nanotubes (nominal diameters of 2–40 nm at 1%, 0.75%, 0.5%, 0.25%, 0.1%, and 0.05% by volume) in a clear cuvette with the same deionized water/surfactant base fluids. Figure 5.7 shows boiling induced in a $\varphi = 0.1\%$ graphite nanofluid heated by an 8-ns pulsed laser, whereas Figure 5.8 presents a series of typical images during a continuous wave laser heating experiment. The lighter areas occur due to transmission of the back-lighting through a region containing a lower concentration (than average) of nanoparticles. The buoyant plume that occurs above the laser spot wavers shows vertical flow in it. The buoyancy that drives this flow may come from laser heating or from microbubbles emerging from the heated region. Temperatures in the plume are not likely to be high enough to affect the optical density of the nanofluid, so the partial transmission associated with the buoyant plume must be due to microbubbles and/or particle depletion in the hot region of the laser column. The brightest regions that occur toward the bottom of the laser beam are probably light transmitted through a vapor bubble. Thus, in this series of

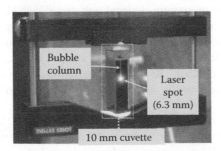

FIGURE 5.7 Boiling is achieved in a $\varphi = 0.1\%$ graphite/H$_2$O nanofluid by Nd:YAG pulsed laser (8 ns ~100 mJ pulses at 15 Hz).

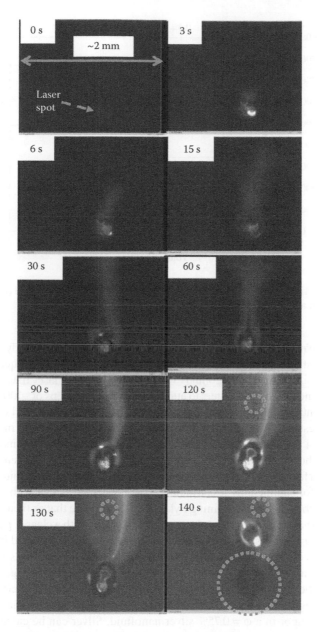

FIGURE 5.8 Bubble generation in a continuous-wave, laser-heated $\varphi = 0.1\%$ graphite/water nanofluid—broken line circles indicate high concentrations of graphite nanoparticles.

images (Figure 5.8), a vapor bubble forms and grows in the heated region. It leaves after the laser is turned off at ~130 s. The last image shows the bubble separated from an area where the nanoparticles are heavily concentrated. Note that laser light is not seen in the images because a cut-off filter is placed in the system to protect the camera (see Figure 5.5).

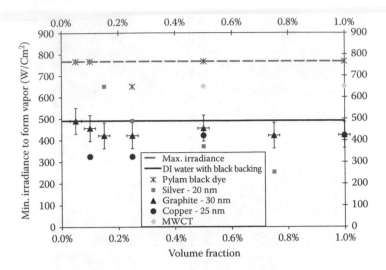

FIGURE 5.9 Minimum irradiance required to generate boiling for deionized (DI) water with a black backing, a black dye solution, and a variety of nanofluids.

Local bubble generation in a subcooled (~25°C) fluid occurs only if the laser irradiance is sufficiently high. The minimum irradiance (in Watt per square centimeter) to cause boiling in these nanofluids was found by varying laser intensity entering the fluid using neutral density filters to attenuate the beam in steps of ~100 W/cm². The nanofluids mentioned above—copper, graphite, silver, and multiwalled carbon nanotubes—were tested to determine the minimum irradiance necessary to initiate boiling as a function of volume fraction. Figure 5.9 shows the results of these measurements, with the solid horizontal line representing the irradiance required to generate boiling for a water/surfactant base fluid with a black backing. Some nanofluids underwent phase change for significantly less irradiance than water/surfactant with a black backing, or for water/surfactant containing Pylam black dye. Note that as there is negligible absorption, no vapor bubbles were observed in pure water with a clear (e.g., nonabsorbing) backing for these laser fluxes—thus, this control fluid is not plotted in Figure 5.9.

These experiments indicate that volumetrically heated nanofluids can undergo boiling more easily than their base fluids exposed to surface heating, that is, than fluids where the heat is introduced at the boundary, such as the "deionized water w/black backing" shown in Figure 5.9. In fact, up to ~50% less irradiance is necessary to create vapor in a φ = 0.75% silver nanofluid. Silver can be calculated (using the independent scattering assumption as given in ref. [37]) to have an order-of-magnitude higher absorption efficiency than graphite at 530 nm. As 20-nm silver nanoparticle/water nanofluids are the best absorbers among the materials tested, it is reasonable that they will generate vapor at the lowest irradiance.

These results also indicated that there may be an optimum volume fraction which minimizes the irradiance necessary to drive phase change for this configuration. This is expected because low particle loadings approach the high transmittance of water and are not effectively heated. Higher particle loadings absorb the light energy close to the wall—approaching conventional surface, or boundary, heating which may lose a significant amount of heat through the wall. The data points for copper in Figure 5.9 appear to follow this trend. Trends as a function of volume fraction for other fluids cannot be inferred from the experiments performed to date.

The experiments exemplified by Figures 5.5 through 5.9 revealed other interesting physics may be at work as well. First, phase change in pure liquid boiling commonly begins in small defects (nucleation sites) on a macroscale *surface*. For light-induced phase change in nanofluids, however, the *particles* become the heating surface. This is an important, and as yet not well-understood, difference, given that the disaggregated particles are smaller than conventional bubble nucleation sites.

Second, as indicated in Figure 5.8, there are several distinct nonuniform spots in the fluid that have high concentrations of dark fluid that must be regions of concentrated particle mass—note that the sample had a uniformly lit background. As graphite melts at ~3850°C and vigorous agitation can break these large regions up, it seems unlikely that these large clumps are molten and/or re-solidified graphite particles. Thus, high-concentration regions are thought to be loosely bonded particle agglomerates. It is unclear whether these dark regions lead or lag vapor formation. Dense collections of particles are expected to absorb light over a shorter path length (i.e., in a smaller volume), which could cause a higher local temperature—that is, driving phase change. Conversely, as vapor forms, particles could be remain behind forming high-concentration regions—that is, nanoparticles are attracted much more strongly to liquid water that to vapor. There is evidence for the latter in that the dark regions appear to grow with exposure time. Also, all resolvable bubbles appear clear—that is, lacking particles.

Third, it is unclear from the images how much vapor is leaving the heated area. Again, in macroscale boiling, vapor bubbles form, grow, and leave the surface. In Figure 5.8, a bubble stays in the same spot even though net buoyancy forces (calculated to be ~0.12 μN) should cause it to rise. As the bubble can grow up to 500 μm in diameter in some cases, the limited 100-μm internal cuvette space could create a restrictive surface tension force on the bubble. Presumably, the vapor would condense in the subcooled surrounding liquid after the laser is turned off. At that point, the reduced restricting force would allow the bubble to rise. Alternatively, tiny (irresolvable) bubbles could be continuously leaving the laser spot. If so, the main bubble could be the generation site for a continuous flow of fluid in as liquid and out as microbubbles. In either case, a larger final volume of vapor is observed to leave just after the laser is turned off. It should be noted that high-temperature gradients should also cause particle migration away from hot regions via thermophoresis.

These results are in contrast to those predicted by recent molecular dynamics simulations [38], and by the experiments those simulations were meant to analyze [39,40]. The experiments involved irradiation by subpicosecond laser pulses, as compared with the nanosecond pulses and steady-state irradiation in our experiments. Those experiments, and the later molecular dynamics simulations, suggested that the heated nanoparticles could reach very high temperatures without initiating boiling. That is, a very high degree of superheat could be achieved, potentially supercritical temperatures. The results in Figures 5.8 and 5.9 appear to contradict these earlier studies. It is possible that the differences are caused by *very* short-time-scale laser heating versus nanosecond-pulsed and steady-state laser heating, but that remains an area of study. Another explanation is that boiling happens around particle agglomerates rather than at individual particles, as will be discussed in the next section.

5.4 MODELING HEAT TRANSFER AND BOILING IN NANOFLUIDS

5.4.1 INDIVIDUAL PARTICLE HEAT TRANSFER

To truly understand heat transfer in a nanofluid, it is necessary to understand heat transfer around an individual particle. In this section, this is done analytically by applying the heat equation on that very localized scale. The diffusive heat flow equation for a homogeneous medium with a source term can be written as [41]

$$kV^2T + q = \rho c_p \frac{\partial T}{\partial t} \tag{5.1}$$

where k is the thermal conductivity, T the temperature field, ρ the density, c_p the specific heat capacity, and t time. The term q is the volumetric power generation term for a single nanoparticle. This term can be estimated from the concentration of light hitting a particle, particle characteristics (material, size, and shape), and the particle's resulting absorption properties. Also, this solution can be significantly simplified by only solving the symmetric one-dimensional case.

The use of this approach, however, needs some justification. An implicit assumption is that heat transfers nearly instantaneously across the characteristic distances involved. More specifically, this equation is only valid if the heat carrier (phonon or electron) has a mean free path smaller than the nanoparticle dimension. In most crystalline solids, mean free paths are on the order of tens of nanometers. Thus, a "lagging" term that accounts for a finite time of heat transfer should be incorporated when modeling small-scale heat transfer in most solids [42]. In contrast, liquids and amorphous solids—due to their lack of a crystal structure—have very short, <1 nm, mean free paths [43]. As we are mostly interested in the surrounding liquid outside of the particle, the use of the heat equation is justified. The assumption of symmetry can be argued through the use of the Biot number. To restate, the Biot number is the ratio of conductive to convective resistance, and is defined as [44]

$$Bi = \frac{hD}{k} \tag{5.2}$$

For a conservative estimate of metallic nanoparticles, the characteristic dimension D is on the order of 100×10^{-9} m, and the thermal conductivity is on the order of 100 W/m/K. Thus, if the convective heat-transfer coefficient is less than 10^8 W/m^2/K, an individual particle can be thought of as a lumped (i.e., symmetric) heat source inside the fluid. In other words, because a metallic nanoparticle is relatively thermally conductive and extremely small, it can be uniformly treated as a homogeneous heat source. This is likely to be the case for all reasonable particle temperatures.

Equation 5.1, assumed to apply to the fluid surrounding a single particle, can be solved by the method of Laplace transforms as is described in ref. [41]. The initial and boundary conditions for this situation are the following:

$$\frac{\partial T}{\partial r}\bigg|_{t=0} = 0 \tag{5.3}$$

$$-k\frac{\partial T}{\partial r}\bigg|_{r=R} = \frac{Q_{particle}}{A} \tag{5.4}$$

$$T_{r=\infty,t=t} = T_\infty \tag{5.5}$$

where the particle surface heat flux $Q_{particle}/A = qV/A$, where A and V are the particle surface area and volume, respectively. These initial and boundary conditions can be described as the following: initial thermodynamic equilibrium, constant heat flux at the particle surface, and constant temperature far away in the infinite medium. Note that "infinite" means from the perspective of the particle medium—that is, generally more than 10 particle diameters between particles.

Using all these assumptions, the heat equation can be solved to yield the following solution [43]:

$$T(r,t) - T_\infty = \frac{Q_{particle}}{4\pi rk}\left[\text{erfc}\left(\frac{r-R}{2\sqrt{\alpha t}}\right) - \exp\left(\frac{r-R}{R} + \frac{\alpha t}{R^2}\right)\text{erfc}\left(\frac{r-R}{2\sqrt{\alpha t}} + \sqrt{\frac{\alpha t}{R^2}}\right)\right]$$
$$\text{for } r > R \tag{5.6}$$

where T represents temperature and the independent variables r and t represent distance in the radial direction and time, respectively. $Q_{particle}$ is the total power generated by the nanoparticle and k is the thermal conductivity of the liquid. The functions denoted erfc() and exp() are the complementary error function and the exponential function. Finally, R and α represent the particle radius and the thermal diffusivity of the liquid, which is defined as [44]

$$\alpha = \frac{k}{\rho c_p} \tag{5.7}$$

where ρ and c_p are the density and the specific heat, respectively (for water, $\alpha \sim 1 \times 10^{-7}$ m^2/s).

There are a couple of interesting limiting conditions which arise from Equation 5.6. For example, as $r \to \infty$ or $t = 0$, the erfc() function approaches 0, and the resulting temperature approaches T_∞, $T \to T_\infty$. This indicates that our final equation checks with the boundary conditions. For $t \to \infty$ at $r = R$—representing the highest steady-state temperature expected—we get the following equation:

$$T(r) = T_\infty + \frac{Q_{particle}}{4\pi kR} \tag{5.8}$$

For this simplified equation, the difficult task is to determine the maximum possible heating rate, $Q_{particle}$. Taking the maximum achievable focused laser flux in the experiments reported here to be 770 W/cm^2 and an absorption cross-section of a nanoparticle to be 5×10^{-14} m^2, it is possible to have a nanoparticle be a generator of up to 4×10^{-7} W. With a 30-nm particle, this will result in a temperature rise around the particle of nearly 3.3°C. It should also be noted that this can be achieved exceptionally quickly. The characteristic time of the above equation can be determined from solving for the following quantity:

$$t_{diffusion} = \frac{R^2}{\alpha} \quad (5.9)$$

As R is less than 50 nm and $\alpha \sim 1 \times 10^{-7}$, the characteristic time is less than 25 ns. After several time constants, say four time constants or ~100 ns, it may be expected that the local region will come close to the steady-state temperature. Thus, this analysis indicates that in a sub-cooled, liquid water surrounding fluid, steady state is reached very quickly and that the localized temperature rise is modest. If, however, the surrounding fluid is at saturated conditions and vapor forms around the particle—which has a much lower thermal conductivity, 1/40th that of liquid water, much higher temperatures can result. As the maximum achievable temperature is inversely related to thermal conductivity, a 1/40th reduction in thermal conductivity could theoretically give a temperature rise around the particle of 130°C. The next section will discuss how vapor may form around a single nanoparticle or a group of nanoparticles and the resultant forces that are theoretically present.

5.4.2 Vapor Nucleation and Kinetics in Nanofluids

Although on the topic of vapor generation in a nanofluid, it is interesting to discuss bubble nucleation and kinetics. As much of this takes place on nano/microtemporal and spatial scales, for the most part these answers lie below the resolution limits of the experimental set-up. Therefore, the discussion of this section is mostly theoretical. Nevertheless, this section will attempt to answer the following questions: what temperatures and pressures are expected inside nanosized bubbles; how do bubbles nucleate; how fast do bubbles grow; and what are the buoyancy forces on the resulting bubbles?

The answer to the first question can be estimated by the Young–Laplace equation simplified for spherical coordinates [45]:

$$P_{vapor} - P_{liquid} = \frac{2\sigma}{R} \quad (5.10)$$

where P_{vapor} and P_{liquid} are the pressures inside and outside of the bubble. The parameter σ is the liquid–vapor surface tension, and R the radius of the bubble. It was found through Lenard–Jones molecular dynamic simulations that Equation 5.10 holds even for nanosized bubbles [46]. Matsumoto and Tanaka [46] also indicated that planar values for surface tension are good approximations even for the very high curvatures of nanosized bubbles. Thus, if $\sigma = 0.04$ N/m (water's surface tension

(a)   (b)

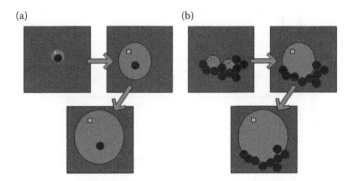

FIGURE 5.10   Bubble formation paths—(a) Bubble formation around a singular particle and (a) bubble formation on particle agglomerates.

at ~300°C) and $R = 10$ nm (the smallest particles we use in the experiments), the vapor pressure inside a bubble is expected to be ~8 MPa (~80 atm) from the Young–Laplace equation. If we assume the vapor inside this bubble is saturated, the saturation temperature at 8 MPa is ~295°C. As the bubble becomes larger, the predicted pressure and temperature inside will go down.

Nanoparticles have a tendency to agglomerate. This always happens to some extent, despite treatment with surfactant and/or other functionalization and stabilization methods. This is, in part, due to the fact that nature seeks to reduce surface energy and nanoparticles have a lot of surface area. This fact also holds for bubble nucleation as well [45]. As a result of this, bubbles tend to nucleate at certain sites—usually microcavities on a solid surface [45]. In the case of a nanofluid, however, the particles themselves are heat sources when irradiated. Therefore, although bubbles may be formed around individual particles, it is more likely that bubbles form in concave sites inside particle agglomerates. Figure 5.10 presents a cartoon of potential bubble formation around either a single nanoparticle or around a nanoparticle agglomerate.

The next question pertains to how fast do bubbles grow? For the answer to this question, a rich, extensive body of research can be drawn upon. This topic can be discussed in terms of two bounding models—inertial-driven bubble growth (fast) versus heat-transfer-driven bubble growth (slow/steady-state growth). There are also several correlations which essentially track between these extremes, but which are based on empirical data for given conditions.

Inertial-driven bubble growth is driven by pressure differences inside and outside of the bubble. Forces which are initially out of equilibrium cause the bubble to expand at a very fast rate. The simple Rayleigh solution to this problem results in the following simplified equation [45]:

$$D_{\text{bubble}}(t) = 2t \sqrt{\frac{2}{3} \frac{T_\infty - T_{\text{sat}}}{T_{\text{sat}}} \frac{\rho_{\text{liquid}} h_{lv}}{\rho_{\text{vapor}}}} \qquad (5.11)$$

where D_{bubble} is the bubble diameter, and t and T are time and temperature, respectively. The constants ρ_{liquid}, ρ_{vapor}, and h_{lv} are the liquid density, vapor density,

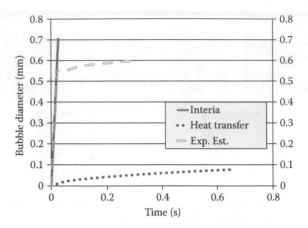

FIGURE 5.11 Bubble size approximations for inertia-induced bubble growth, heat-transfer-induced bubble growth, and an estimate of Zuber's correlation for nucleate bubble growth.

and latent heat of phase change from liquid to vapor. In this equation, the bubble grows linearly with time. This growth is shown as the upper curve in Figure 5.11.

The other extreme, heat-transfer-induced growth, is governed by the following equation [45]:

$$D_{\text{bubble}}(t) = 4Ja\sqrt{3\pi \frac{k_{\text{liquid}}c_{\text{p,liquid}}}{T_{\text{sat}}}t} \tag{5.12}$$

where the diameter, D_{bubble}, of the bubble is a function of the square root of time, INLINE. The properties k_{liquid}, $c_{\text{p,liquid}}$ are the liquid thermal conductivity and the specific heat of the liquid, respectively. The nondimensional variable Ja is the Jakob number [45]:

$$Ja = \frac{\rho_{\text{liquid}}c_{\text{p,liquid}}}{\rho_{\text{vapor}}h_{\text{lv}}T_{\text{sat}}}(T_{\infty} - T_{\text{sat}}) \tag{5.13}$$

From this equation, it can be seen that the bubble size is dependent on the square root of time. Thus, heat-transfer-induced growth is a much slower mechanism of growth than inertia-driven growth.

It follows that bubble growth is initially inertia-controlled, whereas at longer times, the growth is heat-transfer controlled. That is, based on the thermophysical properties of the fluid and the boiling surface, it can be expected to see fast initial growth and slow steady-state growth. It has been argued by a number of researchers that there should be a smooth transition between these two extremes [45,47,48]. One simple approach to determine this transition is to find a transitional radius below which growth is inertia-controlled and above which is heat-transfer controlled. This transitional radius is given by [45] equation:

$$R_{\text{transition}} = \frac{(12k_{\text{liquid}}/\pi\rho_{\text{liquid}}c_{\text{p,liquid}})Ja^2}{\left\{(2[T_\infty - T_{\text{sat}}(P_\infty)]c_{\text{p,liquid}}\rho_{\text{liquid}})/(\rho_{\text{vapor}}h_{lv})\right\}^{1/2}} \tag{5.14}$$

That is, it can be estimated that bubble growth follows inertia-induced bubble growth until the transition size, whereafter it then follows heat-transfer growth. This estimate, based on photos from the experiments, is shown as the middle (dashed) curve in Figure 5.11. Experimental observations indicate that this prediction is a reasonable estimate of the real situation—that is, the first few frames of the video represented in Figure 5.8.

Another question mentioned above is what are the forces acting on a vapor bubble?

In order to observe more easily the responses to the forces, a larger bubble that encompasses the larger heated area is examined. In this region, assume that the forces are balanced and that there is a restricting force in the downward direction which balances with the buoyancy/thermal convective forces pulling the bubble upwards. Experiments have shown a cylindrical bubble of roughly 0.4 mm diameter and a 0.1 mm depth. For this situation, the total force in the positive y-direction is calculated to be ~0.12 µN. A free body diagram of the forces is shown in Figure 5.12.

It is proposed that the force of friction between the bubble and the front and back surfaces of the cuvette provides the balancing force. Van der Waals forces holding the particles together or other particle/bubble interactions could also provide the necessary restricting force in the negative y-direction. Either way, the buoyancy and/or restrictive forces on a single, relatively large bubble are very small. We expect the forces to be orders of magnitude smaller on the nanoscale bubbles shown in Figure 5.10.

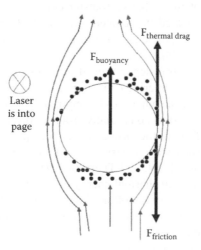

FIGURE 5.12   Free body diagram of bubble during the laser-heating experiments.

5.4.3 Superposition of Particles

Another analysis that can be performed without too much difficulty is that of heat diffusion surrounding a group of heat-generating nanoparticles. That is, if each particle absorbs some heat, then the total heat input will be dependent on the particle absorption cross-section and the density of absorbers. The following equation, which is slightly modified from ΔT_{global} presented in [43], gives an estimate for the steady-state temperature rise of a group of particles over the surrounding medium, in a light-absorbing nanofluid:

$$\Delta T = \frac{\rho_N R_{\text{heat}}^2 I \sigma_a}{2k} \tag{5.15}$$

where ρ_N is the particle density, R_{heat} the radius of the heated region, I the irradiance, σ_a the particle absorption cross-section, and k the thermal conductivity of the surrounding fluid. It should be noted that this equation applies to steady state in an infinite medium. For parameter values representative of the experimental conditions described here, Equation 5.15 predicts ΔT at least two orders of magnitude greater than that for a single nanoparticle (Equation 5.8). That is, much higher temperatures can be obtained by heating many nanoparticles, rather than a single nanoparticle. This is the same conclusion originally reached in ref. [43].

5.5 RADIATIVE PROPERTIES OF NANOFLUIDS

Fundamental to light-induced boiling of nanofluids is an understanding of how light is absorbed by the nanoparticles as well as by the base fluid. Here, a brief listing of the authors' ongoing work in this area will be made to refer the interested reader to the cited publications.

A coupled solution of the radiative transfer equation and the energy equation for a flowing nanofluid yielded the temperature rise of the nanofluid as a result of absorbed radiation and the efficiency of a direct-absorption nanofluid-based solar collector or receiver [8,9,12,14].

An analysis of radiative transfer in typical nanofluids showed that the common assumption of independent scattering cannot always be applied, but rather that multiple dependent scattering can sometimes be important [7,10]. The small scale of the nanoparticles leads to size effects on radiative transfer through nanofluids. This was analyzed by applying a form of the Drude model [10,14].

The radiative properties were measured for four common base fluids, water, propylene glycol, ethylene glycol, and Therminol VP-1, which are likely to be utilized in direct-absorption solar thermal energy conversion systems. In addition, the optical properties of a variety of nanofluids have been investigated for use in direct absorption solar collectors [16,49–52]. The results of such studies indicate the potential to "tune" the spectral radiative properties of nanofluids, as represented by the extinction coefficient, through appropriate choice of nanoparticle materials, shape, volume fraction, or a combination of all these variables [13,14].

The volumetric radiative transfer provided by a nanofluid can lead to higher effective emissivities and/or absorptivities, relative to conventional area-based surface radiative transfer. For example, $\varphi = 1\%$, 30-nm graphite nanofluid is calculated to exhibit a solar absorptance of ~97%, compared with ~90% for bulk graphite [13,14].

5.6 FUTURE DIRECTIONS

Nanofluids remain a fascinating subject, as they have been for several years in the heat-transfer literature and increasingly so in the solar energy literature. A number of areas for further work have already been pointed out above, but some of the most interesting subjects are

- How can nanofluids be utilized for solar-powered direct-steam generation?
- How can the radiative properties be optimized for maximum solar energy conversion efficiency?
- What roles do enhanced convective heat transfer and thermal conductivity, as has been often observed in nanofluids, play in solar energy conversion?

ACKNOWLEDGMENTS

The authors gratefully acknowledge the support of the National Science Foundation through awards CBET-0812778 and CBET-1066705.

NOMENCLATURE

A	Surface area of a single nanoparticle (m^2)
Bi	Biot number for a single nanoparticle
c_p	heat capacity of nanofluid [J/kg/K]
D	Nanoparticle diameter (m)
D_{bubble}	Bubble diameter (m)
h	Convective heat transfer coefficient for a single nanoparticle ($W/m^2/K^2$)
h_{lv}	Latent heat (J/kg)
I	Irradiance (W/m^2)
Ja	Jakob number, defined by Equation 5.13
k	Thermal conductivity (W/m/K)
q	Volumetric power generation term for a single nanoparticle (W/m^3)
$Q_{particle}$	Heat absorbed by a single nanoparticle (W)
r	Radial coordinate (m)
R	Nanoparticle radius (m)
R_{heat}	Radius of a heated (irradiated) region (m)
t	Time (s)
T	Temperature (°C)
V	Volume of a single nanoparticle (m^3)

GREEK SYMBOLS

α Nanofluid thermal diffusivity (m²/s)
φ Volume fraction of nanoparticles (m⁻³)
ρ Density (kg/m³)
ρ_n Number of nanoparticles per unit volume (m⁻³)
σ Surface tension (N/m)
σ_a Absorption cross-section of a single nanoparticle (m²)

SUBSCRIPTS

bubble   Pertaining to a single vapor bubble
particle   Pertaining to a single nanoparticle
sat   Saturation conditions
liquid   Liquid conditions
vapor   Vapor conditions
∞   Fluid conditions far from a nanoparticle

REFERENCES

1. S. Das, S. Choi and H. Patel, Heat transfer in nanofluids—a review, *Heat Transfer Engineering*, 27, 3–19, 2006.
2. V. Trisaksri and S. Wongwises, Critical review of heat transfer characteristics of nanofluids, *Renewable and Sustainable Energy Reviews*, 11, 512–523, 2007
3. W. Daungthongsuk and S. Wongwises, A critical review of convective heat transfer of nanofluids, *Renewable and Sustainable Energy Reviews*, 11, 797–817, 2007.
4. S. Murshed, K. Leong and C. Yang, Thermophysical and electrokinetic properties of nanofluids—a critical review, *Applied Thermal Engineering*, 28, 2109–2125, 2008.
5. W. Yu, D. France, J. Routbort and S. Choi, Review and comparison of nanofluid thermal conductivity and heat transfer enhancements, *Heat Transfer Engineering*, 29, 432–460, 2008.
6. J. e. a. Buongiorno, A benchmark study on the thermal conductivity of nanofluids, *Journal of Applied Physics,* 106, 094312, 2009.
7. R. Prasher and P. Phelan, Modeling of radiative and optical behavior of nanofluids based on multiple and dependent scattering theories, in *Proceedings of IMECE2005, ASME International Mechanical Engineering Congress and Exposition*, Orlando, Florida, 2005.
8. H. Tyagi, P. Phelan and R. Prasher, Predicted efficiency of a low-temperature nanofluid-based direct absorption solar collector, *Journal of Solar Energy Engineering*, 131, 041004, 2009.
9. T. Otanicar, P. Phelan, G. Rosengarten and R. Prasher, Experimental testing and modeling of a micro solar thermal collector with direct absorption nanofluids, in *Proceedings of the Inaugural US-EU-China Thermophysics Conference*, Beijing, China, 2009.
10. T. Otanicar, R. Taylor, P. Phelan and R. Prasher, Impact of size and scattering mode on the optimal solar absorbing nanofluid, in *Proceedings of the 3rd International Conference on Energy Sustainability*, San Francisco, California, 2009.
11. R. Taylor, P. Phelan, T. Otanicar, R. Adrian and R. Prasher, Vapor generation in a nanoparticle liquid suspension using a focused, continuous laser, *Applied Physics Letters*, 95, 161907, 2009.
12. R. Taylor, P. Phelan, T. Otanicar, H. Tyagi and S. Trimble, Applicability of nanofluids in concentrated solar energy harvesting, in *ASME 2010 4th International Conference on Energy Sustainability*, Phoenix, Arizona, 2010.

13. T. Otanicar, P. Phelan, R. Taylor and H. Tyagi, Tuning the extinction coefficient for direct absorption solar thermal collector optimization, in *Proceedings of the 4th International Conference on Energy Sustainability*, Phoenix, Arizona, 2010.

14. T. Otanicar, P. Phelan, R. Prasher, G. Rosengarten and R. Taylor, Nanofluid-based direct-absorption solar collector, *Journal of Renewable and Sustainable Energy*, 2, 033102, 2010.

15. R. Taylor, P. Phelan, T. Otanicar, C. Walker, M. Nguyen, S. Trimble and R. Prasher, Applicability of nanofluids in high flux solar collectors, *Journal of Renewable and Sustainable Energy*, 3, 023104, 2011.

16. R. Taylor, P. Phelan, T. Otanicar, R. Adrian and R. Prasher, Nanofluid optical property characterization: Towards efficient direct-absorption solar collectors, *Nanoscale Research Letters*, 6, 225, 2011.

17. M. Eck and W.-D. Steinmann, Modelling and design of direct solar steam generating collector fields, *Journal of Solar Energy Engineering*, 127, 371–380, 2005.

18. V. Morisson, M. Rady, E. Palomo and E. Arquis, Thermal energy storage systems for electricity production using solar energy direct steam generation technology, *Chemical Engineering and Processing*, 47, 499–507, 2008.

19. M. Montes, A. Abanades and J. Martinez-Val, Performance of a direct steam generation solar thermal power plant for electricity production as a function of the solar multiple, *Solar Energy*, 83, 679–689, 2009.

20. M. Gupta and S. Kaushik, Exergy analysis and investigation for various feed water heaters of direct steam generation solar–thermal power plant, *Renewable Energy*, 35, 1228–1235, 2010.

21. J. Minardi and H. Chuang, Performance of a black liquid flat-plate solar collector, *Solar Energy*, 17, 179–183, 1975.

22. N. Arai, Y. Itaya and N. Hasatani, Development of a volume-heat-trap type solar collector using a fine-particle semitransparent liquid suspension (FPSS) as a heat vehicle and a heat storage medium, *Solar Energy*, 32, 49–56, 1984.

23. G. Camera-Roda and M. Bertelà, A model of black-liquid solar collector, *Solar Energy*, 40, 197–209, 1988.

24. S. Kumar and C. Tien, Analysis of combined radiation and convection in a particulate laden liquid film, *Journal of Solar Energy Engineering*, 112, 293–300, 1990.

25. F. Miller and R. Koenigsdorff, Theoretical analysis of a high-temperature small-particle solar receiver, *Solar Energy Materials*, 24, 210–221, 1991.

26. J. Oman and P. Novak, Volumetric absorption in gas— Properties of particles and particle-gas suspensions, *Solar Energy*, 56, 597–606, 1996.

27. R. Bertocchi, J. Karni and A. Kribus, Experimental evaluation of a non-isothermal high temperature solar particle receiver, *Energy*, 29, 687–700, 2004.

28. H. Klein, R. Rubin and J. Karni, Experimental evaluation of particle consumption in a particle-seeded solar receiver, *Journal of Solar Energy Engineering*, 130, 011012, 2008.

29. A. Z'Graggena and A. Steinfeld, Heat and mass transfer analysis of a suspension of reacting particles subjected to concentrated solar radiation—Application to the steam-gasification of carbonaceous materials, *International Journal of Heat and Mass Transfer*, 52, 385–395, 2009.

30. M. Romero, R. Buck and J. Pacheco, An update on solar central receiver systems, projects, and technologies, *Journal of Solar Energy Engineering*, 124, 98–108, 2002.

31. A. Solar, Jan 2010. [Online]. Available: http://www.abengoasolar.com/corp/web/en/our_projects/solucar/ps10/index.html.

32. D. Wen and Y. Ding, Experimental investigation into the pool boiling heat transfer of aqueous based gamma-alumina nanofluids, *Journal of Nanoparticle Research*, 7, 265–274, 2005.

33. H. Xue, J. Fan, R. Hong and Y. Hu, Characteristic boiling curve of carbon nanotube nanofluid as determined by the transient calorimeter technique, *Applied Physics Letters*, 90, 184107, 2007.

34. D. Wen, G. Lin, S. Vafaei and K. Zhang, Review of nanofluids for heat transfer applications, *Particuology,* 7, 141–150, 2009.
35. R. Taylor and P. Phelan, Pool boiling of nanofluids: Comprehensive review of existing data and limited new data, *International Journal of Heat and Mass Transfer*, 52, 5339–5347, 2009.
36. J. Kim, S. You and J. Pak, Effects of heater size and working fluids on nucleate boiling heat transfer, *International Journal of Heat and Mass Transfer,* 49, 122–131, 2006.
37. C. Bohren and D. Huffman, *Absorption and Scattering of Light by Small Particles*, New York, NY: John Wiley and Sons, 1998, pp. 287–324.
38. S. Merabia, P. Keblinski, L. Joly, L. Lewis and J.-L. Barrat, Critical heat flux around strongly heated nanoparticles, *Physical Review E*, 79, 021404, 2009.
39. M. Hu, H. Petrova and G. Hartland, Investigation of the properties of gold nanoparticles in aqueous solution at extremely high lattice temperatures, *Chemical Physics Letters*, 391, 220–225, 2004.
40. A. Plech, V. Kotaidis, S. Grésillon, C. Dahmen and G. von Plessen, Laser-induced heating and melting of gold nanoparticles studied by time-resolved x-ray scattering, *Physical Review B*, 70, 195423, 2004.
41. H. Carslaw and J. Jaeger, *Conduction of Heat in Solids*, Oxford, UK: Oxford University Press, 1986, 520.
42. D. Tzou, *Macro- To Micro-Scale Heat Transfer: The Lagging Behavior*, Washington, DC: Taylor & Francis, 1997.
43. P. Keblinski, D. Cahill, A. Bodapati, C. Sullivan and T. Taton, Limits of localized heating by electromagnetically excited nanoparticles, *Journal of Applied Physics,* 100, 054305, 2006.
44. F. Incropera, D. DeWitt, T. Bergman and A. Lavine, *Introduction to Heat Transfer*, New York, NY: Wiley, 2006, 912.
45. V. Carey, *Liquid-Vapor Phase-Change Phenomena: An Introduction to the Thermophysics of Vaporization and Condensation Processes in Heat Transfer Equipment*, 2 ed., Washington, DC: Taylor & Francis, 2007, p. 600.
46. M. Matsumoto and K. Tanaka, Nano bubble—size dependence of surface tension and inside pressure, *Fluid Dynamics Research,* 40, 546–553, 2008.
47. P. Griffith, *Bubble Growth Rates in Boiling*, Cambridge, MA, 1956.
48. B. Mikic, W. Rohsenow and P. Griffith, On bubble growth rates, *International Journal of Heat and Mass Transfer,* 13, 657–666, 1970.
49. L. Mercatelli, E. Z. G. Sani, F. Martelli, P. Di Ninni, S. Barison, C. Pagura, F. Agresti and D. Jafrancesco, Absorption and scattering properties of carbon nanohorn-based nanofluids for direct sunlight absorbers, *Nanoscale Research Letters,* 6, 282, 2011.
50. D. Han, Z. Meng, D. Wu, C. Zhang and H. Zhu, Thermal properties of carbon black aqueous nanofluids for solar absorption, *Nanoscale Research Letters,* 6, 457, 2011.
51. Y. Kameya and K. Hanamura, Enhancement of solar radiation absorption using nanoparticle suspension, *Solar Energy,* 85, 299–307, 2011.
52. T. Otanicar, P. Phelan and J. Golden, Optical properties of liquids for direct absorption solar thermal energy systems, *Solar Energy,* 83, 969–977, 2009.

6 Radiative Properties of Micro/Nanoscale Particles in Dispersions for Photothermal Energy Conversion

Qunzhi Zhu and Zhuomin M. Zhang

CONTENTS

6.1   INTRODUCTION

Solar radiation is a vast and renewable energy source. The amount of solar energy absorbed by the earth's atmosphere, lands, and oceans is approximately 4×10^{24} J/year. The solar energy captured by the earth in just 1 h is more than the total energy consumed by the whole world in 1 year [1]. Harness of the solar energy by human started from ancient times. In the 1970s, significant research efforts were devoted to the study of solar energy utilization because of the oil crisis. In the subsequent two decades, solar energy research was slowed down as the crisis was apparently alleviated. In recent years, due to the general concern of environment and global climate change, solar energy utilization has again become an attractive research area.

There are three major approaches in actively exploiting the solar energy, that is, solar thermal technology [2], electricity generation technology [3], and solar chemical technology [4–6]. Solar thermal systems usually provide water heating, space heating and cooling, and process heating. Electricity generation systems include solar photovoltaic (PV) applications which directly convert solar radiation into electricity and concentrated solar power plants which convert solar radiation first into thermal energy and then to electricity. A solar PV panel and a thermal receiver can be combined into one collector. Fluid, such as water or air, is used to cool solar PV cells and the thermal energy gained by the fluid can be utilized. This kind of hybrid systems is called as photovoltaic–thermal system [7–9]. Solar energy can also be applied in chemical processes to drive chemical reactions. The solar chemical technologies can be divided into thermochemical and photochemical technologies [10]. As many chemical reactions are endothermic, solar energy can be used to boost the temperature of chemical reactants for hydrogen generation from water splitting, natural gas cracking, and reforming. Through the analysis of the major solar energy utilization technology, it is clear that photothermal conversion is important to not only the solar–thermal utilization, but also the electricity generation and solar chemical technology.

A solar thermal collector is the most important part in the photothermal conversion. Figure 6.1 categorizes solar thermal collectors according to their working temperatures and designs. Low-temperature collectors can provide fluids at temperatures up to 80°C while medium-temperature ones can supply fluids at temperatures ranging from 80°C to 250°C [11]. Flat-plate collectors and evacuated-tube collectors are common in the low-temperature region. In order to obtain fluids at high temperatures, collectors with low sun concentration ratios, such as compound parabolic concentrator (CPC) collectors and parabolic trough collectors, are required. Furthermore, collectors with high sun concentration ratios, such as parabolic dish collectors and central tower collectors, are necessary to obtain high-temperature working fluids for the generation of electricity.

In many solar thermal collectors, surfaces of the flat plate or the evacuated tube are heated by the solar radiation and then the thermal energy is conveyed away by fluids through convective heat transfer. The thermal resistant in converting solar energy into thermal energy of fluids is large especially when gas or vapor is applied as the working fluid. Thus, the temperature of the surface that absorbs the solar radiation is high. In order to reduce the heat loss from the black surface to the ambient, glazing or vacuum is necessary. Furthermore, a wavelength-selective surface is preferred to decrease thermal emission in the long wavelength region from the heated surface.

Besides the surface absorber, there is another type of solar absorber in which the solar radiation is absorbed through a much longer path inside a volume rather than in a very thin layer at the surface. Figure 6.2 illustrates three typical volumetric

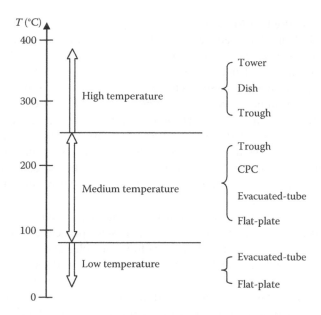

FIGURE 6.1 Types and operation temperature ranges of solar thermal collectors.

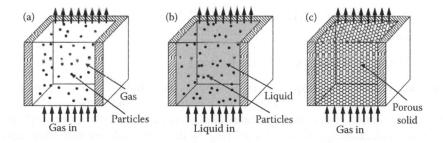

FIGURE 6.2 Volumetric absorber: (a) particles in gas, (b) particles in liquid, and (c) porous solid.

absorbers. In the first design, solar radiation is absorbed by particles dispersed in a gas [12,13]. In the second design, solar radiation is absorbed by a liquid seeded with particles [14–17]. In the third design, a porous solid absorbs the incoming radiation, while a gas flows through the porous absorber and gets heated [18,19]. Compared with surface absorbers, volumetric absorbers have several prominent advantages. Because the particles are in contact with the fluid, the solar energy absorbed by particles can be efficiently transferred to the surrounding fluid. Particles dispersed in gas or liquid have large surface areas that can enhance absorption as well as convective heat transfer. Similarly, a porous solid can increase the surface area for absorbing solar radiation and enhance convection. Therefore, volumetric absorbers can efficiently operate under a high solar flux.

Volumetric solar absorbers have been studied by many researchers for more than 30 years. The idea of seeding a liquid with black particles to absorb solar radiation was initially proposed by Minardi and Chuang [15]. The black liquid was formed by dispersing Indian ink into ethylene glycol–water mixture. The concept of adding solar-absorbing particles into gas came up a few years afterward [12,20]. Extensive research has been carried out on the volumetric absorber for the application of solar thermal power plants and solar chemical reactors since the oil crisis. Owing to the advancement in nanotechnology during the past 20 years or so, applications of nanoparticles in volumetric solar absorbers have also been proposed. Compared with a medium dispersed with micrometer-scale or larger particles, the suspension with nanoparticles may be more stable. In addition, the clogging and fouling would be less significant for suspension with nanoparticles. Furthermore, due to the small size effect, radiative properties of nanoparticle dispersions can be tuned to efficiently absorb the solar radiation.

This chapter reviews the recent advances in volumetric solar absorbers with an emphasis on the radiative properties of micro/nanoscale particle dispersions and the relevant radiative heat transfer within the volumetric solar absorber. Fundamental theories of light scattering and radiative transport, as well as size effects and dependent scattering, are presented in Section 6.2. Radiative properties of particles suspended liquids or gases are summarized in Section 6.3. Applications of particulate media in volumetric receivers are discussed in Section 6.4. A brief summary is provided in Section 6.5 with an outlook of future research directions.

6.2 THEORIES OF LIGHT SCATTERING AND RADIATIVE TRANSPORT

6.2.1 LIGHT SCATTERING BY A SINGLE PARTICLE

The study of light scattering by small particles is important in astronomy, meteorology, biophysics and biomedicine, remote sensing, fire and flame, combustion, thermal insulation, color, and appearance, etc. As early as 1881, Lord Rayleigh formulated electromagnetic wave scattering by dielectric spheres with diameters much smaller than the wavelength. In 1908, Gustav Mie derived the general solution for scattering by spheres of any size with or without absorption. Mie scattering theory reduces to the Rayleigh limit for small particles and the geometric-optics limit for particles with diameters much greater than the wavelength. Detailed discussions of light scattering by individual particles can be found from classical textbooks [21,22]. The Rayleigh scattering theory has been extended by other researchers for particles of irregular shapes under the conditions that the refractive index of the particle is close to that of the free space and that the particle is relatively small so that the phase shift can be neglected. This type of formulation is often called the Rayleigh–Debye–Gans theory, along with other names. The Mie scattering theory originally was developed for a sphere with uniform composition; however, it has been extended to nonspherical and irregularly shaped particles [23]. Numerous methods for rigorous modeling of light scattering by nonspherical particles were reviewed by Kahnert [24,25]. A brief summary of formulations of the Rayleigh scattering and Mie scattering is presented in the following.

For a particle with a complex refractive index (n_p, κ_p) in a nonabsorbing medium with an index of refraction n_m, the normalized refractive index with respect to the medium is

$$m = \frac{n_p + i\kappa_p}{n_m} \tag{6.1}$$

The size parameter, essential to determine the regime of scattering by particles, is

$$\chi = \frac{\pi D}{\lambda} \tag{6.2}$$

where D is the diameter of the spherical particle, and λ is the wavelength in the medium. If $\chi \ll 1$ and $|m|\chi \ll 1$, where $|m|$ stands for the modulus of m, the scattering of a single particle is within the regime of the Rayleigh scattering.

The scattering efficiency factor, absorption efficiency factor, and extinction efficiency factor can be determined as [21,22,26,27]

$$Q_s = \frac{8}{3}\chi^4 \left|\frac{m^2 - 1}{m^2 + 2}\right|^2 \tag{6.3a}$$

$$Q_a = 4\chi \, \mathrm{Im}\left(\frac{m^2 - 1}{m^2 + 2}\right) \tag{6.3b}$$

$$Q_e = Q_a + Q_s \tag{6.3c}$$

The scattering efficiency factor is proportional to the fourth power of the particle size while the absorption efficiency factor is predominantly proportional to the particle size. The phase function for the Rayleigh scattering is

$$\Phi_p(\theta) = \frac{3}{4}(1 + \cos^2 \theta) \tag{6.4}$$

Compared with the Rayleigh scattering, the Mie scattering is more general and can be applied for a spherical particle with an arbitrary diameter. The scattering efficiency factor and extinction efficiency factor can be determined as

$$Q_s = \frac{2}{\chi^2} \sum_{n=1}^{\infty} (2n + 1)(a_n a_n^* + b_n b_n^*) \tag{6.5a}$$

$$Q_e = \frac{2}{\chi^2} \sum_{n=1}^{\infty} (2n + 1)\mathrm{Re}(a_n + b_n) \tag{6.5b}$$

where superscript * refers to the complex conjugate, and a_n and b_n are Mie scattering coefficients, which can be found from refs [21,22,27]. The phase function for the Mie scattering is

$$\Phi_p(\theta) = \frac{1}{Q_s\chi^2}(S_1 S_1^* + S_2 S_2^*)$$

(6.6)

where S_1 and S_2 are complex amplitude functions [21,22,27].

6.2.2 LIGHT SCATTERING IN A PARTICULATE MEDIUM

In order to solve the radiative transfer equation (RTE) of any particulate medium, knowledge of the temperature- and wavelength-dependent absorption coefficient, scattering coefficient, and scattering phase function is required [28–30]. As there are numerous particles in a medium, the scattered wave from one particle may interact with those from nearby particles. For example, a light beam can be scattered more than once and the scattered light can be scattered again by other particles. In addition, the scattered fields can be strongly correlated (i.e., interfere with each other coherently). In order to make the formulation simple, the independent scattering approximation is commonly applied. This approximation states that the interaction between waves from different particles can be neglected. Consequently, the incident field on one particle is the same as the original incident field as if there was no other particle. Furthermore, the scattered waves do not interfere in the far field. On the basis of these assumptions, the extinction cross-section of a particle cloud comprising of M groups of particles with the same size is the sum of the cross-sections of each particle,

$$\beta = \sum_{i=1}^{M} N_i C_{e,i} = \sum_{i=1}^{M} N_i \frac{\pi D_i^2}{4} Q_{e,i} = 1.5 \sum_{i=1}^{M} \frac{f_{v,i}}{D_i} Q_{e,i}$$

(6.7)

where N_i, $C_{e,i}$, and $f_{v,i}$ are the number density, extinction cross-section, and volume fraction of particles in the ith group.

Independent scattering is applicable for cases where interparticle clearance b is sufficiently large compared with the wavelength λ of thermal radiation and the particle diameter D. In contrast, when the interparticle clearance is small in a dense medium, the scattered waves by nearby particles can interact with each other. Figure 6.3 illustrates the interaction of scattered fields. Dependent scattering happens when the scattered fields interfere in the far field. Multiple scattering happens when the scattered wave by one particle is incident onto another particle and then is scattered again. Much research has been performed on multiple scattering and dependent scattering, mainly for the radiative heat transfer in fluidized beds, packed beds, and microsphere insulations.

Generally, independent scattering is applicable for a dilute medium. It may result in good agreement with experimental data. However, the effect of dependent scattering

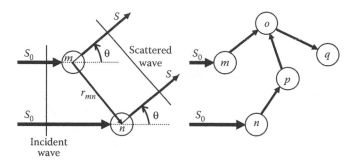

FIGURE 6.3 Schematics of (a) dependent scattering and (b) multiple scattering. Here, m, n, o, p, and q refer to different particles, S_0 and S denote the incidence and scattered waves, respectively.

should be considered if the medium is dense. Initially, it was believed that dependent effects were determined by particle separation only, occurring at spacing between particles on the order of a diameter or less. Hottel et al. [31] realized that interparticle clearance measured in wavelength, that is, b/λ, was a key parameter and considered dependent effect as a function of volume fraction f_v and b/λ. The values of $f_v = 0.27$ and $b/\lambda = 0.3$ were suggested as the simultaneous limits of independent scattering [31,32]. Tein and coworkers did some pioneering research on dependent scattering [33–36]. Experimental results displayed that b/λ was the most important parameter to gauge the onset of dependent scattering and that a high value f_v alone would not result in dependent scattering. As important step forward, Tien et al. [34–36] suggested that the parameter of b/λ alone delineates dependent scattering and independent scattering if $f_v > 0.006$. The regime map for demarcating dependent scattering and independent scattering shown in Figure 6.4 is extensively used by researchers in the study of radiative transfer in particle dispersions. In order for independent scattering theory to be valid, either $f_v < 0.006$ or $b/\lambda > 0.5$ should be satisfied [36]. Although it is easy to calculate optical properties of a particle cloud under the independent scattering approximation, it remains a daunting task to accurately determine the scattering phase function for dependent scattering. In most cases, approximate or semi-empirical phase functions are commonly employed.

Dependent scattering can modify the dispersion relation and Planck's blackbody emissive power in a particulate medium. Multiple scattering changes the effective field in the particulate medium, and the dispersion relation is defined as [37]

$$k^2 = k_1^2 + 4\pi N f^+ \tag{6.8}$$

where k is the effective wave vector of the particulate medium, and k_1 is the wave vector of the matrix in which particles are dispersed, N is number of particles per unit volume, and f^+ is the amplitude of the forwardly scattered electromagnetic wave. The effective field approximation takes into account only multiple scattering while considers that scattering is independent. As multiple scattering changes the dispersion relation of the particulate medium, the effective field approximation is good at

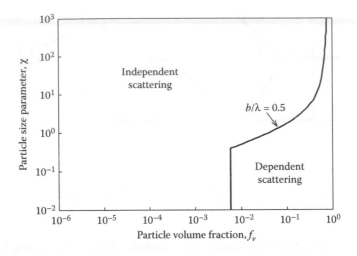

FIGURE 6.4 Regime map of particle scattering. (From C. L. Tien and B. L. Drolen, Thermal radiation in particulate media with dependent and independent scattering. In *Annual Review of Numerical Fluid Mechanics and Heat Transfer*, Hemisphere Publishing Corp, New York, pp. 1–32, 1987.)

capturing the phase velocity and refractive index of the particulate medium, but it is not accurate in modeling the scattering and absorption efficiencies [38,39]. The quasi-crystalline approximation, in contrast, takes account of multiple scattering and dependent scattering simultaneously. Under this approximation, Equation 6.8 is modified to [37]

$$k^2 = k_I^2 + 4\pi N\beta f^+ \qquad (6.9)$$

where β is a correction factor defined as the ratio of the effective field to the coherent field. In particulate media, the effective refractive index of the medium and Planck's blackbody emissive power is mainly affected by multiple scattering [38,39]. In contrast, the scattering and absorption cross-sections are mainly affected by dependent scattering [38,39].

The effect of dependent scattering on scattering efficiency and absorption efficiency in various particulate media has been studied. Dependent scattering efficiency is unusually less than the corresponding independent scattering efficiency [34]. The difference is more significant when the particle volume fraction is higher. However, dependent absorption efficiency is higher than independent absorption efficiency, and the difference increases as the particle volume fraction increases [40]. Therefore, when dependent scattering occurs, the extinction coefficient exhibits a net increase for highly absorbing particles and a net decrease for nonabsorbing particles [40]. The hemispherical transmittance for dependent scattering may be either larger or smaller than that for independent scattering, depending on the optical thickness, whereas the hemispherical reflectance for dependent scattering is generally smaller than that for independent scattering regardless of the optical thickness [35].

Cartigny et al. [34] applied the Rayleigh–Gans–Debye scattering theory to study dependent scattering by particles. The effect of scattering by other particles on the incident and scattered waves were neglected (i.e., multiple scattering). Only the phase difference in the far field of waves scattered by different particles was considered. The Rayleigh–Gans–Debye theory was extended to particles with somewhat larger size parameters, χ, by using the Mie scattering instead of the Rayleigh scattering for individual particles [34]. However, the results may not be very accurate for large particles. A form factor approach, commonly used in x-ray optics, has been employed to calculate dependent scattering efficiency [34,41]. Pair distribution functions have been introduced to correlate the relative positions of particles in a medium. The pair distribution function represents the likelihood of finding a neighboring particle at some distance from the central particle. Gas model and packed-sphere model are simple forms of pair distribution functions. In both models, no neighbor exists within one particle diameter of the scatterer and the likelihood of finding a neighbor beyond the particle diameter is uniform. The difference between two models is that the packed-sphere model includes a Dirac-delta function at the particle diameter. The gas model can be applied to packed beds with small f_v while the packed-sphere model to packed beds with large f_v [34,41].

The Rayleigh–Gans–Debye approximation neglects the near-field interaction of scattered waves, that is, multiple-scattering effect, and assumes that the internal field of each particle is proportional to the incident field. Kumar and Tien [40] considered both the near-filed effect and far-field interference of scattered waves and developed a model for dependent scattering and absorption. Discrete dipole approximation has also been used to study the dependent and independent scattering regimes, including scattering by nanoparticles [42,43].

Although the criterions for delineating dependent scattering and independent scattering have been widely applied by many researchers, some controversy still exists. Ishimaru and Kuga [44] noted dependent effect even at much higher values of b/λ. Singh and Kaviany [45] showed that the independent theory can fail even though the criterion of b/λ is satisfied. There are a few drawbacks in the theoretical model by Cartigny et al. [34] besides the experimental uncertainties in the data reported by various researchers. First, near-field effects were neglected and only the far-field interference effects were taken into consideration. Second, the Rayleigh–Gans–Debye approximation may not accurately predict the scattering efficiency for relatively large particles even though the expansion is based on the Mie theory. Third, nonabsorbing particles were assumed in obtaining the criterions that distinguish dependent scattering and independent scattering. Therefore, further work is needed to expand our knowledge on dependent scattering and to improve the widely used regime map.

6.2.3 Size Effect

Generally, the optical constants for bulk materials are used to determine the absorption and scattering efficiency factors for micro/nanoscale particles. However, as the size of nanoparticles is smaller than the mean free path of free electrons in bulk materials, the optical properties of nanoparticles are modified due to boundary scattering as in the case of thermal conductivity [46]. The dielectric function of metal in the infrared can be modeled by the Drude free-electron model [46]

$$\varepsilon(\omega) = (n + i\kappa)^2 = \varepsilon_\infty - \frac{\omega_p^2}{\omega^2 + i\gamma\omega} \tag{6.10}$$

where ω is the angular frequency of the electromagnetic wave, ε_∞ is high-frequency constant, ω_p is the plasmon frequency, and γ is the damping coefficient. It should be noted that in the visible and ultraviolet region, interband absorption must also be considered [47]. The damping coefficient for nanoparticles is different from that for the bulk materials due to the scattering of electrons by the boundary. Otanicar et al. [48] suggested that

$$\gamma_D = \gamma_b + g\frac{v_f}{D} \tag{6.11}$$

where γ_b is the damping coefficient for the bulk material, g is a constant usually set to 1, and v_f is the Fermi velocity. In the visible and ultraviolet region, the contribution of interband transitions can be modeled with either the Lorentz model or the critical-points model [49]. The dielectric function for the Drude-critical-points model can be expressed as

$$\varepsilon(\omega) = \varepsilon_\infty - \frac{\omega_p^2}{\omega^2 + i\gamma_D\omega} + \sum_p A_p\Omega_p\left(\frac{e^{i\phi_p}}{\Omega_p - \omega - i\Gamma_p} - \frac{e^{-i\phi_p}}{\Omega_p + \omega + i\Gamma_p}\right) \tag{6.12}$$

where Ω, Γ, A, and ϕ are parameters associated with each critical point. For Au and Ag, the parameters are fitted to the experimental data by Vial and Laroche [49].

Nanospheres and nanoshells can be plasmonic particles, whose plasmonic frequencies are dependent on the size and geometry. The plasmonic frequency for metal nanospheres usually falls in the visible and UV regions of the solar spectrum. The resonant frequency of nanoshells, such as a dielectric core coated by a thin layer of metal, is sensitive to the relative diameter of the core and the shell. By combining nanospheres and nanoshells of different materials and geometries, it is possible to absorb solar radiation at high efficiency.

Cole and Halas [50] studied absorptance of solar radiation by a surface deposited with nanospheres and nanoshells. They theoretically investigated optical properties of mixtures of different Au nanospheres and nanoshells with Au as the shell and silica as the core. Figure 6.5 shows an optimal nanoparticle mixture from theoretical simulations and the corresponding air mass 1.5 solar spectrum absorption. It was determined that the optimal mixture consists of three species: species 1 is a nanosphere with a radius of 32 nm, species 2 is a nanoshell with core radius of 28 and shell radius of 42 nm, and species 3 is a nanoshell with core radius of 47 nm and shell radius of 58 nm. The fractions of three species are 35.9%, 22.8%, and 41.3%, respectively. With a total surface coverage of only 50%, the nanospheres and

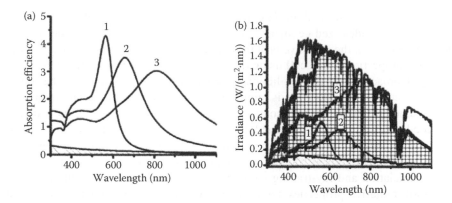

FIGURE 6.5 Illustration of optimal nanoparticle mixtures for absorbing solar energy. (a) Absorption efficiency of three selected nanoparticle geometries. (b) Predicted optimal absorption of the air mass 1.5 solar spectrum for the mixture of nanoparticle species from (a). (From J. R. Cole and N. J. Halas, *Applied Physics Letters*, 89, 153120–1/3, 2006. With permission.)

nanoshells on a silicon substrate in air could efficiently absorb 84% of 805 W/m² of solar radiation.

6.2.4 RADIATIVE TRANSPORT EQUATION FOR PARTICULATE MEDIA

Radiative transfer in particulate and porous media has been widely studied in the last 50 years or so. The steady-state RTE is given as [26,27]

$$\frac{dI_\lambda}{dS} = -a_\lambda I_\lambda(S) - \sigma_{s,\lambda} I_\lambda(S) + a_\lambda I_{\lambda,b}(S,T_m) + \frac{\sigma_{s,\lambda}}{4\pi} \int_{4\pi} I_\lambda(S,\omega_i)\Phi(\lambda,\omega,\omega_i)d\omega_i$$

(6.13)

where I is the radiation intensity in an arbitrary direction S, and T_m is the medium temperature. The terms on the right-hand side of Equation 6.13 can be understood as follows. The first term stands for the loss by absorption, the second term for the loss by out-scattering, the third term for the gain by emission that depends on the medium temperature, and the fourth term for the gain by in-scattering within the particulate medium. Besides the scalar RTE, the vector RTE has been also applied for various applications [51–53]. Mishchenko [52,53] derived the RTE directly from the macroscopic Maxwell equations using a microphysical approach rather than a phenomenological approach.

Fluidized beds and pack beds are important for many engineering applications, such as chemical reactors. In these applications, radiative heat transfer is essential because the temperatures are usually high enough for radiative transfer to be the dominant mode of heat transport. Methods in modeling radiative transfer in fluidized and pack beds can be divided into two groups: discrete models and pseudocontinuous models [54]. The discrete model is a natural approach. In this model, the particulate medium is considered as a regular assemble of cells and these cells have idealized geometry. First, the radiative properties of one cell are determined. Then, the radiative heat transfer in the medium is solved based on the radiative properties of one cell using the standard resistance network or the layer theory approaches. As the prediction is sensitive to the geometry of the idealized cell, careful consideration is required to mimic the particulate medium with the assumed geometry [54]. In contrast, in pseudocontinuous model, the particulate medium is considered as a random assembly of particles. The radiative heat transfer through the medium can be calculated by solving the integrodifferential equation of radiative heat transfer, or simple versions of RTE, or by the Monte Carlo simulation and ray-tracing method. Geometrical optics is often employed to obtain radiative properties for particles whose sizes are considerably large compared with the radiation wavelength. The spatial average theorem and Monte Carlo methods have been applied to determine the average radiative properties and radiative heat transfer in a two-phase medium [55–58]. Three-dimensional geometry of complex structures in porous media has been obtained by means of computer tomography and the digital geometry has been used to calculate effective radiative properties [55,56]. Discrepancy between local radiation intensities in different phases of a heterogeneous medium was observed and a correction term considering the exchange of radiation between different phases was introduced by Gusarov [59].

6.3 SMALL PARTICLES IN DISPERSIONS

6.3.1 Radiative Properties of Particles in Liquids

6.3.1.1 Materials and Morphology of Particles

Materials used as particles include noble metals, carbon-related materials, oxides, and ceramics. The commonly used metal particles include silver, gold, copper, etc. Carbon-related materials are promising as carbon has a broadband absorption in the visible and near-IR spectrum. This category includes graphite, carbon nanotubes, and carbon nanohorns [60,61]. Particles made of composite materials, for example, carbon-coated cobalt particles [62] and carbon-coated copper particles [63] are also interesting as the carbon shell can protect the metal core from the environments. Otherwise, the chemical reaction between the metal and the surrounding could deteriorate the metal nanoparticle. Metal or nonmetal oxides include SiO_2, TiO_2, ZnO, CuO, and Co_3O_4 [64–66]. Ceramic materials include ZrC, TiN, and AlN [65,67].

Nanofluids are promising heat-transfer media. Owing to their anomalous thermal conductivity and enhanced heat transfer, nanofluids have been extensively investigated by various researchers in thermal engineering [68–70]. Water and ethylene glycol are generally used as the base fluid [69] and oil is also explored for higher-temperature applications [63,64]. Recently nanofluids have been incorporated into direct solar absorbers. These nanofluids are usually prepared in two steps. Firstly, nanoparticles and dispersants are added into a certain amount of base fluid. Secondly, the mixture is oscillated by a tip sonicator or a bath sonicator. As the commercially available nanoparticles often come in the form of powder, sonication is necessary to uniformly disperse the powder into the base fluid. Generally, dispersants are used to acquire a stable dispersion. Common dispersants includes sodium hexametaphosphate, Arabic gum, and E80. Zhu et al. [67] investigated the influence of dispersants on the optical properties of water-dispersant solutions and found that the transmittance decrease of the solutions is less than 5% in the visible spectrum and nearly negligible in the near-IR spectrum.

Scanning electron microscopy (SEM) and transmission electron microscopy (TEM) characterizations are frequently applied to determine the geometry and morphology of particles in dispersions. Drotning [64] studied black metallic oxide particles in a molten salt. SEM analysis revealed that the diameters are 60–150 nm for the CuO particles and 30–100 nm for Co_3O_4. Figure 6.6 shows a dahlia-like type of carbon nanohorn aggregate [61]. Carbon nanohorns are similar to carbon nanotubes. Both are made of a single sheet of graphite. However, carbon nanohorns have the shape of an irregular tubule with a cone tip. The diameter of a carbon nanohorn is 2–5 nm and the length is about 30–50 nm. The carbon nanohorns form roughly spherical aggregates of three types. The aggregate shown in Figure 6.6 is of a dahlia-like structure and its dimension ranges from 50 to 100 nm.

Figure 6.7 shows SEM and TEM images of carbon-coated cobalt nanoparticles dispersed in Therminol® VP-1, a synthetic heat transfer fluid [63]. It can be seen that the diameters of most particles are between 20 and 50 nm and some clusters of nanoparticles exceed 100 nm. From the photos of pure VP1 and particle-dispersed VP1 shown in Figure 6.7, it can be seen that pure VP1 is white (scattering) but the particle-dispersed VP1 is black (absorbing). Figure 6.8 shows a TEM image of SiO_2 nanoparticles dispersed in water. The nominal diameter of the nanoparticles is 20 nm. It can be seen that most nanoparticles are around 15–25 nm in diameter. Aggregates with size of about 50 nm can also be observed. It can be inferred from Figures 6.7 and 6.8 that the nanoparticles are of irregular shape and particle diameters cover a

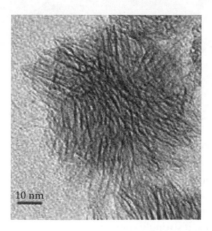

FIGURE 6.6 TEM micrograph of a dahlia-like aggregate consisting of single wall carbon nanohorns. (Adapted from E. Sani et al., *Optics Express*, 18, 5179–5187, 2010. With permission.)

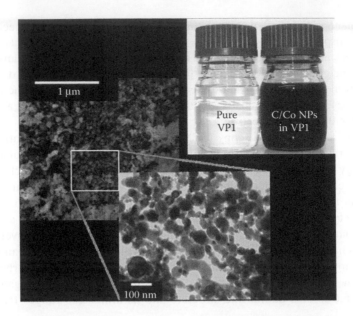

FIGURE 6.7 SEM and TEM images of carbon-coated cobalt nanoparticle in Therminol VP1. (Adapted from A. Lenert, Y. S. P. Zuniga, and E. N. Wang, Nanofluid-based absorbers for high temperature direct solar collectors. In *Proceedings of the 14th International Heat Transfer Conference*, Washington DC, August 8–13, 2010. With permission.)

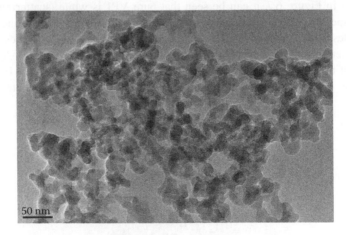

FIGURE 6.8 A TEM image of SiO_2 nanoparticles dispersed in water.

certain range. Furthermore, nanoparticles can form complex aggregates, and some particles are very close to each other with a small interparticle clearance. Although the knowledge acquired through the ex situ characterization might be different from that in undisturbed dispersions, it strongly suggests the independent scattering approximate may be inaccurate to model the scattering of thermal radiation by nanoparticles.

6.3.1.2 Radiative Properties of Liquid Suspensions

As the main purpose of applying nanofluids in solar energy systems is to absorb the solar radiation, the absorptance of nanofluids in the solar spectrum is of particular interest. Although nanofluids also emit thermal radiation, the emissivity of nanofluids is not important in most cases. Although nanofluids have been applied as solar-absorbing media, temperatures of nanofluids are generally low. In addition, nanofluids are contained in a layer sandwiched between two slabs or in a transparent tube within a solar collector. The glazing in the solar collector can effectively block the thermal emission of nanofluids. However, for high-temperature applications, the emissivity of nanofluids may need to be considered.

The typical instruments for measuring the radiative properties of nanofluids are UV/visible/near-IR spectrophotometers. The wavelength range of 300–2500 nm accounts for over 95% of the radiative energy in the solar spectrum. Typical spectrophotometers are capable of measuring regular transmittance. The extinction coefficient of dispersed nanoparticles can be calculated from the regular transmittance using Beer's law through a differential measurement which uses the base fluid as the reference. Some instruments can measure hemispherical transmittance and hemispherical reflectance by means of an integrating sphere (see, e.g., [71,72]). Thus, the absorption of nanofluids can be determined.

Drotning [64] measured hemispherical transmission and reflection for a high-temperature molten salt around 200°C. The absorptance of solar radiation was obtained to be 8% for a 1-cm-thick slab of the molten salt. However, when a 0.1% of hydrated $Co(NO_3)_2$ was added into the fluid, the absorptance could be improved dramatically to 90%. Figure 6.9 shows spectral transmittances of several nanofluids dispersed with dioxide and ceramic nanoparticles [65]. These nanofluids contain different nanoparticles with nearly the same mass fraction. The external appearances of these nanofluids are quite different. The SiO_2–water nanofluid looks semitransparent, the TiO_2–water nanofluid milky, and the ZrC–water nanofluid dark. The transmittance of SiO_2–water nanofluid increases gradually in the wavelength range of 300 to 900 nm. The absorption peaks around 960 and 1160 nm are due to the absorption by water. As silica does not absorb in the solar spectrum, the extinction of the incidence is mainly caused by scattering. The TiO_2–water nanofluid and the ZrC–water nanofluid are almost opaque with a transmittance less than 0.1%. The TiO_2 nanoparticles strongly scatter the incident radiation in the visible spectrum. This is the reason why it appears white. In addition, the TiO_2 nanoparticles can absorb in the wavelength range of $\lambda < 450$ nm, because of a noticeable absorption index (κ). The low transmittance of the ZrC–water nanofluid is attributed to the absorption of the ZrC nanoparticles.

Figure 6.10 displays regular transmittance spectra of four different nanofluids with the same mass fraction of nanoparticles [67]. Again, different characteristics in absorbing solar radiation can be seen. The ZnO–water nanofluid and AlN–water nanofluid are semitransparent while the TiN–water nanofluid is essentially opaque. From Figures 6.9 and 6.10, it can be inferred that TiN and ZrC nanoparticles are preferred over ZnO, AlN, and SiO_2 nanoparticles for the purpose of absorbing solar radiation.

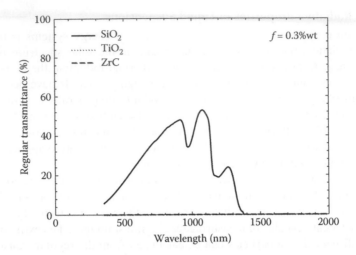

FIGURE 6.9 Spectral transmittance of several aqueous nanofluids with different types of nanoparticles and the same weight fraction of 0.3%. Note that except for the SiO$_2$–water nanofluid, the transmittance is essentially zero. (Adapted from L. Mu, Q. Zhu, and L. Si, Radiative properties of nanofluids and performance of a direct solar absorber using nanofluids. In *Proceedings of the 2nd ASME Micro/Nanoscale Heat and Mass Transfer International Conference*, Shanghai, China, December 18–21, 2009.)

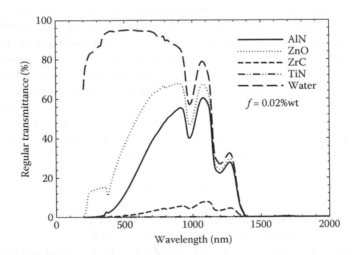

FIGURE 6.10 Regular transmittance of different nanofluids with the same mass fraction. Note that the transmittance of TiN nanofluid is negligible. (Adapted from Q. Zhu, Y. Cui, and L. Mu, Characterization of thermal radiative properties of nanofluids for selectively absorption of solar radiation. In *Proceeding of the 9th Asian Thermophysical Properties Conference*, Beijing, China, Oct. 19–22, 2010.)

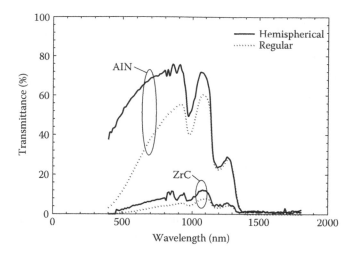

FIGURE 6.11 Comparison of regular and hemispherical transmittance of AlN and ZrC nanofluids. (Adapted from Q. Zhu, Y. Cui, and L. Mu, Characterization of thermal radiative properties of nanofluids for selectively absorption of solar radiation. In *Proceeding of the 9th Asian Thermophysical Properties Conference*, Beijing, China, Oct. 19–22, 2010.)

Figure 6.11 compares the regular transmittance and hemispherical transmittance of AlN and ZrC nanofluids [67]. For the semitransparent AlN–water nanofluid, the hemispherical transmittance is about 10–30% higher than the regular transmittance when $\lambda < 1130$ nm. At $\lambda > 1130$ nm the difference between the regular transmittance and hemispherical transmittance is negligible. This may be due to the weak scattering at long wavelengths. For the optically thick ZrC–water nanofluid, the hemispherical transmittance is noticeably greater than the regular transmittance although absorption dominates the extinction of thermal radiation and thus the transmittance of the ZrC–water nanofluid is relatively low. It indicates that the scattering of solar radiation by nanoparticles also needs to be considered in the analysis of radiative transfer.

Figure 6.12 reveals the effect of mass fraction on the spectral transmittance [65]. As the weight or mass fraction of the SiO_2 nanoparticles increases from 0.2% to 0.7%, the magnitude of the regular transmittances decreases. This is because of the increased total scattering cross-section associated with the increased number of particles per unit volume of the base fluid. Consequently, the extinction coefficient becomes larger and more radiation is scattered off the normal direction. Therefore, the intensity of the radiation propagating in the normal direction is greatly reduced. When the mass fraction of SiO_2 nanoparticles increases, the hemispherical transmittance of nanofluids decreases [66]. The reason is that backscattering becomes stronger as the mass fraction increases. As a consequence, the hemispherical transmittance is also reduced.

Figure 6.13 displays transmittances of carbon nanohorn–water nanofluids with different nanoparticle concentrations [61]. The concentration increases from 0.001 g/L for A1 to 0.050 g/L for A7. As expected, the transmittance of nanofluids decreases with increasing concentration (mass fraction). Sample A7 is nearly opaque

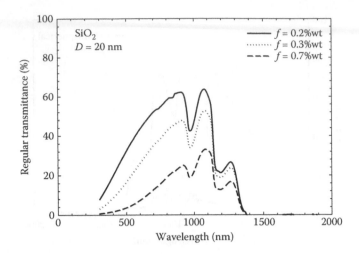

FIGURE 6.12 Transmittance of SiO_2–water nanofluids with different mass fractions. (Adapted from L. Mu, Q. Zhu, and L. Si, Radiative properties of nanofluids and performance of a direct solar absorber using nanofluids. In *Proceedings of the 2nd ASME Micro/ Nanoscale Heat and Mass Transfer International Conference*, Shanghai, China, December 18–21, 2009.)

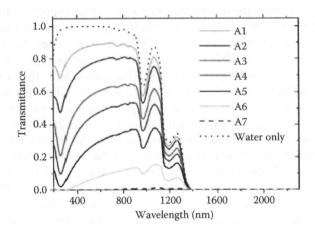

FIGURE 6.13 Transmittance spectra of carbon nanohorn–water nanofluids with different concentrations. (Adapted from E. Sanie et al., *Optics Express*, 18, 5179–5187, 2010. With permission.)

and samples with concentrations higher than 0.050 g/L are too dark with essentially zero transmittance.

Wang et al. [66] studied the effect of particle size on the radiative properties of nanofluids. They found that the hemispherical transmittance decreases if the diameter of the SiO_2 nanoparticles is increased. This can be understood by the enhanced backscattering for larger particles. The extinction coefficient, that is, the scattering coefficient, is proportional to the third power of the particle diameter if the nanoparticles are in the regime of Rayleigh scattering. As the particle diameter increases, the backscattering of thermal radiation by nanoparticles increases dramatically, resulting in a significant decrease in the hemispherical transmittance. The applicability of Beer's law in the study of optical properties of nanofluids is also examined by Wang et al. [66]. It was found that the measurements at different path lengths deviate from the predictions using Beer's law and the deviation is more obvious for larger particles. As the scattering is intense for large particles, Beer's law cannot give an accurate prediction in such a case. It can be inferred that care must be taken in determining the extinction coefficient using Beer's law for a particulate medium with strong scattering. In such cases, the RTE needs to be solved to obtain the radiative properties.

Figure 6.14 illustrates the extinction coefficients of two typical nanofluids, the SiO_2–water nanofluid and the graphite–water nanofluid. The former appears

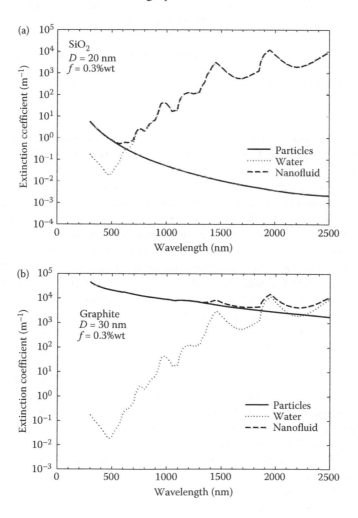

FIGURE 6.14 Extinction coefficients of (a) SiO_2–water and (b) graphite–water nanofluids.

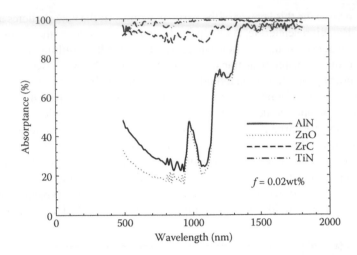

FIGURE 6.15 Absorptances of nanofluids with the same mass fraction. (Adapted from Q. Zhu, Y. Cui, and L. Mu, Characterization of thermal radiative properties of nanofluids for selectively absorption of solar radiation. In *Proceeding of the 9th Asian Thermophysical Properties Conference*, Beijing, China, Oct. 19–22, 2010.)

semitransparent and scattering dominates the extinction due to SiO$_2$ particles. The latter is optically thick and hence absorption dominates. As shown in Figure 6.14, as the wavelength increases, extinction coefficient of particles decreases and the absorption by water becomes essential. Owing to the high absorption index of graphite, the extinction coefficient of graphite nanoparticles is several orders larger than that of silica nanoparticles. For the SiO$_2$–water nanofluid, the absorption of water contributes significantly to the total extinction coefficient for $\lambda > 560$ nm. When $\lambda > 650$ nm, water absorption is much higher than the extinction caused by nanoparticles and, subsequently, absorption of the base fluid starts to dominate. For the graphite–water nanofluid, the absorption of water becomes noticeable when $\lambda > 1330$ nm. Beyond $\lambda > 1850$ nm, the absorption coefficient of the nanofluid is a few times greater than the extinction coefficient due to graphite nanoparticles. It can be concluded that the absorption of solar radiation by the base fluid should be taken into account in the thermal analysis.

Figure 6.15 presents the spectral absorptance of various nanofluids [67]. The mass fraction is 0.02% and the path length of the quartz cell is 1 cm for all nanofluids. The absorptance of AlN–water nanofluid and ZnO–water nanofluid is lower than 50% in a wide spectral region ($\lambda < 1170$ nm). Owing to the base fluid absorption, the absorptance of these two nanofluids exceeds 90% in the spectral region of $\lambda > 1330$ nm. The absorptance of the ZrC–water nanofluid is greater than 90% at most wavelengths and that of the TiN–water nanofluid is greater than 95% in the whole spectrum. From the measurements, it is obvious that the AlN–water nanofluid and ZnO–water nanofluid are not suitable for use as the direct solar absorption fluid, whereas the ZrC–water nanofluid and TiN–water nanofluid are excellent candidates.

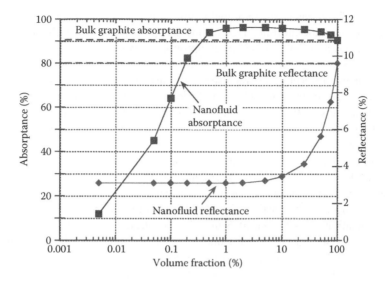

FIGURE 6.16 Illustration of the advantages of a nanofluid volumetric absorber over a black-surface absorber. (Adapted from T. Otanicar et al., *Journal of Renewable and Sustainable Energy*, 2, 033102–1/13, 2010.)

In order to demonstrate advantages of a nanofluid volumetric absorber, a simplified analysis was carried out by Otanicar et al. [60], and their results are given in Figure 6.16. Reflectance and absorptance of a graphite–water nanofluid film are compared with those of a bulk graphite absorber, respectively. The thickness of the nanofluid layer is 0.15 mm. The graphite–water nanofluid reflects about 3% of the incoming solar radiation if the volume fraction is less than 2%, whereas the bulk graphite reflects about 9% of the incident solar radiation. The absorptance of graphite–water nanofluids surpasses that of the bulk graphite if the volume fraction of graphite nanoparticles is higher than 0.5%. The results partly explain that the nanofluid volumetric absorber is advantageous over the conventional black-surface absorber. Beside the high absorptance of nanofluids, there are some other factors. Thermal resistance in converting solar energy into heat is smaller as solar radiation is absorbed directly by the nanofluid. Furthermore, the heat loss from the nanofluid volumetric absorber is smaller because the temperature of glazing is lower than the nanofluid. These will be further discussed in Section 6.4.

6.3.2 Particles in Gases

Solving a RTE requires an input of absorption and scattering coefficients of the participating medium. Generally, the Mie theory is applied to calculate optical properties of spherical particles. After scattering coefficients are calculated by the Mie scattering, transmission of particle dispersions can be predicted by Beer's law, a simplified version of RTE. Generally speaking, computational fluid dynamics and radiative transfer models are extremely time consuming. In order to reduce the demanding computational time, a curve-fitting approach to the Mie theory was proposed by Caldas and Semiao [73] to compromise between accuracy and computational economy.

Radiation and absorption characteristics of particle-laden gases can be modified by particle type, size, and concentration. Abdelrahman et al. [12] investigated effects of absorption index on the absorption of particulate media. It was found that a small value of κ may result in a high transmission while a large value of κ may result in a high reflection.

Particle size can significantly affect the scattering and absorption of particle-laden gases. Klein et al. [13] found that carbon black particles with effective radius less than 100 nm are inefficient in absorbing solar radiation. In order to achieve sufficient absorption, the particle size should be in the same range with the wavelength of radiation (100–1000 nm). Similarly, Abdelrahman et al. [12] observed that size of particles should be on the same order of magnitude of the wavelength in order to have high absorption. It was found that graphite particles with a diameter $D = 0.5$ μm are good candidate for solar radiation absorption.

Griffin and Stahl [74] measured the optical properties of Masterbead, an iron-doped alumina spheroid. The dimension of beads is around 250 μm. A particle curtain was produced for the optical measurements. The phase function and extinction coefficient of submillimeter beads were determined from measurement results.

Klein et al. [13] analyzed the radiative properties of a cloud of carbon black particles. Figure 6.17 shows SEM images of carbon black particles used in the solar receiver. Asymmetric agglomerates can be observed from images at two different magnifications (500× for the top image and 40,000× for the bottom). The equivalent radius of aggregates varies from 0.015 to 12 μm. Although 98.8% of particles have an equivalent radius less than 1 μm, they only contribute 35% of the total projected surface.

Bertocchi et al. [75] measured the physical and optical properties of a carbon black particle cloud. In this study, 99.7% of particles in the cloud are primary particles with diameters ranging from 45 to 570 nm and 0.3% of particles are irregularly

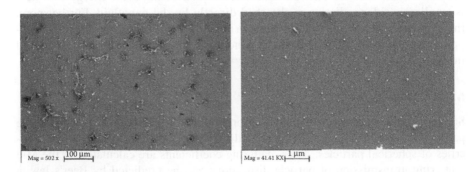

FIGURE 6.17 SEM images of carbon black particles in the solar receiver. (Adapted from H. H. Klein et al., *Solar Energy*, 81, 1227–1239, 2007. With permission.)

shaped agglomerates with equivalent diameters of 1.2–7.25 μm. The particles display strong forward scattering, and 62% of scatted energy is within a forward lobe of 15° at 532 nm and 48% at 1064 nm. It was found that neglecting agglomerates in the theoretical modeling using the Mie theory results in discrepancies with measurement observations. It was also noted that the Mie theory can be applied to estimate the optical properties of a partially agglomerated carbon black particle cloud. The present of a small fraction of large agglomerates can significantly change the particle cloud's scattering pattern. Aggregates at a 0.3% fraction account for 60% of the scattering cross-section. Therefore, detailed knowledge of the particle size distribution is required to accurately predict the optical properties of a particle cloud.

6.4　VOLUMETRIC SOLAR ABSORBERS

6.4.1　Particles-in-Liquid Solar Absorbers

Both theoretical analysis and experimental measurement have been performed to evaluate the performance of solar collectors using particle-laden liquids including nanofluids to absorb the solar radiation [65,76,77]. Otanicar and Golden [78] performed an environmental and economic analysis for nanofluid-based direct absorption solar collectors and conventional solar collectors. It was shown that although nanofluid-based collectors have a lightly longer payback period, it has the same economic saving as the conventional solar collector at the end of its life.

Mao et al. [62] measured the thermal efficiency of a nanofluid-based solar collector. The nanofluid was prepared by dispersing carbon-coated Cu nanoparticles into ethylene glycol aqueous solution. They found that the collector efficiency, defined as the ratio of the useful thermal gain to the amount of solar energy incident on the collector surface, is remarkably high and the maximum efficiency can reach 74%.

Mu et al. [65] compared the temperature increase of four fluids in a volumetric absorber under solar irradiation. These fluids are water, SiO_2–water nanofluid, TiO_2–water nanofluid, and ZrC–water nanofluid. Their measurements demonstrate that the inclusion of ZrC nanoparticles in the base fluid can significantly enhance the solar absorption. It was shown that the ZrC–water nanofluid is the best solar-absorbing medium among the nanofluids studied. Conclusions drawn from the thermal performance measurement are in good agreement with the findings from the radiative property measurements [65].

Tyagi et al. [77] theoretically evaluated the thermal performance of a nonconcentrating direct absorption solar collector. The absorbing medium is water dispersed with aluminum nanoparticles. They found that the collector efficiency first increases rapidly with the volume fraction when $f_v < 0.5\%$, and then gradually reaches a maximum value of around 80%. Compared with the conventional flat-plate collector, the efficiency of the direct solar absorption collector can be up to 10% (absolute value) higher under normal operating conditions.

Otanicar et al. [60] built a micro solar absorber and tested nanofluids made from graphite nanoparticles, carbon nanotubes, and silver nanoparticles. They found that, compared with a conventional black-surface collector, an improvement of 5% can be achieved if 20 nm silver nanoparticles are used and that the difference in the thermal efficiency is small between carbon nanotube-based and graphite sphere-based nanofluids. Both theoretical analysis and experimental results demonstrate that the thermal efficiency of the solar collector initially increases rapidly with the volume fraction of nanoparticles and then levels off after the volume fraction reaches about 0.5%.

Lenert et al. [63] developed a one-dimensional transient heat-transfer model for a volumetric receiver using Thermol VP1 seeded with carbon-coated cobalt nanoparticles. A 7-cm-thick nanofluid layer with a volume fraction of 3.7 ppm nanoparticles can absorb 98% of the solar radiation. They investigated thermal performance of one-pass receivers, two-pass receivers, and idealized surface receivers. The two-pass receiver refers to a receiver with a perfectly reflective bottom surface underneath the nanofluid. An idealized surface receiver refers to a receiver with an absorbing surface whose emissivity is arbitrarily set to unity at $\lambda < 2$ μm and zero beyond. The numerical results indicate that the two-pass receiver runs more efficiently than the one-pass receiver. The comparison between the two-pass receiver and the idealized surface receiver reveals that the former is more efficient at high solar concentration ratios than the latter.

The impact of nanoparticle size on the thermal behaviors has been theoretically investigated [48]. It was found that the prediction using size-dependent optical properties results in higher collector efficiency than that using the bulk optical properties. The difference is more significant when the particle size decreases. This is attributed to the broadening of the absorption peak and enhancement of thermal conductivity due to the dispersed nanoparticles.

Application of nanofluids in chemical reactors utilizing solar energy has also been proposed. Tyagi et al. [79] evaluated the applicability of nanofluids in a biomass gasifier. The nanofluid is a mixture of molten salts and copper nanoparticles. Besides, vapor generation within a nanofluid under illumination of a focused continuous laser has been investigated [80]. Their findings may inspire the development of novel solar collectors in which nanofluids absorb solar radiation and undergo phase change in a single step.

Although reasonable agreements have been observed between the numerical predictions and measurement results, discrepancy exists in some cases. The predicted and measured thermal efficiencies for silver–water nanofluids differ on the overall magnitude [48]. The calculated temperature increase for the SiO_2–water nanofluid is significantly lower than the measured one [81]. Several reasons may be related to the discrepancy. For example, although the nanoparticle is very small so that Rayleigh scattering can be applied, nanoparticles may form aggregates and clusters whose dimension can be on the order of a few hundred nanometers. Therefore, the scattering coefficients calculated assuming individual nanoparticles in the dispersion probably are different from the real case. These aggregates and clusters may be too big to for the Rayleigh scattering to be applicable. Instead, the Mie scattering should be a preferable choice but nonspherical geometries must be considered. In addition, scattering contributes to the extinction of thermal radiation within the dispersion, even though it is not dominant in strongly absorbing nanofluids. Therefore, it is not precise to use Beer's law to determine the radiative intensity within the particulate medium. The radiative transport equation should be resorted to obtain more accurate evaluations.

6.4.2 PARTICLES-IN-GAS SOLAR ABSORBERS

In this kind of absorbers, solar radiation is absorbed by particles suspended in gases. The heated particles transfer thermal energy to the surrounding gas mainly through convection. The cleaned gas can be used in gas turbines and power plants [82]. In some designs, carbon particles are consumed up just after the gas leaves the absorber so that there is no need to clean the gas. Solar thermochemical processes require high-temperature solar reactors operation at 1000–1500°C [83]. The high-temperature reactors can be used in many applications, such as solar gasification of coal, hydrogen production from natural gas cracking, thermal reduction of metal oxides as part of a two-step water splitting cycle [4–6,83,84]. Theoretical modeling and experimental measurements have been performed to investigate the performance of solar absorbers where particles dispersed gas is applied as the moving solar absorber. The effects of particle size and particle concentration have also been studied.

Sasse and Ingel [85] evaluated optical properties of oil shale particles and modeled radiative heat transfer within a fluidized bed acting as a solar gasifier. They found that the temperature profile is sensitive to the optical properties of particles. Temperature differences between the simulation results using either the calculated optical properties or the measured optical properties would be as large as 300°C.

Klein et al. [13] found that particle radius has a small influence on the temperatures of receiver walls and gas; furthermore, increasing particle concentration can improve the heat transfer up to a threshold. Once the threshold is exceeded, further increase of the concentration only has a marginal influence.

Miller and Koenigsdorff [86] applied a modified six-flux model to obtain radiant flux and temperature distribution in a direct solar absorber seeded with submicron carbon particles. It was estimated that the receiver can withstand an incident solar flux such as 5 MW/m^2, producing high-temperature gas of 1400 K at an efficiency as high as 80%.

Bertocchi et al. [87] designed a high-temperature concentrated solar particle receiver. The mass fraction of the submicrometer carbon particles is about 0.2–0.5%. The typical extinction coefficient of an undiluted cloud is around 40 m^{-1} at a volumetric flow rate of 100 L/min. Gas temperature can exceed 2000 K with air as the carrier gas. Owing to the careful design, the wall temperature at the exit plan is at least 100 K lower than that of the gas at steady conditions. Solar radiation to thermal energy conversion efficiency was estimated to surpass 80%.

Various designs of particle solar absorbers have been proposed. In one design, sands fall down from the top to the bottom of a cavity because of gravity, forming a sand curtain [88]. The sand curtain absorbs solar energy and the heated sands can be applied to generate hot fluids and serve as a heat storage medium. Candidate materials that may replace the sands include alumina, silica, silicon carbide, and zircon.

6.4.3 POROUS-SOLID SOLAR ABSORBERS

Porous structures are used to absorb solar energy in this kind of absorbers. The heated solid is cooled by a gaseous medium and then the thermal energy is conveyed away by the gas. The porous structures include wire mesh, corrugated foil, honeycomb structure, and ceramic foams. As the heat-transfer surface is large, the porous structure can withstand high solar flux without reaching the temperature limit of the absorber material. Solar receivers applying both open loop and closed loop have been built and tested. Lansing et al. [19] found that a porous flat-plate solar collector can greatly surpass the performance of a solar absorber with a nonporous opaque surface. Grald and Kuehn [89] analyzed the thermal performance of a parabolic trough solar collector using a porous absorber and observed that the absorber efficiency is higher than a conventional parabolic trough absorber using a selective surface. Recent research has been focused on the design of glass windows in the closed loop and the development of modular structures of the receiver in a concentrated solar power system. Overheating and flow instability can occur in the absorber because of the inhomogeneous flux distribution [18]. Pitz-Paal et al. [90] found that the temperature field is strongly affected by the inhomogeneous irradiation. Nevertheless, owing to radial heat transfer, the flow instability within the receiver can be partially prevented.

6.5 SUMMARY AND OUTLOOK

Knowledge of radiative properties of micro/nanoscale particles in dispersions and understanding radiative heat transport in particulate media are very important in many applications such as solar thermal absorbers, combustors, packed and fluidized beds, and so on. This review summarizes radiative properties of particle-laden liquids and gases, especially in photothermal conversion systems. Recent advances in radiative properties of nanoparticles and dependent scattering and size effect are also presented.

Nanofluids are promising solar absorbers as their radiative properties can be tuned by the size and materials of the nanoparticles dispersed in the base fluid. The enhanced thermal conductivity of nanofluids could be an extra advantage for nanofluids used as the working fluid in a direct solar absorption collector. Nanofluids dispersed with carbon, metal, and ceramic nanoparticles show high absorptance in the solar spectrum. Solar collectors using nanofluids with a small mass fraction of nanoparticles may outperform the conventional collector with a selective surface.

Particles of carbon-related materials are commonly applied to produce a particle-laden gas to absorb solar radiation. Size of particles should be in the same order of wavelengths in the solar spectrum so as to efficiently absorb the solar radiation. Aggregates, even of a small amount, could introduce discrepancies between the model predictions and experimental results. The generalized Mie theory should be applied to estimate the optical properties of a partially agglomerated carbon black particle cloud.

Optimal choice of particle material, size, and concentration in volumetric solar absorbers need to be investigated in future studies. Size distribution of particles should be characterized to precisely predict radiative properties of a particulate medium. The applicability of nanofluids in medium-temperature collectors or high-temperature collectors should also be explored. Effects of dependent scattering on radiative properties of aggregates deserve systematic theoretical and experimental studies. A thorough understanding of radiative heat transfer in photothermal conversion systems is desired to evaluate their performance and to optimize their designs for various applications.

ACKNOWLEDGMENTS

Q. Zhu acknowledges the support from the National Natural Science Foundation of China (Grant No. 50976093) and the Shu Guang Project by Shanghai Municipal Education Commission and Shanghai Education Development Foundation. Z.M. Zhang thanks the support from the US Department of Energy (Contract No. DE-FG02-06ER46343).

NOMENCLATURE

a_n, b_n	Mie scattering coefficients
b	Interparticle clearance
C	Cross-section
D	Particle diameter
f	Volume fraction
g	Proportional factor
I	Radiation intensity
k	Wave vector
m	Relative complex refractive index
n	Real part of the refractive index
Q	Efficiency factor
S	Direction of propagating
S_1, S_2	Complex amplitude functions

GREEK SYMBOLS

α, σ, β	Absorption, scattering, extinction coefficient
χ	Size parameter
ε	Emissivity, dielectric constant
Φ	Phase function
γ	Damping coefficient
κ	Absorption index
λ	Wavelength
θ	Polar angle
ω	Angular frequency

SUBSCRIPTS

a	Absorb
b	Blackbody, bulk

e	Extinction
p	Plasmon, particle
s	Scattering
v	Volume
λ	Wavelength

REFERENCES

1. O. Mortan, Solar energy: A new day dawning? *Nature*, 443, 19–22, 2006.
2. J. A. Duffie and W. A. Beckman, *Solar Engineering of Thermal Processes*, 2nd ed., John Wiley & Sons, New York, 1991.
3. M. A. Green, *Solar Cells: Operation Principles, Technology, and System Applications*, Prentice Hall, Englewood Cliffs, NJ, 1981.
4. S. Abanades and G. Flamant, High-temperature solar chemical reactors for hydrogen production from natural gas cracking, *Chemical Engineering Communications*, 195, 1159–1175, 2008.
5. S. Abanades, P. Charvin, F. Lemont, and G. Flamant, Novel two-step SnO_2/SnO water-splitting cycle for solar thermochemical production of hydrogen, *International Journal of Hydrogen Energy*, 33, 6021–6030, 2008.
6. S. Rodat, S. Abanades, J.-L. Sans, and G. Flamant, A pilot-scale solar reactor for the production of hydrogen and carbon black from methane splitting, *International Journal of Hydrogen Energy*, 35, 7748–7758, 2010.
7. P. Charalambous, G. Maidment, G. Kalogirou, and K. Yiakoumetti, Photovoltaic Thermal (PV/T) collectors: A review, *Applied Thermal Engineering*, 27, 275–286, 2007.
8. J. Ji, J.-P. Lu, T.-T. Chow, W. He, and G. Pei, A sensitivity study of a hybrid photovoltaic/thermal water-heating system with natural circulation, *Applied Energy*, 84, 222–237, 2007.
9. H. A. Zondag, Flat-plate PV-thermal collectors and systems: A review, *Renewable and Sustainable Energy Reviews*, 12, 891–959, 2008.
10. P. V. Kamat, Meeting the clean energy demand: Nanostructure architectures for solar energy conversion, *Journal of Physical Chemistry C*, 111, 2834–2860, 2007.
11. W. Weiss and M. Rommel, Medium temperature collectors—state of the art within task 33/IV, International Energy Agency, Gleisdorf, Austria: AEE INTEC 2005.
12. M. Abdelrahman, P. Fumeaux, and P. Suter, Study of solid-gas-suspensions used for direct absorption of concentrated solar radiation, *Solar Energy*, 22, 45–48, 1979.
13. H. H. Klein, J. Karni, R. Ben-Zvi, and R. Bertocchi, Heat transfer in a directly irradiated solar receiver/reactor for solid-gas reactions, *Solar Energy*, 81, 1227–1239, 2007.
14. N. Arai, Y. Itaya, and M. Hasatani, Development of a "volume heat-trap" type solar collector using a fine-particle semitransparent liquid suspension (FPSS) as a heat vehicle and heat storage medium, *Solar Energy*, 32, 49–56, 1984.
15. J. E. Minardi and H. N. Chuang, Performance of a "black" liquid flat-plate solar collector, *Solar Energy*, 17, 179–183, 1975.
16. M. S. Bohn and H. J. Green, Heat transfer in molten salt direct absorption receivers, *Solar Energy*, 42, 57–66, 1989.
17. M. N. Islam, B. S. Negi, and T. C. Kandpal, Thermal performance of a linear Fresnel reflector solar concentrator using a black liquid, *Renewable Energy*, 2, 533–535, 1992.
18. T. Fend, B. Hoffschmidt, R. Pitz-Paal, O. Reutter, and P. Rietbrock, Porous materials as open volumetric solar receivers: Experimental determination of thermophysical and heat transfer properties, *Energy*, 29, 823–833, 2004.
19. F. L. Lansing, V. Clarke, and R. Reynolds, A high performance porous flat-plate solar collector, *Solar Energy*, 4, 385–694, 1979.

20. A. J. Hunt, *Small Particle Heat Exchangers*, Lawrence Berkeley National Laboratory, University of California, Berkeley, CA, 1978.

21. H. C. Van de Hulst, *Light Scattering by Small Particles*, Dover, New York, 1981.

22. C. F. Bohren and D. R. Huffman, *Absorption and Scattering of Light by Small Particles*, Wiley, New York, 1983.

23. M. I. Mishchenko, J. W. Hovenier, and L. D. Travis, *Light Scattering by Nonspherical Particles: Theory, Measurements, and Applications*, Academic Press, San Diego, CA, 2000.

24. F. M. Kahnert, Numerical methods in electromagnetic scattering theory, *Journal of Quantitative Spectroscopy and Radiative Transfer*, 79–80, 775–824, 2003.

25. M. Kahnert, Electromagnetic scattering by nonspherical particles: Recent advances, *Journal of Quantitative Spectroscopy and Radiative Transfer*, 111, 1788–1790, 2010.

26. R. Siegel and J. R. Howell, *Thermal Radiation Heat Transfer*, 4th ed., Taylor & Francis, New York, 2002.

27. M. F. Modest, *Radiative Heat Transfer*, 2nd ed., Academic Press, San Diego, CA, 2003.

28. J. F. Sacadura and D. Baillis, Experimental characterization of thermal radiation properties of dispersed media, *International Journal of Thermal Sciences*, 41, 699–707, 2002.

29. Z. M. Zhang, C. J. Fu, and Q. Z. Zhu, Optical and thermal radiative properties of semiconductors related to micro/nanotechnology, *Advances in Heat Transfer*, 37, 179–296, 2003.

30. Q. Zhu, H. Lee, and Z. M. Zhang, Radiative properties of materials with surface scattering or volume scattering: A review, *Frontiers of Energy and Power Engineering in China*, 3, 60–79, 2009.

31. H. C. Hottel, A. F. Sarofim, W. H. Dalzell, and I. A. Vasalos, Optical properties of coatings: Effect of pigment concentration, *AIAA Journal*, 9, 1895–1898, 1971.

32. H. C. Hottel, A. F. Sarofim, I. A. Vasalos, and W. H. Dalzell, Multiple scatter: Comparisons of theory with experiment, *Journal of Heat Transfer*, 92, 285–291, 1970.

33. M. Q. Brewster and C. L. Tien, Radiative transfer in packed fluidized beds: Dependent versus independent scattering, *Journal of Heat Transfer*, 104, 573–579, 1982.

34. J. Q. Cartigny, Y. Yamada, and C. L. Tien, Radiative transfer with dependent scattering by particle: Part 1—Theoretical investigation, *Journal of Heat Transfer*, 108, 608–613, 1986.

35. Y. Yamada, J. Q. Cartigny, and C. L. Tien, Radiative transfer with dependent scattering by particle: Part 2—Experimental investigation, *Journal of Heat Transfer*, 108, 614–618, 1986.

36. C. L. Tien and B. L. Drolen, Thermal radiation in particulate media with dependent and independent scattering. In *Annual Review of Numerical Fluid Mechanics and Heat Transfer* T. C. Chawla, ed., Hemisphere Publishing Corp, New York, pp. 1–32, 1987.

37. M. Lax, Multiple scattering of waves. II. The effective field in dense system, *Physical Review*, 85, 621–629, 1952.

38. R. S. Prasher and P. E. Phelan, Modeling of radiative and optical behavior of nanofluids based on multiple and dependent scattering theories. In *Proceeding of ASME International Mechanical Engineering Congress and Exposition*, Orlando, FL, November 5–11, 2005.

39. R. Prasher, Thermal radiation in dense nano- and microparticulate media, *Journal of Applied Physics*, 102, 074316–1/9, 2007.

40. S. Kumar and C. L. Tien, Dependent absorption and extinction of radiation by small particles, *Journal of Solar Energy Engineering*, 112, 178–185, 1990.

41. B. L. Drolen and C. L. Tien, Independent and dependent scattering in packed-sphere systems, *Journal of Thermophysics*, 1, 63–68, 1987.

42. Z. Ivezic and M. P. Mengüç, An investigation of dependent/independent scattering regimes using a discrete dipole approximation, *International Journal of Heat and Mass Transfer*, 39, 811–822, 1996.

43. F. N. Dönmezer, M. P. Mengüç, and T. Okutucu, Dependent absorption and scattering by interacting nanoparticles. In *Proceedings of 6th International Symposium on Radiative Transfer*, Antalya, Turkey, June 13–19, 2010.

44. A. Ishimaru and Y. Kuga, Attenuation constant of a coherent field in a dense distribution of particles, *Journal of Optical Society of America*, 72, 1317–1320, 1982.

45. B. P. Singh and M. Kaviany, Independent theory versus direct simulation of radiation transfer in packed beds, *International Journal of Heat and mass Transfer*, 34, 2869–2882, 1991.

46. Z. M. Zhang, *Nano/Microscale Heat Transfer*, McGraw-Hill, New York, 2007.

47. P. B. Johnson and R. W. Christy, Optical constants of the noble metals, *Physical Review B*, 6, 4370–4379, 1972.

48. T. Otanicar, R. A. Taylor, P. E. Phelan, and R. Prasher, Impact of size and scattering mode on the optimal solar absorbing nanofluid. In *Proceeding of the ASME 3rd International Conference on Energy Sustainability*, San Francisco, CA, July 19–23, 2009.

49. A. Vial and T. Laroche, Comparison of gold and silver dispersion laws suitable for FDTD simulations, *Applied Physics B*, 93, 139–143, 2008.

50. J. R. Cole and N. J. Halas, Optimized plasmonic nanoparticles distributions for solar spectrum harvesting, *Applied Physics Letters*, 89, 153120–1/3, 2006.

51. L. Tsang, J. A. Kong, and K.-H Ding, *Scattering of Electromagnetic Waves: Theories and Applications*, John Wiley & Sons, New York, 2000.

52. M. I. Mishchenko, Multiple scattering by particles embedded in an absorbing medium. 2. Radiative transfer equation, *Journal of Quantitative Spectroscopy and Radiative Transfer*, 109, 2386–2390, 2008.

53. M. I. Mishchenko, Multiple scattering, radiative transfer, and weak localization in discrete random media: Unified microphysical approach, *Reviews of Geophysics*, 46, RG2003–1/33, 2008.

54. C. L. Tien, Thermal radiation in packed and fluidized beds, *Journal of Heat Transfer*, 110, 1230–1242, 1988.

55. S. Haussener, P. Coray, W. Lipiński, P. Wyss, and A. Steinfeld, Tomography-based heat and mass transfer characterization of reticular porous ceramics of high-temperature processing, *Journal of Heat Transfer*, 132, 023305–1/9, 2010.

56. S. Haussener, W. Lipiński, J. Petrasch, P. Wyss, and A. Steinfeld, Tomography characterization of a semitransparent-particle packed bed and determination of its thermal radiative properties, *Journal of Heat Transfer*, 131, 072701–1/11, 2009.

57. W. Lipiński, J. Petrasch, and S. Haussener, Application of the spatial averaging theorem to radiative heat transfer in two-phase media, *Journal of Quantitative Spectroscopy and Radiative Transfer*, 111, 253–258, 2010.

58. M. Tancrez and J. Taine, Direct identification of absorption and scattering coefficients and phase function of a porous medium by a Monte Carlo technique, *International Journal of Heat and mass Transfer*, 47, 373–383, 2004.

59. A. V. Gusarov, Homogenization of radiation transfer in two-phase media with irregular phase boundaries, *Physical Review B*, 77, 144201–1/14, 2008.

60. T. Otanicar, P. E. Phelan, R. Prasher, G. Rosengarten, and R. A. Taylor, Nanofluid-based direct absorption solar collector, *Journal of Renewable and Sustainable Energy*, 2, 033102–1/13, 2010.

61. E. Sani, S. Barison, C. Pagura, L. Mercatelli, P. Sansoni, D. Fontani, D. Jafrancesco, and F. Francini, Carbon nanohorns-based nanofluids as direct sunlight absorbers, *Optics Express*, 18, 5179–5187, 2010.

62. L. Mao, R. Zhang, X. Ke, and Z. Liu, Photo-thermal properties of nanofluid-based solar collector, *Acta Energiae Solaris Sinica*, 30, 1647–1652, 2009.

63. A. Lenert, Y. S. P. Zuniga, and E. N. Wang, Nanofluid-based absorbers for high temperature direct solar collectors. In *Proceedings of the 14th International Heat Transfer Conference*, Washington DC, August 8–13, 2010.

64. W. D. Drotning, Optical properties of solar-absorbing oxide particles suspended in a molten salt heat transfer fluid, *Solar Energy*, 20, 313–319, 1978.
65. L. Mu, Q. Zhu, and L. Si, Radiative properties of nanofluids and performance of a direct solar absorber using nanofluids. In *Proceedings of the 2nd ASME Micro/Nanoscale Heat and Mass Transfer International Conference*, Shanghai, China, December 18–21, 2009.
66. H. Wang, Z.-Y. Luo, J. Cai, T. Wang, J. F. Zhao, and M. J. Ni, Experimental study of influencing factors on transmissivity of SiO_2 nanofluids, *Journal of Zhejiang University (Engineering Science)*, 44, 1143–1148, 2010.
67. Q. Zhu, Y. Cui, and L. Mu, Characterization of thermal radiative properties of nanofluids for selectively absorption of solar radiation. In *Proceeding of the 9th Asian Thermophysical Properties Conference*, Beijing, China, October 19–22, 2010.
68. S. K. Das, S. U. S. Choi, W. Yu, and T. Pradeep, *Nanofluids: Science and Technology*, John Wiley & Sons, New Jersey, 2008.
69. J. A. Eastman, S. U. S. Choi, S. Li, W. Yu, and L. J. Thompson, Anomalously increased effective thermal conductivities of ethylene glycol based nanofluids containing copper nanoparticles, *Applied Physics Letters*, 78, 718–720, 2001.
70. J. Buongiorno, Convective transport in nanofluids, *Journal of Heat Transfer*, 128, 240–250, 2006.
71. Q. Li, B. J. Lee, Z. M. Zhang, and D. W. Allen, Light scattering of semitransparent sintered polytetrafluoroethylene films, *Journal of Biomedical Optics*, 13, 054064–1/12, 2008.
72. X. J. Wang, J. D. Flicker, B. J. Lee, W. J. Ready, and Z. M. Zhang, Visible and near-infrared radiative properties of vertically aligned multi-walled carbon nanotubes, *Nanotechnology*, 20, 215704–1/9, 2009.
73. M. Caldas and V. Semiao, Modeling of scattering and absorption coefficients for a poly-dispersion, *International Journal of Heat and Mass Transfer*, 42, 4535–4548, 1999.
74. J. W. Griffin and K. A. Stahl, Optical properties of solid particle receiver materials, *Solar Energy Materials*, 14, 395–416, 1986.
75. R. Bertocchi, A. Kribus, and J. Karni, Experimentally determined optical properties of a polydisperse carbon black cloud for a solar particle receiver, *Journal of Solar Energy Engineering*, 126, 833–841, 2004.
76. S. Kumar and C. L. Tien, Analysis of combined radiation and convection in a particulate-laden liquid film, *Journal of Solar Energy Engineering*, 112, 293–300, 1990.
77. H. Tyagi, P. Phelan, and R. Prasher, Predicted efficiency of a low-temperature nanofluid-based direct absorption solar collector, *Journal of Solar Energy Engineering*, 131, 041004–1/7, 2009.
78. T. P. Otanicar and J. S. Golden, Comparative environmental and economic analysis of conventional and nanofluid solar hot water technologies, *Environmental Science and Technology*, 43, 6082–6087, 2009.
79. H. Tyagi, P. E. Phelan, and R. S. Prasher, Thermochemical conversion of biomass using solar energy: Use of nanoparticle-laden molten salt as the working fluid. In *Proceedings of the ASME 3rd International Conference on Energy Sustainability*, San Francisco, CA, July 19–23, 2009.
80. R. A. Taylor, P. E. Phelan, T. Otanicar, R. J. Adrian, and R. S. Prasher, Vapor generation in a nanoparticle liquid suspension using a focused, continuous laser, *Applied Physics Letters*, 95, 161907–1/3, 2009.
81. Q. Zhu, Y. Li, L. Mu, and Y. Cui, Theoretical investigation of radiative transport and heat transfer of nanofluids in a direct solar absorption collector. In *Proceedings of the 14th International Heat Transfer Conference*, Washington DC, August 8–13, 2010.
82. J. Oman and P. Novak, Volumetric absorption in gas-properties of particles and particle-gas suspensions, *Solar Energy*, 56, pp. 597–606, 1996.

83. T. Kodama, S. Enomoto, T. Hatamachi, and N. Gokon, Application of an internally circulating fluidized bed for windowed solar chemical reactor with direct irradiation of reacting particles, *Journal of Solar Energy Engineering*, 130, 014504–1/4, 2008.
84. G. Maag, W. Lipiński, and A. Steinfeld, Particle-gas reacting flow under concentrated solar irradiation, *International Journal of Heat and Mass Transfer*, 52, 4997–5004, 2009.
85. C. Sasse and G. Ingel, The role of the optical properties of solids in solar direct absorption pocess, *Solar Energy Materials and Solar Cells*, 31, 61–73, 1993.
86. F. J. Miller and R. W. Koenigsdorff, Thermal modeling of a small-particle solar central receiver, *Journal of Solar Energy Engineering*, 122, 23–29, 2000.
87. R. Bertocchi, J. Karni, and A. Kribus, Experimental evaluation of a non-isothermal high temperature solar particle receiver, *Energy*, 29, 687–700, 2004.
88. T. Tan and Y. Chen, Review of study on solid particle solar receivers, *Renewable and Sustainable Energy Reviews*, 14, pp. 265–266, 2010.
89. E. W. Grald and T. H. Kuehn, Performance analysis of a parabolic trough solar collector with a porous absorber receiver, *Solar Energy*, 42, 281–292, 1989.
90. R. Pitz-Paal, B. Hoffschmidt, M. Bohmer, and M. Becker, Experimental and numerical evaluation of the performance and flow stability of different types of open volumetric absorbers under non-homogeneous irradiation, *Solar Energy*, 60, 135–150, 1997.

7 On the Thermophysical Properties of Suspensions of Highly Anisotropic Nanoparticles with and without Field-Induced Microstructure

Jerry W. Shan, Anna S. Cherkasova, Chen Lin, and Corinne S. Baresich

CONTENTS

7.1   INTRODUCTION

Advances in the large-scale synthesis of a variety of nanoparticles, nanowires, and nanotubes over the last decade have led to new interest in the design of nanoparticle suspensions, or nanofluids, having tailored thermophysical properties. Such engineered, two-phase systems may have utility as enhanced heat-transfer fluids, for example, having greater thermal conductivities and heat-transfer coefficients than conventional fluids. Experimental and theoretical investigations of nanofluids in the heat-transfer community were stimulated by initial reports of large increases in the thermal conductivity of very-low-volume-fraction nanoparticle and nanotube suspensions, for example, copper nanoparticles in ethylene glycol [1], multiwalled carbon nanotubes (MWNTs) in oil [2], and silver nanoparticles in water and toluene [3], among others. Nanofluids were reported to deviate from classical effective medium theory for the thermal conductivity of composite materials in terms of (1) anomalously large increases in thermal conductivity, (2) possible dependence on particle size and shape, and (3) dependence on temperature. With respect to the first two reported anomalies, Nan et al. were among the first to point out that the reported increases in thermal conductivity were within the bounds of existing theory when the aspect ratio of highly anisotropic nanoparticles such as carbon nanotubes (CNTs) is taken into account [4]. For suspensions of spherical nanoparticles, Prasher et al. show the importance of particle–cluster morphology, which, with the formation of large-aspect-ratio particle clusters, can substantially increasing the effective thermal conductivity of a suspension [5]. For several types of well-dispersed nanoparticles (alumina nanorods and nanoparticles, gold nanoparticles, silica nanoparticles, and Mn–Zn ferrite nanoparticles) in oil and water, recent experimental results from the International Nanofluid Property Benchmark Exercise (INPBE) indicate that the measured thermal conductivities are in agreement with classical effective medium theory (EMT) [6].

In light of such evidence suggesting that the thermophysical properties of most nanofluids are adequately described by existing theory, it is reasonable to ask what, if anything, can be truly different about nanoparticle suspensions. Compared with suspensions of micron-sized and larger particles, nanoparticle suspensions can have enhanced stability against gravitational sedimentation or buoyancy, provided the particles are well-dispersed and prevented from aggregation by steric or electrostatic stabilization. They may also exhibit less potential for erosion and clogging compared with suspensions of larger particles. In some cases, nanoparticles may have unique size-dependent physical properties, such as the quantum-confinement-induced differences in the electronic and optical properties of nanoscale semiconductors [7–13], or the unique mechanical properties of CNTs [14–17]. In terms of thermal properties, the ballistic-phonon-dominated thermal conductivity of nanotubes can be extremely high, as in the case of carbon [18–19] or boron-nitride nanotubes [20]. On the other hand, silicon nanowires can have thermal conductivities two order of magnitude lower than bulk silicon, making them attractive for high-efficiency thermoelectric devices [21]. Finally, nanoparticles can also have unique morphologies, including extreme anisotropy (e.g., particle aspect ratios exceeding 10^3 in the case of CNTs) and large surface-area-per-unit-volume, that are not readily accessible with conventional particles. The former can lead to enhanced thermal conductivity of nanofluids, while the latter necessitates the consideration of interfacial resistance between the particle and fluid in the modeling of nanofluid effective thermal conductivities.

In this chapter, we focus on the issue of particle morphology and microstructure on the effective thermophysical properties of suspensions of highly anisotropic particles. Classical effective-medium models for thermal conductivity and viscosity of anisotropic particles are briefly reviewed. In the case of thermal conductivity, quantitative comparison is made between classical effective medium theory and recent experimental results on well-dispersed, randomly oriented suspensions of MWNTs. We further consider the effect of particle microstructure, namely particle aggregation and alignment, on the effective thermophysical properties of the fluid suspension. We discuss how particle alignment might be achieved with external electromagnetic fields, and report on progress toward the experimental demonstration of carbon-nanotube-based nanofluids having anisotropic, actively controllable thermal conductivity and viscosity.

7.1.1 CONTROLLING NANOWIRE/NANOTUBE MICROSTRUCTURE

An experimentally realizable means of controlling the orientation and microstructure of nanowires and nanotubes is the application of external electromagnetic fields. We focus on nanowires/nanotubes under electric fields in the following discussion, although the behavior of ferromagnetic particles under external magnetic fields is largely analogous.

Owing to their one-dimensional shape, nanowires and nanotubes are highly polarizable particles and develop strong induced dipole moments under external electric fields. The interfacial polarization and dipole moments arise from differences in electrical properties between the particle and the fluid in which it is immersed. For prolate-spheroidal particles with semimajor axis, a, larger than the semiminor axis, b, the (complex) effective moment is [22]

$$\vec{P}_{\text{eff}} = 4\pi\varepsilon_1 ab^2 \underline{\underline{K}}\vec{E} \tag{7.1}$$

where ε_1 is the (real) permittivity of the fluid, $\vec{E}$ is the electric field, and $\underline{\underline{K}}$ is the tensor form of the Clausius-Mossotti factor. The Clausius-Mossotti factor, which determines the strength of the induced dipole moment, depends on the complex permittivities (The complex permittivity is defined as $\underline{\varepsilon} \equiv \varepsilon + \sigma/i\omega$, where ε is the real permittivity, σ is the conductivity, and ω the frequency of the applied field.) of the particle and solvent, as well as the geometry of the particle and the frequency of the applied field.

In the low-frequency limit, the Clausius–Mossotti factor (and the resultant induced dipole moment) is determined primarily by the difference in electrical conductivity between the particle and the solvent. In the high-frequency limit, the strength of the induced dipole depends on the (real) permittivity contrast between the particle and fluid. Since some difference in the electric conductivities or permittivities of the particle and fluid is unavoidable, a dipole moment can always be induced by a properly chosen applied field. Moreover, highly anisotropic particles have much stronger induced dipole moments, and are thus easier to manipulate, than spherical particles of the same volume and electric properties.

This induced dipole moment, along with the effective potential (the ζ-potential) at the particle shear surface, makes it possible to induce nanowire rotation and translation with external electric fields. We briefly summarize the lowest-order (neglecting higher-order multipoles) electric-field induced forces and torques acting on a particle in suspension:

1. *Electrophoresis*: Under an electric field, a particle experiences a force proportional to the ζ-potential and the field strength. This mechanism is significant primarily for DC and low-frequency AC fields.
2. *Electroorientation*: A torque is exerted on the nanowire which tends to align it with the electric field [22]. This time-averaged torque

$$\langle T \rangle = \frac{1}{2} \mathrm{Re} \left[\vec{P}_{\mathrm{eff}} \times \vec{E} \,^* \right] \tag{7.2}$$

is proportional to the field squared since the induced dipole moment, $\vec{P}_{\mathrm{eff}}$, is itself proportional to $\vec{E}$. This mechanism, which works for both DC and AC fields, can be used to align nanowires and nanotubes in a liquid solvent, irrespective of the spatial uniformity of the electric field. For our puposes, it will allow precise control of the state of alignment of nanowires in liquid suspension by applying an external field of known magnitude and direction.
3. *Dielectrophoresis (DEP)*: The induced dipole moment of the particle interacts with a nonuniform field to give rise to a frequency-dependent DEP force which is proportional to the polarizability of the particle and the gradient of the field squared [22]:

$$\langle F_{\mathrm{DEP}} \rangle = \frac{1}{2} \mathrm{Re} \left[\vec{P}_{\mathrm{eff}} \cdot \nabla \vec{E} \,^* \right]. \tag{7.3}$$

In general, a particle can be either repelled or attracted to field concentrations depending on the sign of the dipole moment arising from the real part of the complex Clausius–Mossotti function. Even in a spatially uniform external field, the local field of a polarized particle can lead to dielectrophoretic interactions between neighboring particles. This effect is responsible for the typical "pearl-chaining" of particles under dipole–dipole interactions when a uniform external field is applied to an electrorheological fluid.

These forces and torques make it possible to tune the microstructure (alignment and chaining) of polarizable nanowires in suspension with an external electric field. However, the particles are also in general affected by Brownian motion and viscous flow. The relative importance of these various effects can be assessed by comparing

TABLE 7.1

Estimates of the Relative Magnitudes of the Orientational, Thermal, and Viscous-Shear Energies for SWNTs (Dimensions 1 μm by 1 nm) in Distilled Water at Room Temperature under Electric Fields of the Range $E = 10–10^3$ V/mm

Parameter	Explicit Form	Typical Values for $E = 10$ V/mm	Typical Values for $E = 10^3$ V/mm
Orientational/thermal	$$\dfrac{\frac{1}{2}\,\mathrm{Re}\left[\vec{P}_{\mathrm{eff}}\cdot\vec{E}\right]}{k_{\mathrm{B}}T}$$	1.0	10^4
Orientational/shear $(0.1 \le \dot{\gamma} \le 10 \text{ s}^{-1})$	$$\dfrac{\frac{1}{2}\,\mathrm{Re}\left[\vec{P}_{\mathrm{eff}}\cdot\vec{E}\right]}{\frac{4\pi}{3}\mu(2a)^3\dot{\gamma}\left[2\ln\left(2\frac{a}{b}\right)-1\right]^{-1}}$$	1.6–160	1.6×10^4– 1.6×10^6
Chaining interaction/ thermal	$$\dfrac{\frac{8\pi}{r^3}\varepsilon_1\,\mathrm{Re}\left[\vec{P}_{\mathrm{eff}}\right]^2}{k_{\mathrm{B}}T}$$	4.4×10^{-3}	44
Chaining interaction/ shear $(0.1 \le \dot{\gamma} \le 10\text{s}^{-1})$	$$\dfrac{\frac{8\pi}{r^3}\varepsilon_1\,\mathrm{Re}\left[\vec{P}_{\mathrm{eff}}\right]^2}{\frac{4\pi}{3}\mu(2a)^3\dot{\gamma}\left[2\ln\left(2\frac{a}{b}\right)-1\right]^{-1}}$$	6.9×10^{-3}–0.69	69–6.9×10^4

Note: The shear rate is assumed to vary from $\dot{\gamma} = 0.1$–10 s^{-1} and the particle volume fraction is $\varphi = 10^{-2}$.

their magnitudes in dimensionless form. Actual values of these dimensionless parameters are summarized in Table 7.1 for single-wall carbon nanotubes (SWNTs) suspended in a deionized water for a typical range of shear rates $(0.1 \le \dot{\gamma} \le 10/s)$ and electric field strengths in the low-frequency limit. For the DEP particle interactions, a particle spacing is assumed that is consistent with evenly dispersed particles at a solid volume fraction of $\varphi = 10^{-2}$.

As seen from the table, the dimensionless ratios of the electric-field energy to the thermal and viscous-shear energies are >1, and thus sufficient to align the nanotubes against thermal and viscous-shear effects, at reasonable field strengths of 10 V/mm or higher. Moreover, particle chaining can be induced at higher field strengths on the order of 10^3 V/mm. Thus, by controlling the magnitude and duration of the applied electric field, the nanowire orientation and chaining can be precisely tuned to investigate the effect of microstructure on the thermophysical properties of the suspension.

In the following section, we first review models for the effective thermal conductivity of suspensions of anisotropic particles, and then discuss measurements of the conductivity with and without field-induced microstructure.

7.2 THERMAL CONDUCTIVITY

7.2.1 EFFECTIVE MEDIUM MODELS

7.2.1.1 Negligible Interfacial Resistance: Maxwell–Garnett and Fricke's Equations

First, we note that the analytical solution of the Laplace equation for a single sphere, placed in an infinite medium and subjected to a uniform heat flux in the far field, exists [23]. If boundary conditions on the surface of a sphere are given in a form of continuity of temperature and normal component of a heat flux, that is, negligible interfacial resistance, the temperature distribution is known at any point inside and outside the sphere:

$$T_{out} \sim r\left(1 + \left(\frac{R}{r}\right)^3 \frac{(1 - \delta)}{2 + \delta}\right)\cos\theta \tag{7.4a}$$

$$T_{in} \sim \frac{3}{2 + \delta}r\cos\theta \tag{7.4b}$$

where $\delta = k_p/k_B$ is ratio of thermal conductivities of dispersed and dispersing media correspondingly, R is the radius of the spherical particle and r, θ are spherical coordinates.

Now let us briefly describe the approached introduced by Maxwell [24] and used to analyze electric conductivity of the composite medium. As the governing equations for thermal transport are essentially the same, we can use a similar approach to obtain the effective thermal conductivity of a composite material. First, consider an ensemble of spherical particles uniformly distributed inside some volume of fluid. The spheres are assumed to not interact thermally, that is, particle volume concentration is low enough to treat every particle as a single sphere subjected to external uniform heat flux. After applying the Equations 7.4a and 7.4b to every sphere, one can replace the thermal conductivity of such particle ensemble dispersed in base medium with effective thermal conductivity of the material contained within the observed volume. As a result, the effective thermal conductivity of a dispersion containing spherical particles is obtained as following:

$$k = k_B\left(1 + \frac{3\varphi(\delta - 1)}{2 + \delta - \varphi(\delta - 1)}\right). \tag{7.5}$$

In the above equation, which is referred as Maxwell [24] or Maxwell–Garnett [25] formula, φ is volumetric particle concentration.

The main assumptions used to derive the above Equation 7.5 are

1. Suspended particles are spherical.
2. Particles are noninteracting.
3. Interfacial resistance between the liquid and solid phases is negligible.

The spherical particle approximation obviously does not hold for CNTs or any other elongated dispersed objects, such as fibers, whiskers, and so on. The next approximation is to treat the nanotubes and nanowires as prolate spheroids. This approximation allows an analytical solution for the Laplace heat equation in prolate spheroidal coordinate system to be found [23]. We denote one of the spheroidal coordinate as ξ. The surfaces of a constant ξ are confocal spheroids, and $\xi = 0$ corresponds to the surface of the particle. The temperatures inside and outside the particle are given as

$$T_{\text{out}} \sim z\left(1 + \frac{Lr^2}{4} \frac{1 - \delta}{f^{(i)}(\delta - 1) + 1} F(\xi)\right) \tag{7.6a}$$

$$T_{\text{in}} \sim \frac{z}{f^{(i)}(\delta - 1) + 1} \tag{7.6b}$$

when the particle is placed in an infinite medium and subject to a uniform heat flux in the far field. Here F is a function of ξ only and it is defined by

$$F(\xi) = \int_{\xi}^{\infty} \frac{ds}{(s + r^2)\sqrt{(s + (L/2)^2)}} \tag{7.7}$$

For spheroids with long axis oriented either along or perpendicular to the direction of the heat flux, $f^{(i)}$ is

$$f^{(z)} = \frac{1 - e^2}{e^3}(\tanh^{-1} e - e) \tag{7.8}$$

or

$$f^{(x)} = \frac{1 - f^{(z)}}{2} \tag{7.9}$$

respectively. The coefficients $f^{(z)}$ and $f^{(x)}$ are the so-called depolarization factors and depend on the geometry of the particle through its eccentricity e, which is always a nonnegative number <1:

$$e = \sqrt{1 - \left(\frac{D}{L}\right)^2} = \sqrt{1 - \frac{1}{a^2}} \tag{7.10}$$

where, as before, $D = 2R$ and L are the semimajor and semiminor axes of the dispersed spheroid and a is the aspect ratio.

Using the analytic solutions given in Equations 7.6a and 7.6b, a model for the effective thermal conductivity of a dilute suspension of spheroidal particles was developed by Fricke [26]. Assuming that randomly dispersed and oriented spheroidal particles are contained within some volume and performing analysis similar to Maxwell's, Fricke [26] obtained the formula for the effective dielectric constant of the composite medium. To make it consistent with the Maxwell-Garnett equation we rewrite Fricke's formula as the following:

$$k = k_B\left(1 + \frac{n\varphi(\delta - 1)}{(n - 1) + \delta - \varphi(\delta - 1)}\right).$$

(7.11)

Here, n is the dimensionless shape factor given by

$$n = \frac{\beta\delta - \beta}{\delta - 1 - \beta}$$

(7.12)

where

$$\beta = \frac{\delta - 1}{3}\left(\frac{2}{1 + (\delta - 1)f^{(x)}} + \frac{1}{1 + (\delta - 1)f^{(z)}}\right).$$

(7.13)

For spheres, all of the depolarization factors are equal to each other, resulting in $n = 3$ and Fricke's formula is equivalent to Maxwell Equation 7.5. For large-aspect ratio particles, such as nanotubes, the Fricke prediction for effective thermal conductivity can greatly exceed that of the Maxwell model. In the situations when long particles are aligned, the resulting thermal conductivity of a dispersion or a composite is anisotropic and given by

$$k_\| = k_B(1 - \varphi) + k_P$$

(7.14)

$$k_\perp = k_B\left(1 + \frac{2\varphi(\delta - 1)}{1 + \delta - \varphi(\delta - 1)}\right).$$

(7.15)

Here, $k_\|$ and $k_\perp$ denote thermal conductivities along and perpendicular to the alignment vector. As intuitively expected, Equation 7.14 is a simple rule corresponding to the parallel configuration of thermal resistors.

7.2.1.2   Effective Medium Theory Incorporating Interfacial Resistance

The thermal transport in suspensions containing nano-sized particles can be significantly affected by scale-dependent mechanisms that enter into consideration at smaller scales. Among these effects, the interfacial resistance at solid–liquid interfaces becomes increasingly important due to the larger number of interfaces for a given volume fraction of solid content as the particle size decreases. For example, for spherical particles, the particle surface area per unit volume scales as r^{-1} for a fixed particle concentration. This effect is not taken into account in conventional macroscopic-scale models like Maxwell's or Fricke's which were derived under assumption that temperature experiences no discontinuity at the particle surface. This assumption is valid at macroscopic scale, but breaks down as the characteristic length decreases to nanometers. In this case, temperature and/or heat flux may no longer be continuous across the interface of two phases. A phenomenological approach to describe a discontinuity in temperature at liquid/solid interface dates back to Kapitza [27], who suggested that temperature at the interface undergoes a change proportional to the normal component of the heat flux through the interface:

$$T_{out} - T_{in} = -R_K \times q_{interface}. \tag{7.16}$$

Here, R_K is Kapitza, or thermal contact, resistance and $q_{interface}$ is a heat flux through the particle surface. The Kapitza resistance appeared first in the context of liquid helium physics, but it can exist on every interface between two phases. The magnitude of the resistance for particular type of interface depends on many parameters, including thermal and physical properties of both phases. A more physically transparent explanation of interfacial phenomena involves Kapitza length, which is the thickness of a liquid or solid layer equivalent to the interface from the thermal point of view. The Kapitza length is defined as following:

$$l_K = k_i R_K \tag{7.17}$$

where k_i is either k_P or k_B. When the Kapitza length is large compared to the problem characteristic dimension, such as particle size, then interfacial contact resistance plays an important role in thermal transport. In the case of nanofluids, especially containing nanotubes, the thermal conductivity of dispersed phase is extremely high, which means some measurable effects due to interface heat transport can be expected. More importantly, the Kapitza length can become comparable to the size of the particle/tube as the particle size decreases. For most liquid/solid interfaces, the Kapitza resistance is typically of the order 10^{-8} m²K/W. While direct measurement of interfacial resistance is difficult, recent experimental data and molecular dynamics (MD) simulations by Huxtable et al. [28] report an interfacial resistance of $R_K = 8.3 \times 10^{-8}$ m²K/W for CNTs dispersed in water with the aid of sodium dodecyl sulfate (SDS). Molecular dynamic simulations on SWNTs in octane [29] obtained thermal contact resistances ranging from 4.5×10^{-8} up to 14.7×10^{-8} m²K/W. Two different approaches, constant heat rate and relaxation simulations, were used in the MD simulations. For the interfacial resistance of nanoparticles other than CNTs in liquid solvents, R_K values above 10^{-7} m²K/W are usually referred only to weakly bonded (such as hydrophobic) interfaces [30]. For strongly bonded interfaces, reported resistances are around 10^{-8} m²K/W and even lower. For instance, R_K was found to be 7.7×10^{-9} m²K/W for citrate-stabilized Pt and 9.5×10^{-9} m²K/W for gold nanoparticles in water [31,32]. This would lead one to expect relatively low interfacial resistance values for stabilized suspensions where surfactant molecules are strongly adsorbed on nanotubes surface.

It is clear that taking into account Kapitza resistance can lower the resulting thermal conductivity of nanofluid. To estimate the magnitude of this effect we can take a look at Kapitza length for a typical suspension made of a base fluid with thermal conductivity of 0.5 W/mK and magnitude of thermal resistance $10^{-9}-10^{-7}$ m^{2}K/W. The Kapitza length calculated using Equation 7.17 is 0.5–50 nm which of the same order as the diameter of typical MWNTs, and larger than the diameter of SWNTs.

To investigate the effect of contact resistance further we follow the Maxwell approach for obtaining the thermal conductivity of suspensions. Now instead of temperature continuity at the liquid/solid interface, we use Kapitza condition (Equation 7.16) as boundary condition. This leads to the following equation for thermal conductivity of composite containing spherical particles:

$$k = k_B \left(1 - 3\varphi \frac{k_B - \left(\dfrac{k_p}{1 + \dfrac{k_p R_k}{R}} \right)}{2k_B + \left(\dfrac{k_p}{1 + \dfrac{k_p R_k}{R}} \right) + \varphi \left(k_B - \left(\dfrac{kp}{1 + \dfrac{k_p R_k}{R}} \right) \right)} \right) \qquad (7.18)$$

which reverts to the regular Maxwell formula with k_P replaced by an effective thermal conductivity of dispersed phase $k_P^{(e)}$ given by

$$k_P^{(e)} = \frac{k_P}{1 + \alpha} \qquad (7.19)$$

Here, α, the dimensionless measure of the interfacial resistance, introduces the effect of the Kapitza resistance on the particle interface and is given by

$$\alpha = \frac{k_B R_K}{R}. \qquad (7.20)$$

A model for the thermal conductivity of suspensions of ellipsoidal inclusions was also generalized to include the effect of interfacial resistance in the form of an EMT approach by Nan [33]. Nan showed that for a dilute suspension of randomly oriented, spheroidal particles with aspect ratio r and interfacial (Kapitza) resistance R_K, the thermal conductivity is given by

$$k = k_B \left(1 + \frac{\varphi(2\beta_x + \beta_z)}{3 - \varphi(2\beta_x f^{(x)} + \beta_z f^{(z)})} \right) \qquad (7.21)$$

where

$$k_P^{(i)} = \frac{k_P k_B}{k_B + \alpha k_P f^{(i)}(2 + 1/r)} \qquad (7.22)$$

$$\beta_i = \frac{k_P^{(i)} - k_B}{k_B + f^{(i)}(k_P^{(i)} - k_B)}.$$ (7.23)

Nan et al. exploited a methodology known as a multiple-scattering approach to derive Equation 7.21. Later Nan et al. [34] introduced alternative expressions replacing Equation 7.22 for the equivalent thermal conductivity of cylindrical inclusions:

$$k_P^{(z)} = \frac{k_P}{1 + (2k_P R_K /L)}$$
$$k_P^{(x)} = \frac{k_P}{1 + (k_P R_K /R)}.$$ (7.24)

Equations 7.24 can be described as a simple scheme where thermal resistances represented by nanotube and it surfaces are arranged in series. It can be easily verified that Equation 7.24 can deviate significantly from Equation 7.22. For long particles the axial effective thermal conductivity defined by both Equations 7.22 and 7.24 collapses and ultimately reduces to k_P. At the same time, we can see that the transverse thermal conductivity of the long spheroidal particle given by Equations 7.22 and 7.24 depend on the radius of particle and interfacial resistance on its interface.

Later Duan et al. derived their own explicit expression for calculating thermal conductivity of composite material containing randomly oriented spheroidal inclusions [35] and then supplemented it for the case of interfacial resistance on the particle surface [36]. According to Duan et al., the thermal conductivity is given by

$$k = k_B \left(1 + \frac{3\varphi(2\beta_x + \beta_z)}{9 - \varphi(2\beta_x + \beta_z)} \right).$$ (7.25)

Duan et al.'s expression for longitudinal and transverse conductivities coincide with those obtained by Nan et al., however, the general equations for resulting thermal conductivity of a composite are not the same. The difference arises from the fact that they used different averaging schemes when deriving their equation for thermal conductivity. Duan et al. showed that at low volume fractions of dispersed particles both models provide similar results. However, at higher volume fractions, Nan's model predicts a smaller thermal conductivity enhancement compared with that of Duan et al.

In both Nan's and Duan's models, the effective thermal conductivity is seen to depend on the aspect ratio of the particles. As shown in Figure 7.1, the predicted effective thermal conductivity of the suspension increases dramatically with aspect ratio for a given particle volume fraction, before finally saturating at very large aspect ratios. For instance, for spheroidal iron particles in oil at 1% vol, the effective thermal conductivity for randomly oriented particles in suspension increases by a factor of three as predicted by Nan's model and by the factor of 7 according to Duan et al. as the aspect ratio increases from 1 to 10^3, neglecting interfacial resistance.

Thus, randomly dispersed and oriented ellipsoidal particles increase the conductivity of a suspension much more than spherical particles of the same type and

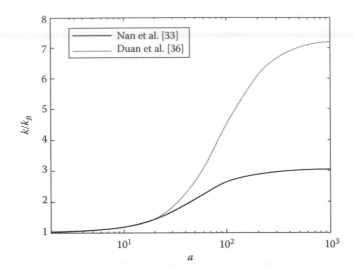

FIGURE 7.1 Predicted thermal conductivity increase in suspension of randomly oriented, 1% vol iron particles of different aspect ratio in oil. Interfacial resistance is neglected, that is, the Kapitza length is negligibly small compared with the size of the particles.

volume fraction. It should be noted that the limiting thermal conductivity calculated with Nan's and Duane's models are different, as seen in Figure 7.1. The difference comes from the fact that these approaches utilize two different averaging schemes to account for particles with uniformly distributed orientations. As will be shown later, Nan's formula for randomly oriented spheroids can be obtained from a composite in which one third of the particles are aligned along each of three orthogonal axes, though it is not the case for Duane's model.

Moreover, the effective medium theory also predicts that the conductivity depends on the orientation of large-aspect ratio particles. For perfectly aligned ellipsoidal particles, Nan's EMT gives the thermal conductivity as

$$k = k_B \left(1 + \frac{\varphi\beta_i}{1 - \varphi\beta_i f^{(i)}} \right) \qquad (7.26)$$

where $i = z$ for the conductivity in the alignment direction and $i = x$ for the conductivity in the direction perpendicular to the alignment. For long aligned fibers, Equation 7.26 ultimately reduces to Equations 7.14 and 7.15, where the thermal conductivity ratio δ is now the ratio of the effective thermal conductivity of the particles and the base fluid thermal conductivity. As was mentioned above, for very long particles, the effective particle thermal conductivity predicted by Equations 7.22 and 7.24 is simply the intrinsic particle conductivity, k_P.

Duan et al. provide the same Equation 7.26 for thermal conductivity for the medium containing perfectly aligned spheroidal particles, which also reduce to Equations 7.14 and 7.15 in the case of ultimately long aligned fibers. This is illustrated in Figure

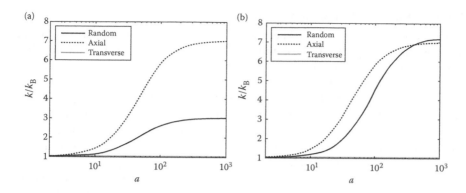

FIGURE 7.2 (a) Nan's effective medium theory [33], and (b) Duan's model [36] predictions of the thermal conductivity of an iron-in-oil suspension with randomly oriented and fully aligned particles, neglecting interfacial resistance.

7.2 for varying-aspect-ratio prolate spheroidal iron particles of 1% volume fraction in oil. One can see that in the suspension of fully aligned long particles, the thermal conductivity increase in the direction of alignment is around three times greater than for the case of randomly oriented particles as predicted by Nan et al. The behavior of the thermal conductivity of randomly oriented particle suspension according to Duan et al. is quite surprising at very high aspect ratios though; it is comparable to and even higher than the thermal conductivity in the direction of alignment of a composite with fully aligned fibers. In both models, the thermal conductivity increase in the direction perpendicular to the alignment is negligible; in fact, it is easy to verify that the increase is smaller than predicted by the Maxwell model for spherical particles. We note that these predictions only account for aligned but noninteracting particles. Particle interaction or aggregation caused by high concentrations, flow, or an external field can further change the effective conductivity of the suspension, as will be discussed later. Finally, in Nan's model, the thermal conductivity of randomly oriented particles (Equation 7.26) can be derived by assuming that three sets of particles with the volume fraction $\varphi/3$ each are aligned along three coordinate axis. The corresponding contributions to the thermal conductivity (the second term of the right hand side of Equation 7.26) can then be added to obtain the effective conductivity for randomly oriented particle suspensions, neglecting second-order terms in volume fraction. The averaging scheme used by Duan et al., however, does not allow such a transformation between the fully aligned and randomly oriented cases.

7.2.2 Experimental Investigation

We next review some recent experimental data on the thermal conductivity of well dispersed randomly oriented nanotube suspensions and make comparison with effective medium theory considering interfacial resistance. We also discuss nanotube suspensions in which the particle microstructure is varied by chemical means as well as external electromagnetic fields.

7.2.2.1 Isotropic Nanotube Suspensions

Before reviewing our own experimental data on the thermal conductivity of MWNT suspensions, we note that a global initiative known as INPBE was launched in 2007 to investigate inconsistencies associated with reported thermal conductivity enhancements in nanofluids [6]. Thirty four scientific organizations from the USA, Belgium, China, France, Germany, India, Italy, Japan, Poland, Puerto Rico, Singapore, South Korea, Switzerland, and the UK participated in this systematic study of thermal conductivity enhancement in nanofluids. All groups performed independent measurements of identical sample nanofluids, including water and poly(α-olefin)-oil (PAO)-based suspensions of alumina, gold, silica, and Mn-Zn ferrite nanoparticles. Very good agreement between the results revealed that thermal conductivity enhancements in nanofluids were consistent between different groups and experimental techniques once identical samples were used. The experimental approaches used in the study [6] included the hot-wire method, which was the method chosen for our studies. The consistency of the hot wire method with other techniques such as steady- state and optical methods, along with recent data by Li et al. showing that hot-wire results are in a good agreement with data obtained with steady-state cut bar for Al_2O_3 aqueous nanofluids at room temperature [37], give us additional confidence that the hot-wire method we use gives accurate and repeatable results.

As previously mentioned, the interfacial resistance between particle and liquid becomes very important in nanofluids, where the diameter of the nanotubes is of the order of tens nanometers. Cherkasova and Shan [38] attempted to compare experimental data for MWNTs aqueous suspension and careful theoretical calculations using Nan's and Duan's models as shown in Figure 7.3. Nanotubes were dispersed in water with the aid of sodium dodecylbenzene sulfonate (NaDDBS). This surfactant was chosen because of its reported effectiveness in dispersing nanotubes in water [39]. The MWNTs used in the experiment are considered relatively short with a nominal length of 0.5–2 μm and corresponding aspect ratios ranging from 10 to 70 to ensure better dispersion characteristics of the suspensions. Volume fraction of nanotubes ranged from 0.01% up to 1% with surfactant to nanotube mass ratio of 5, 10, 20, and 100. It can be seen that measured thermal conductivities are underestimated by both Nan's and also Duan's models if Kapitza resistance is within previously reported range [28,29] of around 10^{-7} m²W/K (Figure 7.3). The experimental data on the MWNT suspensions fit theoretical predictions for interfacial resistance of around 10^{-9} m²K/W for Nan's (later) model and 10^{-8} m²K/W for Duan's model as seen from Figure 7.4. Even though data on interfacial resistance for CNTs suspended in water with the aid of NaDDBS is not available at the moment, such a difference in R_K probably cannot be attributed to the choice of the surfactant. As previously mentioned, Huxtable et al. report a value of $R_K = 8.3 \times 10^{-8}$ m²K/W for CNT suspended in a SDS micelle [28], so R_K for NaDDBS would be expected to be somewhat close to this number.

As was discussed earlier, the thermal conductivity enhancement calculated using Duan et al. [36] and Nan et al. [34] both result in higher heat-transfer augmentation than the later model introduced by Nan in 2004 [34]. However, experimental data on thermal conductivity of nanofluid showed an agreement with all three effective medium theories [33,34,36] provided a lower magnitude of the interfacial resistance

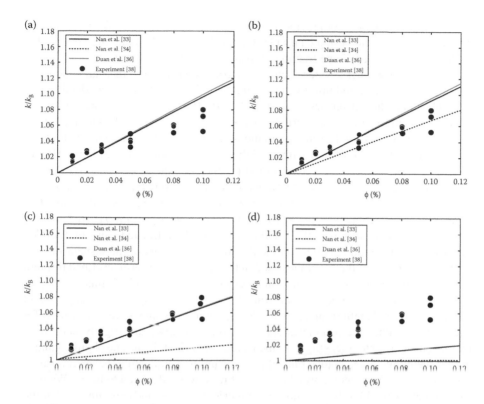

FIGURE 7.3 Measured thermal conductivity enhancements in aqueous MWNTs suspensions [38] compared with theoretical calculations [33], [34], and [38] at different magnitudes of interfacial resistance: (a) $R_K = 0$ m²K/W, (b) $R_K = 10^{-9}$ m²K/W, (c) $R_K = 10^{-8}$ m²K/W, (d) $R_K = 10^{-7}$ m²K/W.

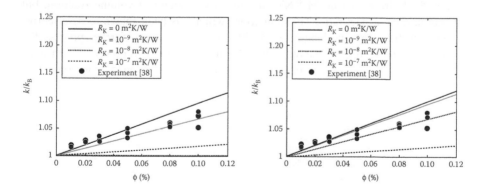

FIGURE 7.4 Measured thermal conductivity enhancements in aqueous MWNT suspensions [38] compared with EMA by Nan [34] and Duane et al. [36] for different magnitudes of interfacial resistance.

is used in calculations. As previously noted, calculations of the effective thermal conductivity calculated using Nan's [33,34] and Duan's [36] models are comparable in this case since the volume fraction of dispersed particle is so low.

Let us analyze other key parameters that could contribute to thermal conductivity of suspension. In addition to the volume fraction, the thermal conductivity of nanotubes and their geometry have a profound effect on thermal conductivity of nanofluid. With regard to the thermal conductivity of nanotubes, k_P, the magnitude of 1000 W/mK was used in all calculations as a reasonable estimate of the thermal conductivity of MWNTs volume-weighted average diameter of 21.6 nm. Higher thermal conductivities were only reported for SWNTs and MWNTs with diameter <10–15 μm [40–42], while lower thermal conductivities were typically only reported for thick nanotubes (with the exception of one of Choi et al.'s results [43] among data obtained with direct methods [44–48]). For the moderate aspect-ratio particles of the described study, variations in the value of k_P about the estimated value of 1000 W/mK would have a minimal effect on the thermal conductivity of the suspension.

The aspect ratio of the nanotubes used in the suspensions was determined through direct measurement from transmission electron microscope (TEM) images (Figure 7.5) of the suspension dried on a substrate. Several TEM images were analyzed and the diameter and length of more than 200 individual nanotubes were measured for each sample. Since the nanotubes deposited on a substrate are slightly bent, both end-to-end and contour lengths were measured for each of them. The obtained diameter of the individual nanotube ranged from 12 up to 35 nm, while the length was as high as 1.4 μm. The volume-weighted aspect ratio was calculated to be 33. We note that a recent paper [38] showed through numerical analysis and experiments that the volume-weighted aspect ratio (rather than other measures such as the mean, peak, or median aspect ratio) is the correct measure to use when particle aspect ratios are not monodisperse but are rather widely distributed. For our MWNT samples, the volume-weighted aspect ratio is higher than the mean aspect ratio of 28, as expected.

We further studied the aspect-ratio effect on the effective thermal conductivity of nanofluids by preparing MWNT suspensions of identical volume fraction, but varying particle aspect ratio. Tip sonication, which has been shown to be effective in reducing the nanotube length [38,39,49,50], was chosen to further process the suspensions and reduce the nanotube length. In contrary to less powerful bath sonication which is usually used for dispersing particles in suspension [51], tip-sonication

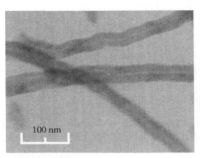
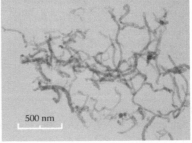

FIGURE 7.5   TEM micrographs of MWNTs.

TABLE 7.2

Measured Aspect Ratio of MWNTs

Tip-Sonication Time (min)	0	5	10	15
Mean aspect ratio	28.04	14.61	11.42	8.92
Volume-averaged aspect ratio	32.94	20.34	15.02	14.47

is much more powerful tool and breaks nanotubes. Cherkasova and Shan used tip-sonication to reduce the aspect ratio of the MWNTs in aqueous dispersion from 28 down to 9 after 15 min [38]. The sonication was carried out by 5 min intervals and full geometric characterization of the particles in suspension was made at every step of the procedure. As can be seen from Table 7.2, the volume-averaged aspect ratio was ultimately reduced by a factor of 2 after a 15 min of sonication, while the mean aspect ratio was reduced by the factor of 3.

Cherkasova and Shan [38] measured thermal conductivities of such nanofluids containing 0.1% vol MWNTs of varying aspect ratio, as presented in Figures 7.6 and 7.7. It can be clearly seen that the thermal conductivity enhancement of the suspension dropped in half, from 8% to 4%, after the first 5 min of sonication. After another 10 min of intense ultra-sonication the thermal conductivity enhancement goes down to 1% and remains essentially unchanged afterwards. Careful comparison with calculations showed that the thermal conductivity drop occurred slightly faster than predicted with decreasing aspect ratio. This effect could result from a decrease in thermal conductivity of MWNTs as their aspect ratio is reduced since nanotube thermal conductivity can depend on length. Secondly, the intense sonication has been shown to damage the nanotube structure, likely causing an additional decrease in conductivity due to the introduction of defects in the nanotube walls [52].

In addition to our own data, experiments by Yang et al. found that the thermal conductivity of MWNT/PAO suspensions at a volume fraction of $\varphi = 0.0021$ decreases with sonication time [49]. Figure 7.8 plots Yang et al.'s measured thermal conductivities as a function of the nanotubes' volume-weighted aspect ratio (obtained using

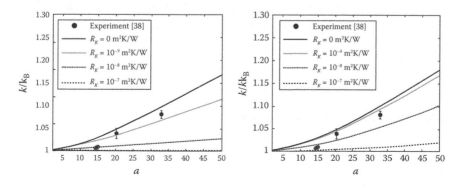

FIGURE 7.6 Measured thermal conductivity increases in aqueous suspension of MWNTs compared to calculations [33] and [36] for different magnitudes of interfacial resistance.

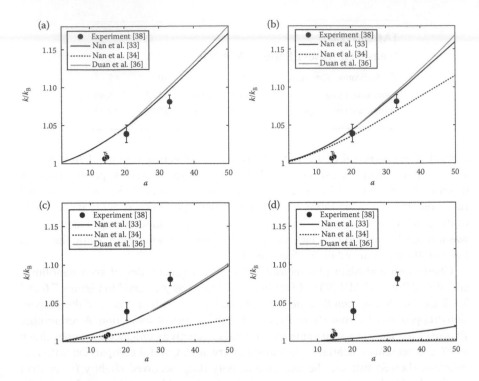

FIGURE 7.7 Measured thermal conductivity increases in aqueous suspension of MWNTs [38] compared to calculations at different magnitudes of interfacial resistance: (a) $R_K = 0$ m²K/W, (b) $R_K = 10^{-9}$ m²K/W, (c) $R_K = 10^{-8}$ m²K/W, (d) $R_K = 10^{-7}$ m²K/W.

their reported aspect-ratio distributions). As seen in the figure, the changes that Yang et al. observe in the thermal conductivity of the MWNT suspension are consistent with the reduction in thermal conductivity brought about by decreasing aspect ratio. Qualitative agreement is seen between the data and Nan's and Duan's models assuming an interfacial resistance of around 10^{-9} m²K/W. Quantitative comparisons with the models are difficult, however, since, as noted by Yang et al., their suspensions contained substantial aggregates and therefore it was not possible to separate the effect of decreasing particle aspect ratio from particle (de)aggregation, which may have competing effects on the thermal conductivity. Nevertheless, it is clear that the effective thermal conductivity of a nanofluid depends strongly on the aspect ratio of the suspended particles.

7.2.2.2 Microstructured Suspensions

Within the context of effective medium theory, the microstructure of suspensions of particles can have a profound effect on the effective thermal conductivity. Flocculation of nanowires on one hand may be expected to decrease the conductivity by reducing the effective aspect ratio of the nanowires through aggregation. On the other hand, induced aggregation of spherical particles into linear chains may increase the effective thermal conductivity. Furthermore, as discussed in Section 7.2.1.2 and

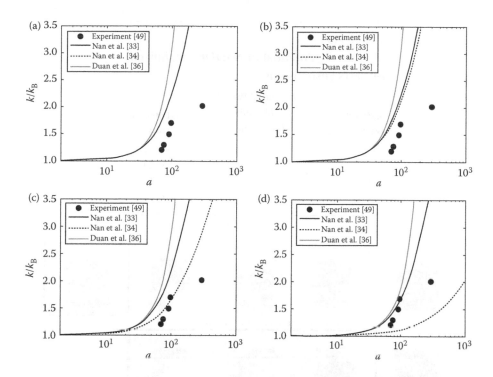

FIGURE 7.8 Measured thermal conductivity increases in aqueous suspension of MWNTs [49] compared to calculations at different magnitudes of interfacial resistance: (a) $R_K = 0$ m²K/W, (b) $R_K = 10^{-9}$ m²K/W, (c) $R_K = 10^{-8}$ m²K/W, (d) $R_K = 10^{-7}$ m²K/W.

shown in Figure 7.2, alignment of highly anisotropic particles is predicted by effective medium theory to increase a suspension's conductivity in the particle-alignment direction, while decreasing it in the two perpendicular directions. In the following, we briefly review our experimental data on the differing effects of particle aggregation on the conductivity of in two types of suspensions: MWNT suspensions in water, and spherical, micron-sized iron particles in oil. We also present preliminary experimental data on the conductivity of suspensions in which MWNTs are aligned by an external electromagnetic field.

Cherkasova and Shan attempted to correlate the thermal conductivity of MWNT suspensions to the agglomeration state of particles. The aggregation state of MWNTs in aqueous suspension was characterized through a combination of visual observation, optical microscopy, and ζ-potential and optical absorbance measurements [38]. As seen in Table 7.3, above a NaDDBS surfactant concentration of 20–50 g/L, the ζ-potential decreases in magnitude with increasing surfactant concentration, going through its isoelectric point (corresponding to minimal colloidal stability of the particles) between 100 and 200 g/L of surfactant. This result is consistent with visual observations which show good stability in MWNT samples with NaDDBS concentration below ~20 g/L (Figure 7.9). No aggregation or sedimentation was detected in these low-surfactant-concentration samples over a period of a week. However,

TABLE 7.3

Measured ζ-Potential of NaDDBS-Stabilized MWNTs Suspensions

NaDDBS Fraction (g/L)	Measured Particle Mobility $\left(\mu s^{-1}/Vcm^{-1}\right)$	ζ-Potential (mV)
20	−4.83	−61.81
50	−4.71	−60.29
100	−1.88	−24.10
200	2.68	34.29

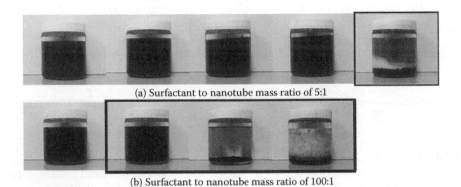

(a) Surfactant to nanotube mass ratio of 5:1

(b) Surfactant to nanotube mass ratio of 100:1

FIGURE 7.9 Nanofluids 1 week after preparation. MWNT volume fractions are 0.01%, 0.02%, 0.05%, 0.1%, and 1%, from left to right. Owing to obvious suspension instability, the highest MWNT concentration of 1% was not prepared for the 100:1 surfactant to nanotube mass ratio.

above a critical NaDDBS concentration of approximately 20–50 g/L, sedimentation occurred regardless of the nanotube volume fraction. This corresponds to the regime of decreasing repulsive interaction between particles, as evidenced by the decreasing magnitude of the ζ-potential. Thus, somewhat counter-intuitively, increasing surfactant concentration can sometimes decrease the stability of the suspension.

Thermal conductivity measurements of the aqueous suspensions of MWNTs showed an initial linear increase in conductivity with nanotube volume fraction only for stable nanofluids. As seen in Figures 7.3 and 7.4, good agreement between the experimental data and theoretical predictions is observed at this stage. This regime holds up to NaDDBS concentrations of around 20–50 g/L and does not depend on surfactant-to-nanotube mass ratio in the 5:1–100:1 range. For suspensions at higher NaDDBS concentrations (regardless of the actual MWNT concentration), the effective conductivities actually decreased due to compromised colloidal stability. This decrease in thermal conductivity can be interpreted as being due to a reduction in the volume fraction of individually dispersed, high-aspect ratio particles that are primarily responsible for the enhanced thermal conductivity of the nanofluid. For this particular MWNT–surfactant system, the optimum surfactant-to-particle mass ratio was found to be 10:1, which yielded the largest thermal conductivity enhancement and good stability in a wide range of MWNT concentrations.

The effect of colloidal stability on the thermal conductivity of MWNTs dispersions was demonstrated further [38] by destabilizing previously stable nanofluids by addition of ethanol, which strips surfactant molecules from the nanotube surface [53]. The volume ratios of nanofluid to added ethanol were 1:1 (E1), 2:1 (E2), and 3:1 (E3). Optical absorbance measurements showed that the addition of ethanol caused MWNT aggregation and sedimentation [38]. For nanofluids of a given (final) MWNT concentration, the effective thermal conductivity enhancement dropped significantly when the suspensions were destabilized with ethanol as seen in Figure 7.10. The thermal conductivity drop occurred immediately upon ethanol addition, suggesting that the initial agglomeration, and not the slow sedimentation, is responsible for the decrease in conductivity. Thus, individually dispersed, high-aspect ratio particles are primarily responsible for the enhanced effective thermal conductivity of these MWNT-based nanofluids.

It should be noted that although aggregation of highly anisotropic nanoparticles can reduce the effective thermal conductivity of the suspension, particle aggregation can in some cases increase the conductivity. This has been simulated and experimentally demonstrated for suspensions of spherical particles induced to aggregate and form elongated chains in the field direction by dipolar particle interactions under an external magnetic field. Simulations by Heine et al. [54] predict that, for large ratios of particle conductivity to fluid conductivity, the effective thermal conductivity of a suspension roughly doubles as particle chains form under an external field. This was experimentally demonstrated by our group [55] in suspensions of micron-sized, spherical iron particles. It was found in these suspensions that the effective thermal conductivity increased in the field direction over that of the unchained suspension by

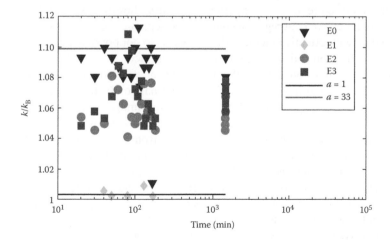

FIGURE 7.10 Thermal conductivities of aqueous suspensions of MWNTs for varying amount of added ethanol. Ethanol strips surfactant from the MWNTs and causes nanotube aggregation and sedimentation over time.

up to a factor of 2, depending on the particle concentration. A factor of 1.25 increase in thermal conductivity was recently found by ref. [56] in a similar magnetorheological (MR) suspension.

To model the effective thermal conductivity of the chained-particle suspensions, we have developed a two-level homogenization model: (1) the effective conductivity within the particle chains is first calculated using a Bruggeman-type model which takes into account thermal interactions between particles and is suitable for high local volume fractions, and (2) then, a Fricke/Nan-type EMT for aligned chains is used to model the overall thermal conductivity of the microstructured suspension. The model, which has only one adjustable parameter (the packing fraction of particles within the chains) has been successfully used by two different groups [55,56] to fit measured thermal conductivities for MR fluids under external fields, using the same value of the adjustable parameter. To summarize, in marked contrast to the aggregated MWNT suspension, induced particle aggregation of spherical particles in this case causes an increase in effective conductivity of suspensions due to an increase in the resultant aspect ratio of the inclusions.

Finally, as discussed in Section 7.2.1.2, effective medium theory also predicts that the alignment of highly conductive, highly anisotropic nanoparticles can increase the thermal conductivity of the suspension in the alignment direction. Suggestive experimental evidence for this increased thermal conductivity is seen in the experiments of Wright et al. on aqueous suspensions of Ni-coated SWNTs [57]. For 0.01 and 0.02 wt% particle loadings, the thermal conductivities in the field direction increased by up to 75% when a magnetic field was applied. Under a continuous magnetic field, the effective thermal conductivities first increased then decreased over time, with the time required to reach the peak thermal conductivity depending inversely on the strength of the magnetic field. This increase in thermal conductivity was hypothesized by Wright et al. to depend on the Ni-coated nanotubes forming aligned chains under the applied magnetic field. These thermal conductivity measurements required relatively long durations (2 min), however, so that the effects of particle alignment and aggregation could not be separated. Quantitative comparison of these data with effective medium theory was also not possible because the nanotube volume fraction and aspect ratio were not reported.

In order to more rigorously test the predictions of effective medium theory on the effect of particle alignment on the effective thermal conductivity, we have investigated magnetically responsive MWNTs suspensions under external magnetic fields. The volume-weighted particle aspect ratio was determined by TEM imaging to be 33, and the particle volume fraction was varied from 0.06% to 0.08%. The particles were made magnetically response by functionalizing them with ferromagnetic nanoparticles of characteristic size 10 nm adapting a procedure originally developed by Korneva et al. [58]. Commercially sourced ferrofluid (Ferrotec EMG 911) containing ferromagnetic nanoparticles in a kerosene carrier liquid was deposited on dry nanotubes and allowed to fill and coat the nanotubes. After drying, the magnetically responsive nanotubes were resuspended in ethanol and filtered twice to remove excess magnetite. Finally, aqueous suspensions of washed, ferromagnetic-nanoparticle-functionalized nanotubes were prepared at the desired volume fractions using NaDDBS as surfactant.

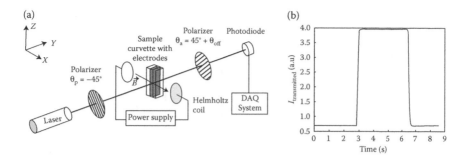

FIGURE 7.11 (a) Schematic of laser polarimetry apparatus, and (b) transmission through the suspension when a magnetic field is applied, indicating alignment of the ferrofluid-functionalized nanotubes.

To confirm that the suspended nanotubes were in fact aligning under an external magnetic field, laser polarimetry was applied to the samples. Transmission of linearly polarized light is affected by the state of alignment of the nanotubes, since large aspect ratio particles absorb anisotropically [59,60]. As shown in Figure 7.11, linear polarized light was transmitted through a sample that was placed between to nearly crossed polarizers. When a magnetic field of 400 Gauss was applied to induce nanotube alignment at 45° to the analyzer polarizer, transmission through the optical train to the photodetector increased, indicating the orientation of MWNTs by the magnetic field. The precise fraction of nanotubes that are aligning is not known, although Korneva et al.'s work with wet filling of larger-diameter MWNTs showed yields of close to 100% [58].

Thermal conductivities of the magnetically-responsive nanotube suspensions were measured with a 3ω method [61] in which a 25-mm length of tantalum wire acting as both a heater and temperature sensor is immersed into a 2-mL sample of fluid. The third harmonic of the electrical heating voltage is extracted using a lock-in amplifier, and the effective thermal conductivity is found from measurements at two frequencies ω_1 and ω_2 as

$$k = C \ln\left(\frac{\omega_2}{\omega_1}\right) \frac{1}{V_{\omega 1} - V_{3\omega 2}} \tag{7.27}$$

where C is a constant that can be found from calibration. It was verified in tests with pure water that the custom-built 3ω apparatus was insensitive to external magnetic fields, with an error of only 1% at an external field of 375 Gauss.

The thermal conductivity of the ferromagnetic nanotube suspensions was measured with a 295-Gauss magnetic field oriented along the 3ω-wire axis, that is, so that the measured conductivity is that transverse to the particle orientation. As shown in Figure 7.12, the transverse conductivity falls significantly below the conductivity of the suspension with random particle orientation. This decrease in conductivity with the magnetic field is believed to be primarily due to alignment rather than chaining, because the entire measurement (and the duration of the magnetic

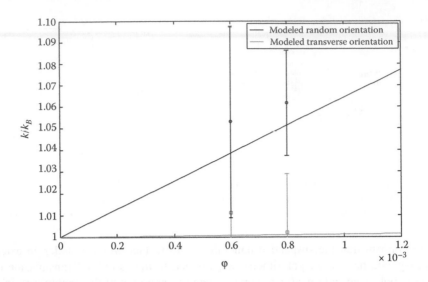

FIGURE 7.12 Measured changes in effective thermal conductivity of ferromagnetic-carbon-nanotube suspensions with and without field-induced alignment. Comparison lines are based on model of Nan [34].

field) is only 7 s, which limits time for particle aggregation at these low volume fractions. Furthermore, the experimentally measured conductivities, with and without particle alignment, are consistent with the predictions of Nan's effective medium theory at volume fractions of both 0.06% and 0.08%. The changes in conductivity with field-induced particle alignment are modest in this case because of the relatively low aspect ratio (33) of the MWNTs used. Effective medium theory predicts that changes of a factor of three are possible for particles of aspect ratio >10^2 (Figure 7.2). We note that our axial 3ω-wire geometry prevents us from directly measuring the conductivity along the direction of particle alignment because of the difficulty in producing a perfectly radial magnetic field. This axial conductivity (in the direction of particle alignment) is the subject of ongoing research. Nonetheless, it is clear that the thermal conductivity of suspensions of anisotropic particles can be actively controlled by inducing particle alignment with external fields.

7.3 EFFECTIVE VISCOSITY

We next turn our attention to the effective viscosity for dilute suspensions of highly anisotropic nanoparticles. For suspensions of noninteracting spherical particles, Einstein found the effective viscosity to be

$$\mu = \mu_B \left(1 + 2.5\varphi\right) \tag{7.28}$$

where μ_B is the viscosity of the fluid and φ is the particle volume fraction [62]. For suspensions of anisotropic particles, the rheological behavior is considerably more complicated and cannot be described by a single viscosity, but is in general non-Newtonian and flow dependent. Even in simple shear flow, $u = \dot{\gamma}\, y$, for instance, a dilute suspensions of large-aspect ratio particles can have a shear-rate-dependent effective viscosity relating shear stress, σ_{xy}, with shear rate, $\dot{\gamma}$,

$$\sigma_{xy} = \mu(\dot{\gamma})\dot{\gamma}. \tag{7.29}$$

Moreover, in such a flow, the suspension of anisotropic particles can generate normal stress differences $\sigma_{xx} - \sigma_{yy}$ and $\sigma_{zz} - \sigma_{yy}$ that also depend on shear rate. Such a suspension tested in a conventional cone-and-plate viscometer would generate both normal and shear forces on the upper and lower plates. However, the normal stress differences are typically an order of magnitude smaller than the shear stress [63], and we focus on the effective shear viscosity in the following discussion.

7.3.1 Models

In a simple shear flow in the absence of Brownian motion or an external electric field, small, anisotropic particles will rotate in periodic Jeffery orbits [64]. The angular velocity is nonuniform, however, and a large-aspect-ratio particle with negligible inertia will spend a majority of its time in the flow direction. Thus, in a suspension of anisotropic particles absent Brownian motion, the orientation distribution is not uniform but strongly peaked in the flow direction. Brownian motion acts to randomize the orientations of particles against the aligning effect of flow. The relative magnitude of these two effects is parameterized by the rotary Peclet number

$$\mathrm{Pe}_r \equiv \frac{\dot{\gamma}}{D_r} \tag{7.30}$$

where the rotary diffusivity, D_r, for rigid, large-aspect-ratio particles can be approximated as [65]

$$D_r \cong \frac{3k_B T}{32\pi\mu_B a^3}\left[2\ln(2r) - 1\right] \tag{7.31}$$

where a is the semimajor axis of the particle and $r = a/b$ is the aspect ratio.

At small Pe_r, Brownian motion dominates and the particle orientation distribution is nearly uniform. At large Pe_r, shear flow causes the particle to rotate in rotate but develop a preferential orientation distribution in the flow direction. Even in the limit of very large Pe_r Brownian motion still has an important effect, acting to render the steady-state particle orientation independent of the initial conditions. Hinch and Leal discuss solutions for the steady-state distribution of particle orientations for various limiting cases of Pe_r [63].

Assuming that (1) the particles are freely suspended and free of external forces, (2) the suspension is sufficiently dilute so that hydrodynamic interactions between the particles can be neglected, and that (3) a separation of length scales exists between

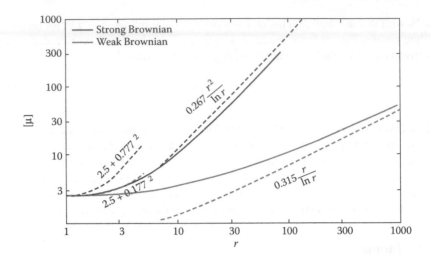

FIGURE 7.13 Intrinsic viscosity of suspensions of spheroidal particles in simple shear flow as a function of particle aspect ratio. Upper (black) line shows the low-shear (small Pe_r) limit while the lower (gray) line shows the high-shear (large Pe_r) limit. Dashed lines indicate asymptotic expressions in the limits of nearly spherical and large-aspect ratio particles. (Adapted from E. J. Hinch and L. G. Leal, *J. Fluid Mech.*, 52, 683, 1972.)

the average spacing of particles and the smallest relevant length scale for the bulk flow, the effective bulk stresses in the suspension of anisotropic particles have been calculated in the high, intermediate, and low Pe_r limits by [63]. At low Pe_r, where Brownian motion is important, the shear stress increases linearly with shear rate, $\dot{\gamma}$, that is, there is a constant shear viscosity. The effective viscosity calculated by ref. [63] in this case is illustrated in Figure 7.13 in terms of the intrinsic viscosity, $[\mu] \equiv 1/\varphi \lim_{\varphi \to 0}(\mu/\mu_B - 1)$. As seen from the figure, in the low Pe_r limit, the particle contribution to the shear stress is approximately $[\mu] \cong 2.5 + 0.777e^2$ for nearly spherical particles ($e \ll 1$), and $[\mu] \cong 0.267r^2/\ln r$ for very long particles ($r \gg 1$). In the strong shear-flow limit (high Pe_r), the shear viscosity is again constant and the intrinsic viscosity goes as $[\mu] \cong 2.5 + 0.177e^2$ for nearly spherical particles and $[\mu] \cong 0.315r/\ln r$ for very long particles. Two conclusions can be drawn about the effective shear viscosity of suspensions of noninteracting anisotropic particles in the absence of external forces: (1) At both high and low shear rates, particle anisotropy causes a greatly enhanced effective viscosity in the suspension compared with the spherical-particle case, where $[\mu] = 2.5$. For particles of aspect ratio $r = 100$ for instance, the intrinsic viscosity is between 230 (in the low-shear case) and 6.8 (in the high-shear case) times higher than for a spherical-particle suspension of the same volume fraction. (2) Moreover, suspensions of large-aspect particles are also very shear thinning, with $[\mu]$ decreasing by a factor of $0.85r$ from the low-shear to high-shear limit.

In the case of shear flow combined with electric fields, both particle orientation and the effective viscosity are affected by the external field. Mason and coworkers have calculated the motion and particle orientation of spheroidal polarizable particles under an electric field, in the absence of Brownian motion [66,67]. By considering the relative magnitudes of electrostatic and shear-flow torques acting on the particle, they find that the particle orientation, neglecting particle interactions (the dilute limit), depends on $E^2/\dot{\gamma}$. Below a critical electric-field strength, prolate spheroidal particles are predicted to rotate in a modified Jeffery orbit, while, for sufficiently large field strengths, the electrostatic and hydrodynamic torques on each particle balance and the particles should reach an equilibrium orientation angle

$$\varphi_\infty = \tan^{-1}\{r[f - (f^2 - 1)^{1/2}]\} \tag{7.32}$$

where f is a dimensionless measure of the relative electrostatic-to-hydrodynamic torques acting on the isolated particle at a particular orientation:

$$f \equiv \frac{-\varepsilon_1 P(r)E^2(r^2 + 1)}{\mu_B r \dot{\gamma}}. \tag{7.33}$$

In the limit of a prolate spheroidal conducting particle such as a SWNT immersed in a dielectric fluid, the function $P(r)$ reduces to [66,67]

$$P(r) = \frac{(3A - 2)Q(r)}{8\pi A(A - 1)} \tag{7.34}$$

where

$$Q(r) = \frac{2r^2 + (1 - 2r^2)A(r)}{4(r^2 + 1)} \tag{7.35}$$

and

$$A(r) = \frac{r^2}{r^2 - 1} - \frac{r\cosh^{-1} r}{(r^2 - 1)^{3/2}}. \tag{7.36}$$

The parameter f, which governs the particle orientation behavior, is a dimensionless measure of the relative electrostatic-to-hydrodynamic torques acting on the isolated particle in a fixed orientation. Above a critical electric-field strength corresponding to $f = 1$, particles are predicted to stop rotating and reach an equilibrium orientation angle that is only a function of the parameter f (which is in turn proportional to $E^2/\dot{\gamma}$) for all shear rates and field strengths. This will be experimentally tested in SWNT suspensions, as discussed later.

The effective viscosity of a suspension of anisotropic, polarizable particles can also be strongly affected by electric fields, that is, such suspensions are electrorheological. For suspensions of very large-aspect ratio particles, the simple change in preferential particle orientation from the flow direction (at high Pe_r) to the field direction can be expected to lead to changes in effective viscosity. In the limit of a strong

($f \gg 1$) electric field turned on in a direction perpendicular to the flow at large Pe_r, the change in viscosity due to particle-orientation-distribution changes alone will exceed the difference between the low and high Pe_r lines shown in Figure 7.13. Thus, particle alignment with the electric field can cause electrorheological behavior.

Moreover, field-induced particle chaining over longer times can also lead to an increase in the effective shear viscosity, as well as a nonzero static yield stress. By analogy to the behavior of conventional electrorheological (ER) suspensions of spherical particles, the anisotropic-particle aggregation kinetics and electrorheological behavior are presumably be governed by a dimensionless ratio of the hydrodynamic to electrostatic dipole–dipole interaction energies [68,69]. This parameter, the Mason number, is scales as $\dot{\gamma}/E^2$, and so is proportional to the inverse of f. Thus, if the so-called polarization model of ER fluids applies to suspensions of highly anisotropic nanowires, one would expect the effective shear viscosity μ/μ_B to collapse with $\dot{\gamma}/E^2$ for different shear rates and field strengths.

In the following, we discuss our recent experiments on the equilibrium particle orientation and electrorheological behavior of dilute suspensions of SWNTs in steady shear flow under an external electric field.

7.3.2   CARBON NANOTUBE SUSPENSIONS UNDER ELECTRIC FIELDS

Dilute suspensions of SWNTs were studied in simple shear flow under an electric field that was oriented perpendicular to the flow direction [70,71]. To prepare the suspensions, SWNTs were dispersed in a dielectric fluid, α-terpineol, which has been reported to be a good solvent for CNTs [72]. Samples at four different volume fractions of SWNTs, $\varphi = 1.5 \times 10^{-6}, 3.7 \times 10^{-6}, 1.5 \times 10^{-5}$, and 2.2×10^{-5}, were prepared at room temperature with the assistance of bath sonication. Visible/near-infrared absorbance measurements and analysis with a Brookhaven disk-centrifuge particle sizer confirmed the presence of individual SNWTs in suspension, and showed that the particles/particle aggregates in suspension were small, with 90% of particles present in the suspension having equivalent hydrodynamic diameters below 160 nm. One sample was also centrifuged after preparation to remove particle aggregates and increase the fraction of individualized (rather than aggregated) SWNTs in suspension.

Simultaneous measurements of the effective shear viscosity and the particle orientation angles were made with a modified concentric-cylinder viscometer, as shown in Figure 7.14. Shear flow was generated between a rotating inner cylinder and a stationary outer one, while a radially oriented electric field was produced by applying a potential difference between the cylinders. The tested shear rates ranged between 1.1 and $11/s^1$, while a 4 kHz AC electric field was varied in strength up to $270\ V_{rms}/mm$. For this range of parameters, Brownian motion is not expected to be dominant since the calculated values of Pe_r are in the range of 2–18, and, furthermore, the electric field is very strong, that is, $f \gg 1$. A laser-polarimetry system, based on the optical rheometry techniques developed by Fuller [73] was used to measure the ensemble-averaged orientation of the individually suspended and small bundles of SWNTs. As seen in Figure 7.12, the rotating half wave plate is used to rapidly modulate the polarization direction of light incident upon the SWNT suspension. Because the nanotubes absorb anisotropically, preferential SWNT

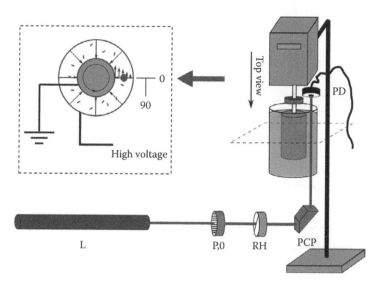

FIGURE 7.14 Schematic of the customized Brookfield concentric-cylinder viscometer used to simultaneously measure apparent viscosity and ensemble-averaged particle-orientation angles. L, HeNe laser; P,0, polarizer; RH, rotating half-wave plate, PCP, polarization-conserving prism; PD, photodetector.

alignment changes the intensity of the linearly polarized light that is transmitted through the sample and detected by a photodetector. The ensemble-averaged orientation angle of SWNTs in the path of the laser can then be extracted from the demodulated photodetector signal using a lock-in amplifier. It is important to note that the nonintrusive optical technique used to measure the ensemble-averaged particle orientation angle is sensitive to individual nanotubes and small bundles only; particles much larger than the wavelength, 632.8 nm, of the transmitted light do not cause linear dichroism and thus do not contribute to the measurement [74].

Representative time traces of the measured apparent viscosity in steady shear at a volume fraction of $\varphi = 1.5 \times 10^{-5}$ are shown in Figure 7.15a for a constant shear rate of 2.81 s^{-1} and various electric-field strengths [71]. Before the field is turned on, the suspension viscosity is indistinguishable from that of the base fluid because of the very low particle concentration. With an electric field, the shear viscosity increases, with the magnitude of the viscosity change increasing (and the time scale of the viscosity change decreasing) with field strength. Despite the diluteness of the suspension, the apparent viscosity nearly triples at the highest electric field for this shear rate. The response time for the viscosity change is very slow, on the order of 10^2 s, however. Although not shown, the suspension is shear thinning, with effective viscosity decreasing as the shear rate is increased at fixed electric field strength. As the electric field is increased beyond a critical strength (266 V_{rms}/mm at $\dot{\gamma} = 6.1/s$), the apparent viscosity can spike suddenly due to the growth of SWNT chains that span the entire gap between the inner and outer cylinders of the viscometer (Figure 7.15b), that is, field-induced percolation. As seen in Figure 7.15c, this secondary jump in apparent viscosity coincides with a jump in the current delivered by the amplifier

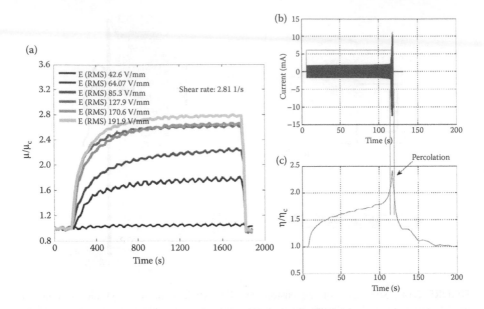

FIGURE 7.15 (a) Representative time traces of normalized apparent viscosity at a shear rate of 2.8 s⁻¹ and varying field strengths. (b) Spike in apparent viscosity at yet higher electric field strengths (280 V_{rms}/mm) when field-induced percolation occurs. (c) Percolating chains are formed, as shown by the simultaneous jump in current delivered by the amplifier.

to the suspension. Because we are limited in both the current that can be supplied by our amplifier and in the torque that the viscometer can measure, we are unable to completely characterize the rheology of the suspension in the case of percolating chains. However, the actual upper limit of the increase in apparent viscosity upon application of the electric field is believed to significantly exceed that shown in Figure 7.15a.

Even in the absence of percolating chains, the magnitude of the ER response for the SWNT suspension is surprisingly large in comparison to a conventional ER suspension of the same particle volume fraction. As shown in Figure 7.16, a suspension of spherical glassy-carbon particles (0.4 μm in diameter) of the same volume fraction as that of the SWNT suspensions shows no measurable ER response, despite the fact that the glassy carbon is conducting and polarizable at low frequencies like the nanotubes. In fact, the glassy-carbon suspension required a particle volume fraction that is three orders-of-magnitude higher than that of the SWNT suspension to achieve similar increases in apparent viscosity, under the same conditions (E_{rms} = 160 V/mm, shear rate of 6.12 s⁻¹). This is due to the high aspect ratio of the SWNTs, which leads to higher polarizability and stronger electrostatic interactions than a spherical particle of the same volume.

The classical polarization model of ER fluids predicts that the steady-state apparent viscosity should collapse with the ratio of viscous to dipole–dipole interaction forces. As shown in Figure 7.17, the measured effective viscosities do appear to collapse with the dimensionless Mn, or, in dimensional form, $\dot{\gamma}/E^2$, for a variety of

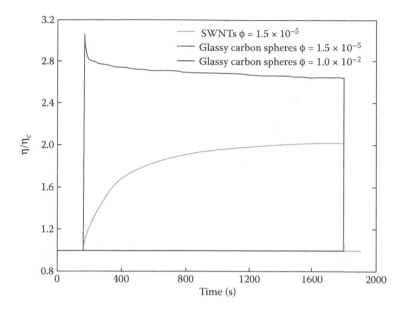

FIGURE 7.16 Comparison of the electrorheological response of SWNT and glass-carbon suspensions of different particle volume fractions [70]. The electric field and shear rate were kept constant (F_{rms} = 160 V/mm, shear rate of 6.12 s^{-1}).

shear rates and field strengths, at least at high Mn. Above a critical field strength of 133 V_{rms}/mm, the viscosities at each shear rate begin to plateau and deviate from one another, however. This is believed to be due to a field-induced conductivity change in which fluid electrical conductivity is no longer constant but increases with field strength beyond a critical value. The dipole–dipole interaction force in this case has been found to depend linearly [75] rather than quadratically on E. Evidence for this explanation for the plateauing viscosities is seen in the inset, which shows the collapse of the high-field data with $\dot{\gamma}/E$. Thus, we believe that the ER behavior of the SWNT/α-terpineol suspension can be interpreted in terms of a electrostatic polarization model, where the governing parameter is the ratio of viscous to dipole–dipole forces. Only the dependence of the electrostatic-interaction force on E changes in the high- and low-field limits.

As discussed previously, the observed increase in effective shear viscosity of the SWNT suspension under the electric field could arise from at least two mechanisms: (1) the orientation of the particles, and (2) the chaining of particles into macroscopic chains which can resist shear. To clarify the relative importance of these two mechanisms in dilute SWNT suspensions, the equilibrium orientation angle of the particles was measured simultaneously with the effective viscosity. Figure 7.18 shows the that the ensemble-averaged particle orientation in the shear flow without an electric field is in the flow direction (90° as defined in our coordinate system), consistent with the predicted preferential orientation at high Per. Upon application of the electric field at t = 160 s, the measured equilibrium orientation angles quickly deviate toward the radial direction, 0°. The alignment of nanotubes in the radial direction is

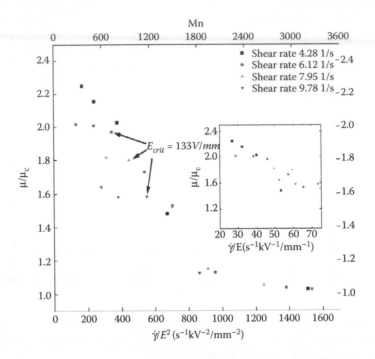

FIGURE 7.17 Collapse of the measured apparent viscosities at different shear rates and field strengths with the ratio of hydrodynamic to electrostatic particle interaction forces. Above critical field strength of 133 V_{rms}/mm, the viscosities at each shear rate begin to plateau and deviate from one another due to field-induced conductivity effects, as discussed in refs. [70,75].

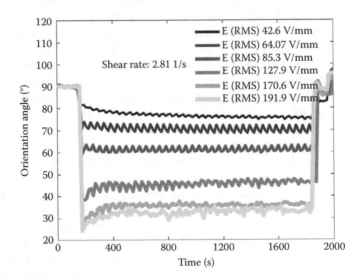

FIGURE 7.18 Equilibrium SWNT orientation angles measured simultaneously with the electrorheological data shown in Figure 7.15a. (Adapted from C. Lin and J. W. Shan. *Phys. Fluids*, 22, 022001, 2010.)

increased with an increase in field strength, reaching 33° at a shear rate of 2.81 s⁻¹ for E_{rms} = 191 V/mm. Once the electric field is turned off at t = 1900 s, the nanotube orientations return to the flow direction, 90°. Of particular note is the much shorter time scale (on the order of 1–10 s) of the particle alignment response, which is one to two orders of magnitude faster than the rheological response shown in Figure 7.15a. Thus, at these low particle concentrations, the alignment of the particles with the electric field has negligible effect on the apparent viscosity of the suspension, despite the very large aspect ratio of the nanotubes. Rather, consistent with the behavior of conventional ER fluids [68,76–78] the chaining of nanotubes under electrostatic particle interactions is the primary mechanism for the change in apparent viscosity under an external electric field. In this light, the slow ER time response observed in these experiments is due to the diluteness of the SWNT suspensions, which increases the mean spacing between particles, and thus the chaining time, in the suspension.

The measured equilibrium orientation angles for various shear rates and electric-field strengths are plotted in Figure 7.19 against the parameter $E^2/\dot{\gamma}$ predicted by Mason's theory to govern the orientation of ellipsoidal polarizable particles in simultaneous shear flow and electric fields, Equation 7.32. As seen in the figure, the measured orientation angles collapse well against $E^2/\dot{\gamma}$ for an order-of-magnitude variation in both electric field strengths (21.3 V_{rms}/mm < E < 383.7 V_{rms}/mm) and shear rates (1.12 s⁻¹ < γ < 11.22 s⁻¹). The collapse of the data indicates that the $E^2/\dot{\gamma}$ dependence for the equilibrium orientation angle predicted by Mason's theory is essentially correct.

However, substantial quantitative differences are seen between the predictions of Equation 7.32 and the measured SWNT orientation angles. The measured angles fall below the theory and are inclined toward the electric-field direction by as much as

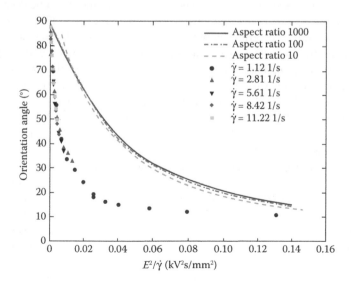

FIGURE 7.19 Measured equilibrium SWNT orientation angles as a function of E2/gamma_dot. Predicted angles (Equation 7.32) are shown for a wide range of particle aspect ratios.

$40°$. This implies a higher aligning torque, either electrostatic or hydrodynamic in the electric-field direction than predicted by classical theory. Thus, while the collapse of the data for various shear rates and electric fields indicates that the particle orientation is indeed governed by the ratio between electric field and shear-flow torques, as parameterized by $E^2/\dot{\gamma}$, or, equivalently, f, Equation 7.32 does not completely predict SWNT orientation in these suspensions. As discussed in more detail in ref. [71], the reason for this discrepancy between theory and experimental results is believed to be due to long-range electrostatic interactions between particles, which are neglected in the classical model. Although the particle concentrations are low, the extremely large aspect ratios of the SWNTs lead to large-induced dipole moments, which in turn causes nonnegligible electrostatic interactions between neighboring particles that tend to align further align particles in the field direction. Experiments with lower particle concentrations (hence a greater average spacing between particles and weaker particle–particle interactions) showed better agreement in the orientation angle with theory [71]. However, at the lowest detectable SWNT concentration ($\phi = 1.5 \times 10^{-6} = 1.5$), substantial differences persist between the data and the model. Thus, even at very dilute volume fractions, the large-aspect ratio particles renders electrostatic particle interactions important under an applied external electric field.

7.4 CONCLUSIONS

In summary, the addition of highly anisotropic particles can have profound effects on the effective thermophysical properties of fluids. For suspensions with randomly oriented particles, the effective thermal conductivities can be predicted by effective medium theories [33–35] which account for particle anisotropy and interfacial resistance, among other factors. Experiments with MWNTs of various aspect ratios and concentrations dispersed in aqueous suspension confirm the theoretical predictions for the dependence of thermal conductivity on those parameters [38]. Although not discussed here, similar careful comparisons have found good agreement between experiment and theory for the suspensions of micron-sized silicon carbide particles in organic solvents [79]. There is uncertainty, however, about the precise value of interfacial resistance for various nanoparticle/fluid systems, and further studies in this area are needed. The rheological behavior of flowing suspensions of anisotropic particles is complex and a single viscosity cannot be defined for such a flow, although, in simple shear flow, the shear stress that is generated is much larger than the normal stress. The effective shear viscosities for suspensions of anistotropic particles for various ranges of Pe_r have been analytically studied by Hinch and Leal, and show a strong dependence on particle aspect ratio, as well as shear-thinning behavior [63]. The latter is due to the dependence of the viscosity on the particle orientation distribution, which changes from a uniform distribution at low shear rates (small Pe_r) to a preferential distribution in the flow direction at high shear rates (large Pe_r).

For microstructured suspensions, which may have nonuniform particle-orientation distributions or particle clustering, the thermophysical properties are again different. This opens an avenue to the development of novel "smart" fluids having anisotropic, actively controllable properties. Experimental demonstrations of suspensions with such controllable thermal conductivities have been done with magnetorheological fluids [55] and magnetically responsive MWNT. In the former, the chaining of spherical particles increases the conductivity in the field direction, while, in the latter, the alignment of MWNTs is shown to decrease conductivity in the directions perpendicular to alignment. At very low particle volume fractions (10^{-5}), the presence of large-aspect ratio particles alone does not cause a significant change in apparent shear viscosity, regardless of the orientation distribution of the particles. However, particle chaining due to dipolar interactions between particles does increase the apparent shear viscosity of suspensions. In the dilute SWNT suspensions studied, both the effective shear viscosity and the particle orientation can be parameterized by the ratio of electrostatic to viscous flow forces [70,71]. A key difference of these SWNT suspensions from conventional electrorheological fluids is the extremely large aspect ratio of the particles, which enhances the polarizability of the particles. This results in both a larger change in viscosity as compared to suspensions of conventional particles, as well as long-range electrostatic interactions between particles which causes differences in particle orientation distributions from that predicted by classical theory. As is apparent from these examples, the control of the microstructure of suspensions of anisotropic nanoparticles offers a rich arena in which to explore basic science, as well as to engineer novel fluids with anisotropic, controllable thermophysical properties.

ACKNOWLEDGMENTS

This work was supported by the National Science Foundation (NER-ECS-0404181 and CAREER-CBET-0644719) and the New Jersey Space Grant Consortium (NJSGC-04–47).

NOMENCLATURE

a	Semimajor axis of spheroidal particle
b	Semiminor axis of spheroidal particle
T	Temperature
t	Time
k	Thermal conductivity of suspension
k_P	Particle thermal conductivity
k_B	Base-fluid thermal conductivity
R	Particle radius
D	Particle diameter
L	Particle length
e	Eccentricity
n	Shape factor
f	Dimensionless ratio of electrostatic to hydrodynamic torques acting on particle
$f^{(i)}$	Depolarization coefficient of a spheroid
r	Aspect ratio of a spheroid
R_K	Kapitza resistance
l_K	Kapitza length

$\overline{P}_{eff}$ Complex effective moment of particle
E Electric field magnitude
V Voltage

GREEK SYMBOLS

$\overline{\varepsilon}$ Complex permittivity
ε Permittivity
σ Electrical conductivity
ω Electrical frequency
φ Volume fraction of particles
δ Thermal conductivity ratio
α Interfacial resistance coefficient
μ Viscosity
$[\mu]$ Intrinsic viscosity
σ Stress
$\dot{\gamma}$ Shear rate

SUBSCRIPTS

P   Particle
b   Base fluid
V   Volume-weighted average
K   Kapitza
mean   Mean value
peak   Peak value
out   Outside
in   Inside

COORDINATES

x, y, z Cartesian coordinates
r, z, θ Spherical coordinates
ξ Ellipsoidal coordinate

REFERENCES

1. J. A. Eastman, S. U. S. Choi, S. Li, W. Yu, and L. J. Thompson, Anomalously increased effective thermal conductivities of ethylene glycol- based nanofluids containing copper nanoparticles, *Appl. Phys. Lett.*, 78, 718–720, 2001.
2. S. U. S. Choi, S. G., Zhang, W. Yu, and E. A. Grulke, Anomalous thermal conductivity enhancement in nanotube suspensions, *Appl. Phys. Lett.*, 79, 2252–2254, 2001.
3. H. Patel, S. Das, T. Sundararajan, N. A. Sreekumaran, B. George, and T. Pradeep, Thermal conductivities of naked and monolayer protected metal nanoparticle based nanofluids: Manifestation of anomalous enhancement and chemical effects, *Appl. Phys. Lett.*, 83(14), 2931–2933, 2003.

4. C.-W. Nan, Z. Shi, and Y. Lin, A simple model for thermal conductivity of carbon nanotube-based composites, *Chem. Phys Lett.*, 376, 666–669, 2003.
5. R. Prasher, W. Evans, P. Meakin, J. Fish, P. Phelan, and P. Keblinski, Effect of aggregation on thermal conduction in colloidal nanofluids, *Appl. Phys. Lett.*, 89, 143119, 2006.
6. J. Buongiorno, D. C. Venerus, N. Prabhat, T. McKrell, J. Townsend, R. Christianson, Y. V. Tolmachev et al., A benchmark study on the thermal conductivity of nanofluids, *J. Appl. Phys.*, 106, 094312, 2009.
7. J. Hu, T.W. Odom, and C.M. Lieber, Chemistry and physics in one dimension: Synthesis and properties of nanowires and nanotubes, *Acc. Chem. Res.*, 32(5), 435–445, 1999.
8. Y. Kasumov, R. Deblock, M. Kociak, B. Reulet, H. Bouchiat, I. I. Khodos, Y. B. Gorbatov, V. T. Volkov, C. Journet, and M. Burghard, Supercurrents through single-walled carbon nanotubes, *Science*, 284(5419), 1508, 1999.
9. P. G. Collins and P. Avouris, Nanotubes for electronics. *Sci. Am.*, 283(6), 62–69, 2000.
10. Y. Cui and C.M. Lieber, Functional nanoscale electronic devices assembled using silicon nanowire building blocks, *Science*, 291(5505), 851, 2001.
11. X. Duan, Y. Huang, Y. Cui, J. Wang, and C. M. Lieber, Indium phosphide nanowires as building blocks for nanoscale electronic and optoelectronic devices, *Nature*, 409(6816), 66–69, 2001.
12. M. H. Huang, S. Mao, H. Feick, H. Yan, Y. Wu, H. Kind, E. Weber, R. Russo, and P. Yang, Room-temperature ultraviolet nanowire nanolasers, *Science*, 292(5523), 1897, 2001.
13. P. Jarillo-Herrero, J. A. Van Dam, and L. P. Kouwenhoven, Quantum supercurrent transistors in carbon nanotubes, *Nature*, 439(7079), 953–956, 2006.
14. J.P. Lu. Elastic properties of carbon nanotubes and nanoropes, *Phys. Rev. Lett.*, 79(7), 1297–1300, 1997.
15. Krishnan, E. Dujardin, T. W. Ebbesen, P. N. Yianilos, and M. M. J. Treacy, Youngs modulus of single-walled nanotubes. *Phys. Rev. B*, 58(20), 14013–14019, 1998.
16. J. P. Salvetat, G. A. D. Briggs, J. M. Bonard, R. R. Bacsa, A. J. Kulik, T. Stockli, N. A. Burnham, and L. Forro, Elastic and shear moduli of single-walled carbon nanotube ropes, *Phys. Rev. Lett.*, 82(5), 944–947, 1999.
17. J. Y. Huang, S. Chen, Z. Q.Wang, K. Kempa, Y. M. Wang, S. H. Jo, G. Chen, M. S. Dresselhaus, and Z. F. Ren, Superplastic carbon nanotubes, *Nature*, 439(7074), 281, 2006.
18. S. Berber, Y. K. Kwon, and D. Tomanek, Unusually high thermal conductivity of carbon nanotubes, *Phys. Rev. Lett.*, 84, 4613, 2000.
19. P. Kim, L. Shi, A. Majumdar, and P. L. McEuen, Thermal transport measurements of individual multiwalled nanotubes, *Phys. Rev. Lett.*, 87, 215502, 2001.
20. T. Ouyang, Y. Chen, Y. Xie, K. Yang, Z. Bao, and J. Zhong, Thermal transport in hexagonal boron nitride nanoribbons, *Nanotechnology* 21(2010), 245701.
21. D. Li, Y. Wu, P. Kim, L. Shi, P. Yang, and A. Majumdar, Thermal conductivity of individual silicon nanowires, *Appl. Phys. Lett.*, 83, 2934, 2003.
22. T. J. Jones. *Electromechanics of Particles*. Cambridge Univ. Press, Cambridge, 1995.
23. L. D. Landau, L. P. Pitaevskii, and E. M. Lifshitz, *Electrodynamics of Continuous Media, Second Edition: Volume 8 (Course of Theoretical Physics)*, Butterworth-Heinemann, Oxford1984.
24. J. C. Maxwell, *A Treatise on Electricity and Magnetism*, Dover Publications, New York, 1954.
25. J. C. M. Garnett, Colours in metal glasses and metal films, *Trans. Royal Soc.*, CCIII, 385–420, 1904.
26. H. Fricke, A mathematical treatment of the electric conductivity and capacity of disperse systems, *Phys. Rev.*, 24(5), 575–587, 1924.

27. P. L. Kapitza, Heat transfer and superfluidity of helium II, *Phys. Rev.*, 60, 354–355, 1941.
28. S. Huxtable, D. Cahill, S. Shenogin, L. Xue, R. Ozisik, P. Barone, M. Usrey et al., Interfacial heat flow in carbon nanotube suspensions, *Nature Mater.*, 2, 731–734, 2003.
29. S. Shenogin, L. Xue, R. Ozisik, P. Keblinski, and D. G. Cahill, Role of thermal boundary resistance on the heat flow in carbon-nanotube composites, *J. Appl. Phys.*, 95(12), 8136–8144, 2004.
30. C. H. Henager and W. Pawlewicz, Thermal conductivities of thin, sputtered optical films, *Appl. Opt.*, 32, 91–101, 1993.
31. A. Plech, V. Kotaidis, S. Gresillon, C. Dahmen, and G. von Plessen, Laser-induced heating and melting of gold nanoparticles studied by timeresolved x-ray scattering, *Phys. Rev. B*, 70, 195423-1–195423-7, 2004.
32. M. W. Wilson, X. Hu, D. G Cahill, and P. V. Braun, Colloidal metal particles as probes of nanoscale thermal transport in fluids, *Phys. Rev. B*, 66(22), 224301, 2002.
33. C.-W. Nan, R. Birringer, D. R. Clarke, and H. Gleiter, Effective thermal conductivity of particulate composites with interfacial thermal resistance, *J. Appl. Phys.*, 81(10), 6692–6699, 1997.
34. C.-W. Nan, G. Liu, Y. Lin, and M. Li, Interface effect on thermal conductivity of carbon nanotube composites, *Appl. Phys. Lett.*, 85(16), 3549–3551, 2004.
35. H. L. Duan, B. L. Karihaloo, J. Wang, and X. Yi, Effective conductivities of heterogeneous media containing multiple inclusions with various spatial distributions, *Phys. Rev. B*, 73, 174203, 2006.
36. H. L. Duan, and B. L. Karihaloo, Effective thermal conductivities of heterogeneous media containing multiple imperfectly bonded inclusions, *Phys. Rev. B*, 75, 064206, 2007.
37. C. H. Li, W. Williams, J. Buongiorno, L.-W. Hu, and G. P. Peterson, Transient and steady-state experimental comparison study of effective thermal conductivity of Al_2O_3/water nanofluids, *J. Heat Transf.*, 130(4), 042407, 2008.
38. A. S. Cherkasova and J. W. Shan, Particle aspect-ratio and agglomeration-state effects on the effective thermal conductivity of aqueous suspensions of multiwalled carbon nanotubes, *J Heat Transf.*, 132, 082402, 2010.
39. M. F. Islam, E. Rojas, D. M. Bergey, A. T. Johnson, and A. G. Yodh, High weight fraction surfactant solubilization of single-wall carbon nanotubes in water, *Nano Lett.*, 3(2), 269–273, 2003.
40. P. Kim, L. Shi, A. Majumdar, and P. L. Mceuen, Thermal transport measurements of individual multiwalled nanotubes, *Phys. Rev. Lett.*, 87(21), 215502, 2001.
41. J. P. Small, L. Shi, and P. Kim, Mesoscopic thermal and thermoelectric measurements of individual carbon nanotubes, *Solid State Commun.*, 127, 181–186, 2003.
42. R. Prasher, Thermal boundary resistance and thermal conductivity of multiwalled carbon nanotubes, *Phys. Rev. B*, 77, 075424, 2008.
43. T.-Y. Choi, D. Poulikakos, J. Tharian, and U. Sennhauser, Measurement of the thermal conductivity of individual carbon nanotubes by the four-point three-ω method, *Nano Lett.*, 6(8), 1589–1593, 2006.
44. Q. Li, C. Liu, X. Wang, and S. Fan, Measuring the thermal conductivity of individual carbon nanotubes by the Raman shift method, *Nanotechnology*, 20, 145702, 2009.
45. M. Fujii, X. Zhang, and K. Takahashi, Measurements of thermal conductivity of individual carbon nanotubes, *Phys. Status Solidi B*, 243(13), 3385–3389, 2006.
46. T.-Y.Choi, D. Poulikakos, J. Tharian, and U. Sennhauser, Measurement of thermal conductivity of individual multiwalled carbon nanotubes by the 3-method, *Appl. Phys. Lett.*, 87, 013108, 2005.
47. H. Shioya, T. Iwai, D. Kondo, M. Nihei, and Y. Awano, Evaluation of thermal conductivity of a multi-walled carbon nanotube using the delta V-Gs method, *Jpn. J. Appl. Phys., Part 1*, 46(5A), 3139–3143, 2007.

48. D. J. Yang, S. G. Wang, Q. Zhang, P. J. Sellin, and G. Chen, Thermal and electrical transport in multi-walled carbon nanotubes, *Phys. Lett. A*, 329(3), 207–213, 2004.
49. Y. Yang, E. A. Grulke, Z. G. Zhang, and G. Wu, Thermal and rheological properties of carbon nanotube-in-oil dispersions, *J. Appl. Phys.*, 99(11), 114307, 2006.
50. S. Wang, R. Liang, B. Wan, and C. Zhang, dispersion and thermal conductivity of carbon nanotube composites, *Carbon*, 47, 53–57, 2009.
51. L. Xue, P. Keblinski, S. R. Phillpot, S. U. S. Choi, and J. Eastman, Two regimes of thermal resistance at a liquid-solid interface, *J. Chem. Phys.*, 118(1), 337–339, 2003.
52. M. Zhang, M. Yudasaka, A. Koshio, C. Jabs, T. Ichihashi, and S. Iijima, S., Structure of single wall carbon nanotubes purified and cut using polymer, *Appl. Phys. A: Mater. Sci. Process.*, 74, 7–10, 2002.
53. Z.-B. Zhang, J. Cardenas, E. E. B. Campbell, and S.-L. Zhang, Reversible surface functionalization of carbon nanotubes for fabrication of field-effect transistors, *Appl. Phys. Lett.*, 87, 043110, 2005.
54. M. C. Heine, J. d. Vicente, and D. J. Klingenberg, Thermal transport in sheared electro- and magnetorheological fluids, *Phys. Fluids*, 18, 023301, 2006.
55. B. N. Reinecke, J. W. Shan, K. K. Suabedissen, and A. S. Cherkasova. On the anisoptric thermal conductivity of magnetorheological suspensions. *J. App. Phys.*, 104, 023507, 2008.
56. G. Cha, Y. S. Ju, L. A. Ahuré, and N. M. Wereley, Experimental characterization of thermal conductance switching in magnetorheological fluids, *J. Appl. Phys.* 107, 09B505, 2010.
57. B. Wright, D. Thomas, H. Hong, L. Groven, J. Puszynski, E. Duke, X. Ye, and S. Jin, Magnetic field enhanced thermal conductivity in heat transfer nanofluids containing Ni coated single wall carbon nanotubes, *Appl. Phys. Lett.*, 91, 173116, 2007.
58. G. Korneva, H. Ye, Y. Gogotsi, D. Halverson, G. Friedman, J.-C. Bradley, and K. G. Kornev, Carbon nanotubes loaded with magnetic particles, *Nano Lett.*, 5, 879–884, 2005.
59. K. Bubke, H. Gnewuch, M. Hempstead, J. Hammer, and M. L. H. Green, Optical anisotropy of dispersed carbon nanotubes induced by an electric field, *Appl. Phys. Lett.*, 71, 1906, 1997.
60. M. Brown, J. W. Shan, C. Lin, and F. M. Zimmermann, Electrical polarizability of carbon nanotubes in liquid suspension, *Appl. Phys. Lett.*, 90, 203108, 2007.
61. D. G. Cahill, Thermal conductivity measurement from 30 to 750 K: The 3ω Method, *Rev. Sci. Instr.*, 61.2(1989), 802–808, 1990.
62. A. Einstein, Eine neue bestimmung der molekuldimensionen. *Ann. Physik.*, 19, 289–306, 1906.
63. E. J. Hinch and L. G. Leal, The effect of Brownian motion on the rheological properties of a suspension of non-spherical particles, *J. Fluid Mech.*, 52, 683, 1972.
64. G. B. Jeffery, The motion of ellipsoidal particles immersed in a viscous fluid, *Proc. R. Soc. London, Ser. A*, 102, 161, 1922.
65. M. Doi and S. F. Edwards, *The Theory of Polymer Dynamics*, Oxford University Press, New York, 1986.
66. A. Okagawa, R. G. Cox, and S. G. Mason, Particle behavior in shear and electric fields. VI. The microrheology of rigid spheroids, *J. Colloid Interface Sci.*, 47, 536, 1974.
67. A. Okagawa and S. G. Mason, Particle behavior in shear and electric fields. VII. Orientation distributions of cylinders, *J. Colloid Interface Sci.*, 47, 568, 1974.
68. M. Parthasarathy and D. J. Klingenberg, Electrorheology: Mechanisms and models, *Mater. Sci. Eng.*, R. 17, 57, 1996.
69. R. G. Larson, *The Structure and Rheology of Complex Fluids*, Oxford University Press, New York, 1999.

70. C. Lin and J. W. Shan. Electrically tunable viscosity of dilute suspensions of carbon nanotubes, *Phys. Fluids*, 19, 121702, 2007.
71. C. Lin and J. W. Shan. Ensemble-averaged particle orientation and shear viscosity of single-wall-carbon-nanotube suspensions under shear and electric fields. *Phys. Fluids*, 22, 022001, 2010.
72. B. Kim, Y. H. Lee, J. H. Ryu, and K. D. Suh, Enhanced colloidal properties of single-wall carbon nanotubes in α-terpineol and texanol, *Colloids Surf.*, A 273, 161, 2006.
73. G. G. Fuller and K. J. Mikkelsen, Optical rheometry using a rotary polarization modulator, *J. Rheol.*, 33, 761, 1989.
74. G. G. Fuller, Optical rheometry, *Annu. Rev. Fluid Mech.*, 22, 387, 1990.
75. N. Foulc, P. Atten, and N. Felici, Macroscopic model of interaction between particles in an electrorheological fluid, *J. Electrost.*, 33, 103, 1994.
76. W. M. Winslow, Induced fibration of suspensions, *J. Appl. Phys.*, 20, 1137, 1949.
77. H. Block and J. P. Kelly, Electrorheology, *J. Phys. D*, 21, 1661, 1988.
78. T. Hao, Electrorheological suspensions, *Adv. Colloid Interface Sci.*, 97, 2002.
79. A. S. Cherkasova and J. W. Shan, Particle aspect-ratio effects on the thermal conductivity of micro- and nanoparticle suspensions, *J. Heat Transf.*, 130, 082406, 2008.

8 Advances in Fluid Dynamic Modeling of Microfiltration Processes

John E. Wentz, Richard E. DeVor, and
Shiv G. Kapoor

CONTENTS

8.1 INTRODUCTION

Membrane filtration was developed as a technology that provides for the removal of fluid contaminants at flow rates that enable its continuous application while the fluid itself is in use. Membranes are sized for applications so that contaminant particles will not pass through the filter while, in principle, those of the fluid constituents will. Over the last 50 years membrane filtration has been developed and applied in the pharmaceutical, chemical, food, and agricultural industries [1–2]. More recently, the use of membrane filtration has been extended to environmental engineering applications as well as in wastewater treatment, pollution prevention, and industrial fluid recycling [1,3–6]. Both the contaminant particles and fluid components, like microemulsions, can become lodged in the membrane pore system, thereby fouling the membrane and greatly reducing the microfiltration system flow rate, referred to as the flux of the membrane.

Membrane filtration, and the fouling that interferes with its application, has seen significant experimental and modeling work in the past with several comprehensive studies describing the transport mechanics of microfiltration in a wide range of applications [1,2,7,8]. Membrane fouling due to the deposition in the pore structure of particulates, macroemulsions, and colloids from both contaminants and the fluid constituents like surfactants and microemulsions has been a major focus of much of this chapter [1,9]. This fouling results in flux decline and accordingly decreased productivity of the microfiltration system. Therefore, it is important to obtain a better understanding of the mechanisms of the fouling that occurs during microfiltration processes. Such will lead to process improvements in terms of the operating parameters of microfiltration systems as well as the creation of tools for the improved design of filtration systems, membranes, and fluids.

One application of membrane filtration that has received considerable recent attention is the purification of metalworking fluids (MWFs). The use of MWFs is critical to the success of machining processes, but contaminants such as tramp oils and bacteria are known to reduce the effectiveness of MWFs over time [4]. Both the acquisition and the disposal costs associated with MWFs are high, particularly given the standards and regulations that exist for the use of these fluids today. Therefore, technologies such as membrane filtration have been sought as a way to remove contaminants without a disruption of the chemical integrity of the fluid.

There have been a number of studies that have focused on experimentally determining the mechanisms for membrane fouling [4,10] including those that focus on how certain MWF additives, like polyoxyalkylenes [11], can exacerbate the problem of membrane fouling. Skerlos et al. [4,10,11] studied the microfiltration of synthetic MWFs and demonstrated how the productivity of the microfiltration process can be lowered by both the uncontaminated fluid constituent particles and by contaminants introduced through the machining process. This research was able to characterize flux decline in uncontaminated MWFs (see Figure 8.1) in terms of the presence or absence of various fluid additives, which helped to better understand how the fluid composition itself contributes to membrane fouling.

Although experimental work has led to important insights regarding membrane fouling, the use of modeling techniques has been found essential to creating a more fundamental understanding of the factors that drive the various mechanisms of

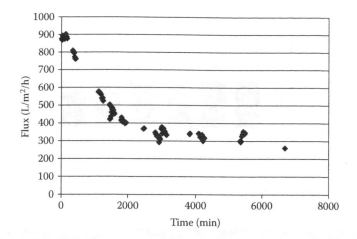

FIGURE 8.1 Flux curve of metalworking fluid test showing flux decline [12]. Results correspond to a 500-nm membrane.

membrane fouling. Such modeling work, both analytical and numerical, has been accomplished via one of two basic approaches, macroscale models that evaluate flux decline over time based on mass-transfer calculations, and modeling of the interaction between individual particles and the membrane pore system at a microscale.

The research presented in this chapter will focus on the use of computational fluid dynamic (CFD) modeling as a way to provide a comprehensive approach to understanding how membrane fouling occurs at the microscale. In so doing, the challenges that are present in modeling particle capture and micro and nanoscale particle forces acting in the flow field will be discussed and the specific mathematical modeling methods used to address these forces will be presented. Most of the modeling work presented here addresses the microfiltration of MWFs.

The next section of this chapter will outline the basics of the microfiltration process, filter morphology, and terminology. This will be followed by a brief history of pore-scale microfiltration modeling. The next section will discuss initial two-dimensional CFD modeling efforts and the introduction of the electrostatic and Brownian motion forces that act on fluid particles. The fifth section will introduce the advancement to three-dimensional membrane CFD modeling for tortuous pore geometries that exist in membranes, such as those made of sintered aluminum oxide, to enable a better understanding of the fouling mechanisms and the prediction of flux decline. The last section of this chapter will summarize the conclusions from the work presented.

8.2   THE MICROFILTRATION PROCESS

The membrane filtration process is based on size exclusion of particles. Membrane filtration systems are designed so that particles (fluid contaminants) larger than the pore size of the membrane are unable to enter the pores and, therefore, are unable to pass from the feed side of the membrane to the clean permeate side. A pressure differential between the feed side of the membrane and the permeate side, called

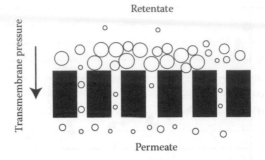

FIGURE 8.2 Size exclusion in membrane filtration. (Adapted from F. Zhao, M. Urbance, and S. Skerlos, *Journal of Manufacturing Science and Engineering*, 126(3), 435–444, 2004.)

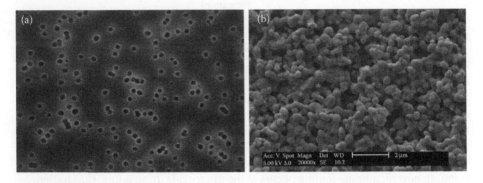

FIGURE 8.3 Track-etched polycarbonate membrane surface (a), and sintered aluminum oxide membrane surface (b). ((a) Adapted from K. L. Tung, Y. L. Chang, and C. J. Chuang, *Tamkang Journal of Science and Engineering*, 4(2), 127–132, 2001. (b) Adapted from J. E. Wentz et al., *Transactions of NAMRI/SME*, 33, 281–288, 2005.)

the transmembrane pressure, drives the filtration of the process flow. Figure 8.2 illustrates the size exclusion of particles in a membrane filtration system.

Although it might be thought that the membrane structure is composed of flat surfaces with cylindrical pores (e.g., Figure 8.2) this is only true for track-etched polycarbonate membranes like the one shown in Figure 8.3a [13]. The majority of membranes, both organic and inorganic, have morphology that is significantly different from this case. Figure 8.3b [14] shows the surface of an inorganic aluminum oxide membrane that is manufactured by a sintering process. This is the type of membrane structure that is the focus of the research described later in this chapter. Cheryan and Rushton both point out that these nonuniform pore geometries can allow membranes to work as depth filters with some particles becoming trapped within the pores [1,8]. The particles, some of which are as small as one fifth of the maximum diameter of the pores, can become stuck in the crevices that are part of the torturous pore geometry until the pores become clogged and flow is reduced. Such is referred to as membrane fouling.

The efficacy of membrane filtration processes is dependent on keeping a high level of permeate flow per membrane unit area, also known as flux and given the units of $L/m^2/h$. The flux level decreases over time due to four mechanisms in which fluid constituents and contaminants restrict fluid flow by fouling the membrane: pore constriction, partial pore blocking, complete pore blocking, and cake formation [9,12]. Pore constriction is the reduction in pore diameter due to adsorption or deposition of molecules in the fluid onto pore walls. Partial pore blocking occurs when particles that are small enough to enter a pore become stuck due to the tortuous nature of the pores and cause a buildup of like particles, often in time leading to complete pore blocking [12]. Complete pore blocking is when an entire pore entrance or pathway becomes essentially sealed due to a physical obstruction such as a fouling particle larger than the pore opening or a buildup of partial pore blocking. Cake formation occurs on the surface of the membrane when particles that do not enter the pores create a layer of particles that allows a decreased flow of fluid through the layer but little or no transfer of other particles.

The remainder of this chapter will focus primarily on membrane fouling and research that has been directed toward better understanding of the mechanisms responsible for fouling. While quite a number of such studies have been conducted, using both experimental and modeling approaches, the bulk of what follows will focus on the use of CFD methods directed toward understanding the interactions that occur between fluid particles and the membrane structure in an effort to better understand the mechanisms of membrane fouling.

8.3   EARLY MODELING APPROACHES

In this section, early approaches to modeling pore and particle interactions on the pore level are discussed. Although a number of papers reviewed in this section talk only about the interactions of particles and pores and not about fouling, the methods used in these studies can be extremely important to the understanding of the way that fouling progresses at the pore level. As membrane fouling is, at its root, an interaction between the particles and between the particles and the membrane/pore surface, the understanding of how these interactions take place is necessary for a full understanding of membrane fouling mechanisms.

8.3.1   ANALYTICAL MODELS

The first models of microfiltration fouling were analytical. Hermia [15] introduced mathematical formulas for each of the fouling mechanisms with the exception of partial pore blocking. These formulas were later unified independently by Koltuniewicz et al. [16] and Song [17]. Experimental validation and improvement of the unified flux decline models were carried out by Zhao et al. [6]. A second approach to mathematical modeling of flux decline examines the individual pores in a probabilistic manner based on the size of the pores and the size of the particles approaching them. This is called the *sieving modeling* approach, and it grows out of experimental investigations into the effect of particle size distributions on microfiltration [18–20]. Chang et al. [21] investigated the impact that particles both smaller and larger than the pore diameter can have on membrane fouling. They found that there was significant flux dependence on particle size with the smaller particles causing a more drastic flux reduction due to a tighter-packed cake layer. The approach of a polydisperse suspension through a membrane was investigated by several researchers who based their approach on filtration of rigid particles through the entrance to ideal cylindrical pores [18–20]. Wentz et al. [12] adapted this approach for use with MWFs, characterized by a polydisperse particle size distribution that changes with usage. However, while analytical methods help to understand fouling and its impact on flux decline in general, they do not illuminate what happens at the pore level.

8.3.2 COMPUTATIONAL FLUID DYNAMIC MODELS

The first computational work in the modeling of particle-pore interactions was undertaken by Bowen et al. [22,23] through the introduction of finite element techniques. They extended analytical approaches to account for the effects of repulsive electrostatic interactions between a given spherical particle and the charged flat surface of a membrane. Through the use of finite elements, they solved the Navier–Stokes equations for a stationary particle placed at arbitrary positions in the flow. This one-way analysis was based on the assumption that the presence of the fluid affects the particle but the particle does not affect the fluid. Later, the same finite element approach was used to solve the nonlinear Poisson–Boltzmann equation to evaluate the electrostatic forces. Particle trajectories were not determined in this chapter, but the equilibrium position of a particle at the mouth of a pore was evaluated from an overall force balance. Bowen and Sharif [24] extended their previous work to determine a critical dead-end filtration velocity below which the particles could not enter the pore and were kept from contacting the membrane due to electrostatic repulsion.

Kim and Zydney [25] took the simulation of particle–pore interactions further through adding the effect of Brownian forces on the particle trajectory. In this chapter, the evaluation of particle trajectory was accomplished through an uncoupled approach, meaning the velocity of the fluid was used to determine the particle trajectory but the movement of the particle did not change the flow of the fluid. The velocity of the flow in the presence of a two-dimensional cross-section of a cylindrical pore with rounded edges was evaluated by solving the steady-state Navier–Stokes equation for an incompressible fluid without the inertial terms. No-slip boundary conditions were applied at all surfaces of the membrane model and the feed velocity and pressure were assumed to be uniform far from the membrane surface. Their results showed that in the absence of electrostatic repulsion, particles followed streamlines, but as the electrostatic repulsion increased the particles were pushed toward the center of the pore and ultimately excluded from entering it at all. Kim and Zydney [26] built on their previous work by evaluating the interaction of two spherical (circular) particles approaching the same cylindrical pore cross-section using the same assumptions and calculations as in their previous work. These simulations also followed an uncoupled approach with the steady-state velocity in the pore area being solved first followed by the trajectories of the two particles. However, the drag effect of one particle on the other was incorporated through expressions developed by Jeffery and Onishi [27]. Kim and Zydney [26] found that if one particle had attained an stationary equilibrium position above the pore mouth, the approach of a second particle was able to push the first particle past the electrostatic repulsive barrier at the pore mouth, but only if electrostatic repulsion between the particles was strong enough.

8.3.3 SUMMARY OF PREVIOUS MODELING

Analytical, computational, and explicit numerical methods have been previously used to analyze membrane fouling. Analytical methods are useful for flux calculations on the entire membrane level but lack the ability to model accurately tortuous pores. CFD methods in the past have focused on ideal cylindrical pores and ideal spherical particles. However, to further research in this area, a more realistic depiction of the tortuous nature of the pore structure and the particles (shape and size distribution) is necessary.

8.4 TWO-DIMENSIONAL MODELING USING REALISTIC PORE MORPHOLOGY

Previous computational modeling approaches all used the assumption that pores were ideal cylindrical holes in a flat plane while particles were perfectly spherical. In order to create a more realistic simulation of how partial and complete pore blocking occur, Wentz et al. [28] developed a two-dimensional model that mimicked actual pore and particle geometries and also included electrostatic and Brownian motion forces. This modeling approach, and most of the work presented throughout the rest of this chapter, was developed specifically for the microfiltration problem of purifying semisynthetic MWFs. These fluids are composed of oil microemulsions ranging in size from 16 to 500 nm in suspension in an aqueous solution. In this specific microfiltration application, it is desired that the microemulsions pass through the membrane without being removed, while extraneous foulants such as tramp oils and bacteria (1–3 μm) are separated out. The membranes typically used in this application are sintered Al_2O_3.

8.4.1 PARTICLE AND PORE MODEL CREATION

The analysis described in ref. [28] considers the microfiltration of an uncontaminated semisynthetic MWF. It focuses on the consequences of having particles that are contained within the uncontaminated fluid fouling the membrane by becoming lodged in the membrane pore structure. CFD simulations of the fluid flow within the membrane were accomplished using the commercially available software Fluent (ANSYS, Inc) and the solid modeling and meshing software Gambit (ANSYS, Inc). These software packages were used to create a two-dimensional model of a typical alumina membrane section that is then used to create a velocity flow field for a modeled pore area.

To create as realistic a simulation as possible, a cross-section of an alumina membrane was captured with a scanning electron microscope (SEM) and used to represent the membrane pore structure within which the two-dimensional CFD model is applied. This portion is highlighted in white in Figure 8.4a. It is important to note that talk of individual "pores" in referring to these membranes is somewhat misleading as the tortuous membrane structure appears to make almost all pores interconnected. Each "pore" has multiple outlets that lead to other "pores" creating a porous network of alumina particles. Figure 8.4b shows the cross-section of the pore area created and

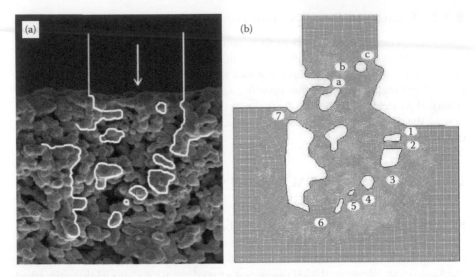

FIGURE 8.4 Cross-section view of alumina membrane surface showing modeled area (a), meshed model of same pore geometry (b). (Adapted from J. E. Wentz et al., *Journal of Manufacturing Science and Engineering*, 130(6), 061015.1–9, 2008.)

meshed in Gambit. The white areas in the mesh are the alumina pore structure highlighted in Figure 8.4a. They are modeled as solids. To characterize the pore entrance in two dimensions, only the membrane solids in the front plane of the cross-section are included (the white outlying sections in Figure 8.4a). Flow enters the model from the top and passes through the open spaces between these membrane solids (white sections in Figure 8.4b). The modeled pore area was scaled so that the primary flow path had a minimum width of 500 nm to match the largest pore size present in the membrane. The inlets range in width from 115 to 534 nm and are lettered a–c while the outlets are numbered 1–7 and range in width from 125 to 500 nm.

The particle sizes used in the CFD simulations are randomly chosen based on a probability density function of the particle diameters present in the fluid being filtered [12]. A logarithmic distribution was used to represent the particle size distribution. The great majority (99.991%) of particles in the flow are smaller than 125 nm. Particles smaller than 125 nm will not be able to block any pore outlets present in the modeled pore structure, therefore, the simulations will only consider the few particles that are able to block the pore outlets. Each time a particle is sent through the pore area the size is randomly chosen from the population of particles, the minimum size of which is larger than 125 nm. Five particle shapes were created based on those seen in SEM images of fouled membranes as shown in Figure 8.5. These shapes were randomly chosen along with particle size for the simulations.

8.4.2 Continuous Phase Flow Simulation

The CFD simulations were conducted using an uncoupled approach, that is, the flow of the continuous phase is solved first and the paths of particles are calculated based

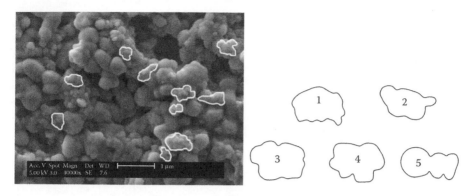

FIGURE 8.5 Particle shapes created from SEM image of fouled membranes. (Adapted from J. E. Wentz et al., *Journal of Manufacturing Science and Engineering*, 130(6), 061015.1–9, 2008.)

on the set flow pattern. This is the same approach taken by Kim and Zydney [25,26] in simulating the approach of a spherical particle to an ideal cylindrical pore with relative sizes similar to those in this research. This approach is well suited to flows in which particles are injected into a continuous phase flow with a well-defined entrance and exit condition [29]. The fluid velocity was evaluated by using Fluent to solve the steady-state Navier–Stokes equation for an incompressible fluid without the inertial terms [25].

Three types of boundaries were present in the pore area: (1) a velocity inlet at the top of Figure 8.4b, (2) unconstrained pressure outlets at the bottom and sides of Figure 8.4b, and (3) no-slip wall treatment for all other surfaces. The velocity inlet was chosen to provide uniform flow into the pore. This is assumed to be an accurate representation of flow into the pore mouth due to the presence of a laminar sublayer on the membrane surface.

The boundary condition imposed at the inlet was a uniform flow rate of 8.5×10^{-5} m/s. The pressure outlet boundary condition at the pore exit was set at −24,821 Pa based on the transmembrane pressure used in the experimental work of Wentz et al. [12]. The use of a pressure outlet boundary allows the flow to take its natural course without imposing a mass transfer rate.

Figure 8.6 shows the flow of the aqueous continuous phase through the pore area as velocity vectors. The inlet to the model is shown in greater detail in the enlargement on the right of Figure 8.6 where arrows indicate the direction of flow. The majority of the flow passes through the center of the pore area and exits by the largest pore outlet at the bottom.

Owing to the uncoupled nature of the CFD simulations the path taken by each particle was impacted by the flow around it but the presence of the particle does not impact the flow of the continuous phase. Particle trajectories were calculated by numerical integration of the Langevin equation [25,29],

$$m_p \frac{d\mathbf{u}_p}{dt} = 6\pi\mu a \left[K_p \mathbf{u}_p - K_f \mathbf{u}_f \right] \tag{8.1}$$

where $\mathbf{u}_p$ is the velocity vector for the particle, $\mathbf{u}_f$ is the unperturbed fluid velocity at the particle center, a is the particle radius, μ is the fluid viscosity, and m_p is the

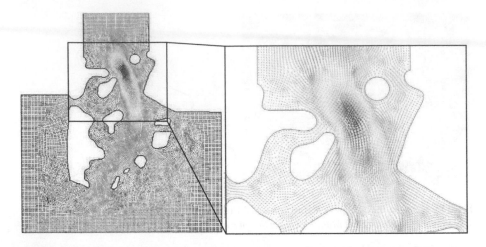

FIGURE 8.6 Flow through pore area (enlargement of pore area on right). (Adapted from W. R. Bowen and A. O. Sharif, *Chemical Engineering Science*, 53(5), 879–890, 1998.)

particle mass. In the summation of forces on particle motion gravitational forces are ignored based on their negligible magnitude when compared with drag forces. Drag forces were calculated based on the diameter of the particle but the particle track was evaluated for a point mass at the center of the particle.

8.4.3 INCLUSION OF ELECTROSTATIC AND BROWNIAN FORCES

In investigating fouling mechanisms, it is important to recognize that they are a function of the motion and interactions of small particles. Small particles in the velocity flow field are subject to not only hydrodynamic forces, but also additional forces including electrostatic forces and those causing Brownian motion (Figure 8.7). These forces may be included in the calculation of particle trajectories through the addition of two terms to Equation 8.1 to give

$$m_p \frac{d\mathbf{u}_p}{dt} = 6\pi\mu a \left[K_p \mathbf{u}_p - K_f \mathbf{u}_f \right] + \mathbf{F}_E + \mathbf{F}_B \qquad (8.2)$$

where $\mathbf{F}_E$ is the electrostatic force acting on the particle, and $\mathbf{F}_B$ is the force due to Brownian motion of the particle.

8.4.3.1 Electrostatic Force

Electrostatic forces play an important role in particle aggregation, microemulsion coalescence, and particle deposition [31]. Increasing electrostatic forces through the use of charged surfactants have been shown to improve the microfiltration performance of semisynthetic MWFs through aggregation reduction [12]. It is therefore important to incorporate electrostatic forces into this study of particle motion within a pore in the presence of other particles.

The electrostatic force acting on a particle can be calculated by integrating the electromagnetic stress acting on the particle over the surface as given by

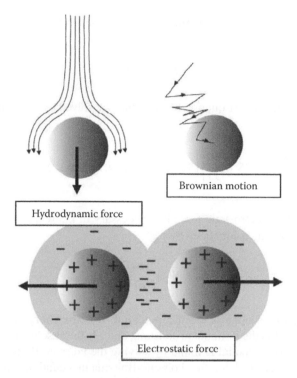

FIGURE 8.7   Hydrodynamic, electrostatic, and Brownian forces. (Adapted from S. Ham et al., *Journal of Manufacturing Science and Engineering*, 132(1), 011006.1–9, 2010.)

$$F_F = \iint_s T \cdot n \, dS \tag{8.3}$$

where T is the total stress tensor acting on the particle and n is the unit normal vector on the surface. The electrostatic force acting on a particle originates from excess osmotic pressure and Maxwell's stress around the particle [31]. The total stress tensor, T, can be expressed as the sum of the excess osmotic pressure and the Maxwell's stress,

$$T = \left(\Pi + \frac{1}{2} \varepsilon_0 \varepsilon_r E \cdot E \right) I - \varepsilon_0 \varepsilon_r E \times E \tag{8.4}$$

where Π is the excess osmotic pressure and all other terms combine to make up Maxwell's stress. The electric field is shown as E, I is the identity tensor, ε_0 is the permittivity of free space, and ε_r is the dielectric constant. The excess osmotic pressure, Π, is related to the electric potential, ψ, by

$$\Pi = \frac{1}{2} \varepsilon_0 \varepsilon_r \kappa^2 \psi^2 \tag{8.5}$$

where κ is the inverse Debye screening length, and the electric potential is related to the electric field by

$$E = -\nabla \psi \tag{8.6}$$

The electrostatic force (Equation 8.3) can then be solved by integrating the stress tensor T.

$$F_E = \iint_S T \cdot n \mathrm{d}S = \iint_S \left[\left(\Pi + \frac{1}{2}\varepsilon_0 \varepsilon_r E \cdot E \right) I - \varepsilon_0 \varepsilon_r E \times E \right] \cdot n \mathrm{d}S \tag{8.7}$$

To calculate the electrostatic force it is necessary to know the electric potential, ψ. This can be nondimensionalized by

$$\Psi = \frac{\upsilon e \psi}{kT} \tag{8.8}$$

where Ψ is the dimensionless electric potential, υ is the charge number, e is the electronic charge, k is the Boltzmann constant, and T is the absolute temperature. The value of the electric potential is found by solving the Poisson–Boltzmann equation. For a symmetric electrolyte, the Poisson–Boltzmann equation over a dielectric continuum separating two charged surfaces [31] is

$$\nabla^2 \Psi = \kappa^2 \sinh{(\Psi)} \tag{8.9}$$

If the electric potential is small the Poisson–Boltzmann equation can be linearized,

$$\nabla^2 \Psi = \kappa^2 \Psi \tag{8.10}$$

The linearized Poisson–Boltzmann equation is traditionally used to calculate the electric potential due to the lack of an analytical solution to the nonlinear Poisson–Boltzmann equation. The Derjanguin approximation [32] of the linearized Poisson–Boltzmann equation is used to calculate the electrostatic force between two spherical particles with thin double layers that are separated from each other at a small distance compared with the radii of the particles. This approximation is described for the scalar electrostatic force between two particles as

$$F_E = \frac{2\pi \kappa \varepsilon_r \varepsilon_0}{g} \left[\frac{2\psi_{Pa}\psi_{Pb}\exp{(\kappa h)} - (\psi_{Pa}^2 + \psi_{Pb}^2)}{\exp{(2\kappa h)} - 1} \right] \tag{8.11}$$

which is the approximated version of Equation 8.7 where h is the distance of closest approach measured between the particle surfaces, and ψ_{Pa} and ψ_{Pb} are the surface potentials of the two particles. The parameter g is a steric factor associated with the specific geometry of the system. For two spherical particles with radii a and b, the steric factor is

$$g = \frac{1}{a} + \frac{1}{b} \tag{8.12}$$

The semisynthetic MWF (Castrol Clearedge 6519) used to provide the particle size distribution in Wentz et al. [28] was tested to determine the surface charge on the particles. Surface potential is not a measureable quantity [31], so its value was estimated from a distribution of values from ζ-potential measurements using laser Doppler electrophoresis. Based on these results, a value of −50 mV was taken as the approximate average for the surface potential of the particles used in the simulations.

In addition to electrostatic repulsive forces between charged particles, there can be a charge on the membrane surface that causes either repulsive or attractive forces, depending on whether it is like or unlike the charge on the particles. To determine the charge on the aluminum oxide membranes used in the earlier study [28], a measurement of their ζ-potential was conducted. It was found that the pH of the MWF (9.3) matched the point of zero charge of the membranes. Therefore, it was assumed that there is no electrostatic interaction between the particles and the membrane under the conditions simulated. The authors recognize that this is not always the case although it is typical for an inorganic ceramic membrane.

8.4.3.2 Brownian Motion

Particles suspended in an aqueous solution are subject to Brownian motion caused by the collisions between the submicron particles and the fluid molecules surrounding them [31]. Brownian motion is included in these simulations due to its ability to change the trajectory of microscopic particles undergoing Stokesian flow (Re < 10^{-3}), which is the case of flow within the modeled membrane pore geometry. The individual Brownian force at each time step, $F_{B,i}$, was modeled after the method of Ounis et al. [33] as a Gaussian white noise process,

$$F_{B,i} = Z_i \sqrt{\frac{2\pi S_0}{\Delta t}} \tag{8.13}$$

where Z_i is a distribution of independent Gaussian random numbers with zero mean and unit variance, S_0 is the spectral intensity of the white noise process, and Δt is the time step used in the simulation. The spectral intensity, S_0, is dependent on fluid parameters and calculated as

$$S_0 = \frac{216\mu kT}{\pi^2 d_P^5 \rho_P^2 C_c} \tag{8.14}$$

where T is the fluid absolute temperature, k is the Boltzmann constant, ρ_p is the particle density and d_p is the diameter of the particle. From Equation 8.14 it can be seen that the factor with the largest impact on the Brownian force is the particle diameter, d_p, which inversely affects the strength of the Brownian force.

8.4.4 SIMULATION METHODOLOGY

Pore fouling simulations were conducted by first simulating the continuous phase velocity flow field within the pore area. A random choice software program was used to pick a particle shape and size as well as its starting location and orientation in the inlet region from assigned distributions. Particles rotate from their initial orientation as they flow through the pore area. This rotation was implemented by having the particle rotate in steps of 15° with the direction of rotation randomly chosen for each step. The initial orientation of each particle was randomly chosen and rotation rate was kept constant through the simulation. Once the particle's shape, size, starting location, and orientation were determined, Equation 8.1 was used in Fluent to calculate the track of the particle. The method used to produce the final particle track was a combination of the movement calculated by the software and the rotation calculated outside of the software and added to the motion. The particle path is then evaluated to see if the particle escapes the simulation area. If the particle escapes the process is repeated with a new particle. If the particle becomes stuck the pore area is remeshed with the inclusion of the stuck particle prior to starting a new particle through the model. This process is repeated until flow through the pore area becomes entirely blocked or it becomes apparent flow is unlikely to become blocked.

8.4.5 PORE MODEL RESTRUCTURING

A particle that has been placed in the flow of the pore area can either become wedged in an outlet as shown in Figure 8.8a or pass through an outlet without becoming stuck. As it passes through the pore area it may come within impact distance of the wall. Impact distance is defined as one half of the hydraulic diameter of the particle. If the particle track, modeled for a point mass, falls within the impact distance then

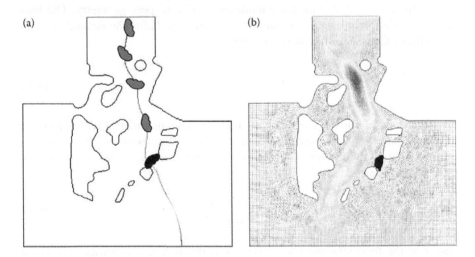

FIGURE 8.8 Particle rotating and becoming stuck in pore (a), new velocity field including stuck particle (b). (Adapted from J. E. Wentz et al., *Journal of Manufacturing Science and Engineering*, 130(6), 061015.1–9, 2008.)

the particle is assumed to roll along the impacted wall until it is either reintroduced into the flow or the particle track passes through an outlet. Reintroduction of the particle into flow occurs when the particle track exceeds impact distance from the wall. When a particle becomes stuck in an outlet the flow of the MWF within the pore area will change. This flow change is brought about through changing the shape of the pore model by incorporating the new stuck particle as an extension of the membrane wall and remeshing the entire pore area. Following the remeshing the new continuous phase flow is calculated using the CFD software. Figure 8.8b shows how the continuous phase flow has changed with respect to Figure 8.6 following the restructuring and remeshing due to the particle trapped in Figure 8.8a. It is also possible for a stuck particle to become unstuck and reenter the flow, in which case the pore area is remeshed without the particle in it. All calculations of impact distance, trapping, particle releasing, and pore remeshing were done independent of the Fuent.

8.4.6   INITIAL SIMULATIONS

The model described in the preceding sections was used to investigate how fouling occurs within a pore and the impact that individual forces have on membrane fouling. While Kim and Zydney [26] previously looked at the use of hydrodynamic, electrostatic, and Brownian motion for cylindrical pore geometries, it was unclear whether the inclusion of nonhydrodynamic forces would cause any change to fouling in a more realistic pore geometry situation such as depicted in Figure 8.8. For this purpose, two sets of simulations were conducted using the model and methodology described above. The first set included only hydrodynamic forces while the second set included hydrodynamic, electrostatic, and Brownian forces. Table 8.1 provides the simulation sequence of particles simulated entering the pore area. In the table "particle shape" refers to which particle shape (recall Figure 8.5) is used, "position" refers to the initial starting location of the particle at the pore entrance, which is divided into 50 equally spaced positions, "size" refers to the particle diameter in nanometers, "rotation" refers to the particle orientation in degrees when it either gets stuck or exits the pore area, "disposition" refers to the fate of the particle. The last column of Table 8.1 provides the results of the simulation when all three forces are included in the model. For this simulation the pore was not completely closed until particle 39 while the pore with only hydrodynamic forces was closed at particle 24. The particle shape, position, size, and rotation for particles 1 through 24 in these simulations are identical in the two simulations. The sequential fouling of the pore for the three-force simulation (Disposition 2) is shown in Figure 8.9.

In Figure 8.9, the status of partial blocking for the three force simulation is shown after each successive particle that becomes stuck in one of the pore inlets or outlets. The change in flow patterns within the model pore can be seen as particles progressively block individual inlets and outlets. For instance, prior to particle 11 becoming stuck the flow splits between two primary flow paths. But after particle 11 becomes stuck the primary flow path on the right is closed and more flow passes to the left and through other outlets. The largest impact on the blocking of the model pore occurs when particle 38 blocks the main entrance, allowing a much smaller particle

TABLE 8.1

Conditions and Results of Simulations with Hydrodynamic Forces Only (Unshaded) and with Hydrodynamic, Electrostatic, and Brownian Forces (Shaded)

Number	Position	Size	Rotation	Disposition 1	Disposition 2
1	21	150	60	Pass through	Pass through
2	41	325	270	Stuck	Stuck
3	16	175	90	Pass through	Pass through
4	11	400	300	Pass through	Pass through
5	17	325	150	Pass through	Pass through
6	12	175	180	Pass through	Pass through
7	20	175	150	Pass through	Pass through
8	19	175	30	Pass through	Pass through
9	49	150	210	Stuck	Stuck
10	26	175	300	Pass through	Pass through
11	45	475	330	Stuck	Stuck
12	19	150	330	Pass through	Pass through
13	39	175	150	Stuck	Pass through
14	33	275	240	Pass through	Stuck
15	26	325	270	Pass through	Pass through
16	38	250	210	Stuck	Stuck
17	32	225	90	Pass through	Pass through
18	29	175	210	Pass through	Pass through
19	9	500	90	Stuck	Stuck
20	37	225	330	Pass through	Pass through
21	34	425	180	Stuck	Pass through
22	1	225	330	Stuck	Stuck
23	40	200	30	Stuck	Pass through
24	32	200	30	Pore closed	Pass through
25	43	450	240		Pass through
26	44	150	210		Pass through
27	7	250	120		Pass through
28	14	475	300		Pass through
29	13	225	150		Pass through
30	10	375	360		Pass through
31	4	200	180		Pass through
32	11	250	360		Pass through
33	31	200	180		Pass through
34	37	150	120		Pass through
35	40	175	330		Pass through
36	33	425	300		Pass through
37	16	150	60		Pass through
38	20	525	240		Stuck
39	46	350	330		Pore closed

Source: Adapted from S. Ham, *Journal of Manufacturing Science and Engineering*, 132(1), 011006.1–9, 2010.

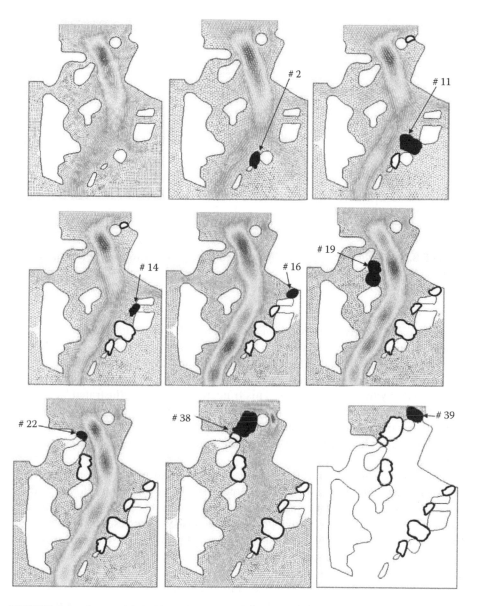

FIGURE 8.9 Sequential fouling of two-dimensional pore for three-force simulation, particle number indicated. (Adapted from S. Ham et al., *Journal of Manufacturing Science and Engineering*, 132(1), 011006.1–9, 2010.)

to totally close off the pore. In viewing Figure 8.9 it should be noted that the mechanism of partial-blocking progresses through each inlet or outlet being blocked by only a single particle.

It is important to note that even though both simulations described in Table 8.1 have identical initial conditions for the first 24 particles the blocking

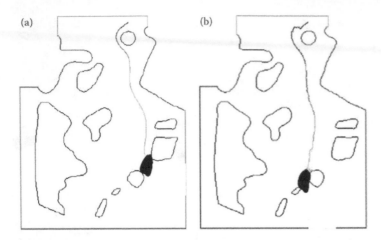

FIGURE 8.10   Particle trajectory: (a) hydrodynamic only; (b) hydrodynamic and Brownian motion. (Adapted from S. Ham et al., *Journal of Manufacturing Science and Engineering*, 132(1), 011006.1–9, 2010.)

progresses differently in each case. The cause for this deviation can be attributed to the inclusion of both electrostatic forces and Brownian motion forces. In the case of particle 2, the added forces caused the trajectory of the particles through the pore area to change and the particle to block a different pore outlet than in the case of only hydrodynamic forces. This trajectory change is illustrated by Figure 8.10. In this case the difference in trajectory is caused solely by the action of Brownian motion because the electrostatic force is assumed only to act between particles and at this point in the simulation there are no other particles stuck in the pore.

The effect of the electrostatic force can be observed by the action of particle 26. In Figure 8.11a, the particle moves under solely hydrodynamic forces and in Figure 8.11b the electrostatic and Brownian forces have been included causing the particle to track further away from previously deposited particles. The difference in the approach of particle 26 to the particles that had previously become stuck shows how the electrostatic repulsion between them caused the particle to deviate from the path it would have otherwise taken.

A significant difference in the way that the two simulations concluded is that the solely hydrodynamic force simulation was progressively clogged through an aggregation of particles combining to fill the outlets (Figure 8.12a). In the second set of simulations the pore area became progressively blocked but particles never aggregated together to form a multiple particle bridge as they did in the solely hydrodynamic simulation. This means that in the presence of a strong electrostatic repulsive force (−50 mV) each inlet or outlet was blocked through the individual action of a large particle. This is caused by the electrostatic repulsive force between stuck particles and particles-in-motion being stronger than the hydrodynamic drag force that tries to push the particles together.

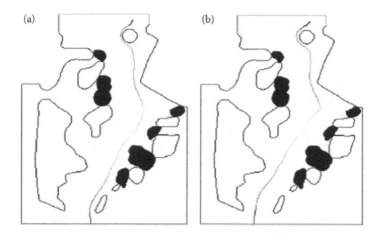

FIGURE 8.11   Particle trajectory: (a) hydrodynamic only; (b) hydrodynamic, Brownian, and electrostatic forces. (Adapted from S. Ham et al., *Journal of Manufacturing Science and Engineering*, 132(1), 011006.1–9, 2010.)

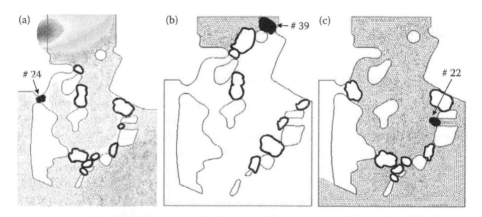

FIGURE 8.12   Final pore-blocking results of (a) hydrodynamic only; (b) hydrodynamic, Brownian, and electrostatic at −50 mV; (c) hydrodynamic, Brownian, and electrostatic at −10 mV. (Adapted from S. Ham et al., *Journal of Manufacturing Science and Engineering*, 132(1), 011006.1–9, 2010.)

8.4.7   EFFECT OF SURFACE POTENTIAL ON PARTIAL BLOCKING

The results of the two simulation sets showed that electric potential is a significant role player in partial blocking. Therefore, a third set of simulations was conducted with a particle surface potential of −10 mV to determine if a lower surface potential would provide the same effect of eliminating particle aggregation within the pore. The same particle feed conditions presented in Table 8.1 were used for the third simulation set. Figure 8.12b shows the final disposition of the second simulation set while Figure 8.12c shows the location of particles stuck within the pore after it became completely blocked in the third simulation set with −10 mV surface potential. The result of this simulation is very similar to that of the first simulation that only took into consideration hydrodynamic forces (Figure 8.12a). The most important insight from this comparison is that both simulations concluded with multiple particles congregating together to block a large outlet. This means that particles within the pore itself could aggregate together and block outlets larger than the individual particles. In the simulation with −10 mV surface potential, the pore became completely blocked after 22 particles were introduced to the flow, compared with 24 particles for the simulation with zero surface potential. This difference may be primarily attributed to the action of Brownian motion changing the flow path of the particles and causing them to become deposited in different pore outlets in the two simulations.

When the results of the −50 mV simulation (Figure 8.12b) are compared with that of the other two simulations, it can be seen that the stronger electrostatic force has eliminated particle aggregation within the pore as a fouling mechanism. In fact, under strong electrostatic forces the partial blocking of the pore is dependent on each pore inlet or outlet finding an individual particle large enough to completely block it on its own. From the different results for the three levels of surface potential, it can be concluded that the level of surface potential, and therefore the electrostatic force, plays a significant role in the nature of partial-blocking progression.

8.4.8   SUMMARY OF TWO-DIMENSIONAL MODELING EFFORTS

The creation of a two-dimensional model that recreates realistic pore and particle morphology represents an improvement on previous efforts that used cylindrical pores and spherical particles. The simulation of particle motion through realistic pores showed for the first time how particle bridging and particle aggregation within the pore can cause the mechanism of partial blocking, sometimes leading to complete pore blocking. The inclusion of electrostatic and Brownian motion forces in the simulations showed how higher electrostatic repulsive forces can eliminate particle aggregation and reduce partial blocking. However, a two-dimensional model is unable to properly model membrane productivity via flux, an important measure of performance. To achieve this, the use of three-dimensional modeling techniques is required.

8.5   THREE-DIMENSIONAL MODELING USING REALISTIC PORE MORPHOLOGY

In the research presented so far, the models have been two dimensional. Although these models provided some understanding about how partial blocking occurs due to their more realistic representation of two-dimensional pore morphology and particle shapes and sizes, they were still lacking in some important areas. Two-dimensional models are limited in their ability to depict flux decline in membranes with tortuous and interconnected pores because all interconnections occurring on planes not included in the model are lost. Two-dimensional models also assume that particles becoming stuck in pore inlets and outlets completely shut off flow through them.

Therefore, the impact of partial blocking on flux decline can be significantly over-estimated using a two-dimensional approach. It is unlikely that three-dimensional aggregates becoming stuck in a three-dimensional outlet will completely seal off flow through those outlets, rather the pore outlets will be partially blocked and flux will be reduced. By introducing the third dimension into the model a realistic flow pattern through a pore network can be simulated and the true nature of particles becoming stuck in tortuous pore paths can be seen. In this section a three-dimensional modeling effort using pore morphology based on a series of membrane scans that move progressively into the pore structure is presented [34].

8.5.1 Pore Geometry Recreation

The same membrane type used for the two-dimensional CFD study was used in this chapter. In order to recreate the geometry in three dimensions a focused ion beam (FIB) was utilized to "slice off" a succession of layers of the membrane to obtain the pore features. The FIB removes material by ionizing the substrate, much like a grinding process. An alumina layer revealed with the FIB is shown in Figure 8.13b. Material layers were removed in the amount of 0.5 μm in a direction deeper into the membrane to attain consecutive layers—see Figure 8.13c—for imaging. Layer dimensions were width 10.00 μm, length 10.00 μm, and depth 0.60 μm. Images of successive FIB-sliced layers were adjusted to match features extant on multiple layers upon layer stacking. Matched features ensured that upon three-dimensional reconstruction, pores existing in one layer passed smoothly to adjacent layers.

There are two types of features that dictate how the layers stack and in what position and orientation: (1) a pore with a distinct geometry in two layers signifies that the pore started or continued from the present layer and continued or ended on the next layer; (2) a distinct membrane geometry that is only found on certain layers signifies that the membrane material had not emerged on the present layer but became extant in the next layer.

The alumina membrane structure is approximated by spheres, as this membrane initially consisted of microscale alumina spheres that are superheated and sintered together. For one layer, the two-dimensional membrane surface area is approximated

FIGURE 8.13 Alumina membrane (a), FIB cut (b), and successive membrane slices (c). (Adapted from B. Yu, S. G. Kapoor, and R. E. DeVor, *Proceedings of ASME 2010 International Manufacturing Science and Engineering Conference,* 2, 447–457, 2010.)

(a)   (b)

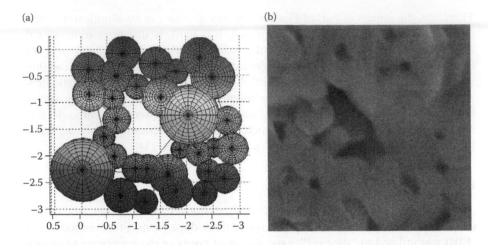

FIGURE 8.14 Comparison of a circle recreated with spheres projected onto a 2D plane (a) and image from the FIB (b). (Adapted from B. Yu, S. G. Kapoor, and R. E. DeVor, *Proceedings of ASME 2010 International Manufacturing Science and Engineering Conference,* 2, 447–457, 2010.)

by circles as this geometry grasps the most flexibility in edge curvature to match the FIB-obtained geometry of the material. Intersecting circles approximate the fused mating surfaces; their locations in two-dimensional space, associated layer marking their z-location, and the radii of h and their locations in the three-dimensional space, and the radii of the circles were recorded. Figure 8.14 shows the similarity of a circle-approximated layer (a) compared to its FIB image complement (b).

The circle-approximation procedure is applied to each consecutive layer. With the circle positions of each layer and the distance between layers known, reconstruction can begin. Spheres were inserted into the location coordinates of the circles and each layer was then combined 0.5 μm apart to form a three-dimensional representation of the tortuous pore structure. Single and multiple layer combinations can be seen in Figure 8.15. A material transition exists between layers because membrane particle sizes are large enough to intersect spheres from other layers and bridge the 0.5-μm layer–layer gap.

The membrane geometry in Figure 8.15b is subtracted from a rectangular block to form a negative of the volume. This generates a space that represents the flowable volume for the fluid. This space is a faithful representation of the actual microfilter membrane and therefore has many advantages over geometries employed in previous works. These advantages are (1) the geometry is three-dimensional enabling more degrees of freedom in particle movement than two-dimensional simulation; (2) the ability to show partial blocking since complex geometry pores are difficult to be completely blocked by spherical particles that become trapped; (3) the pores are tortuous and their cross-sectional areas change in shape and size throughout their paths allowing particles to be trapped in multiple locations; (4) the ability to visualize and predict particles agglomerating on the membrane surface.

A flow volume above the surface leading into the pores was introduced to allow particles to come to steady-downward movement after being injected and prior to

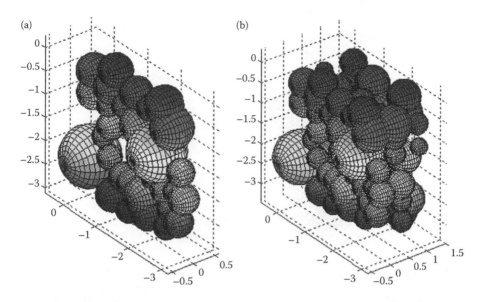

FIGURE 8.15 (a) One layer and (b) three layers of spherical structure (units in micrometers). (Adapted from B. Yu, S. G. Kapoor, and R. E. DeVor, *Proceedings of ASME 2010 International Manufacturing Science and Engineering Conference,* 2, 447–457, 2010.)

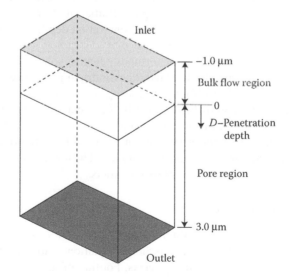

FIGURE 8.16 Schematic of the particle penetration depth within the bulk flow region and the pore region. (Adapted from B. Yu, S. G. Kapoor, and R. E. DeVor, *Proceedings of ASME 2010 International Manufacturing Science and Engineering Conference,* 2, 447–457, 2010.)

entering the tortuous pore region. This pore volume, also referred to as the bulk flow region is shown in Figure 8.16. If the particles are able to pass through the tortuous pore region, they exit at the outlet. Particles that do not exit the pore will stop either at or above the interface of the two regions. Therefore, it is useful to define a

FIGURE 8.17 Isometric views of the three-dimensional mesh of the modeled pore structure. (Adapted from B. Yu, S. G. Kapoor, and R. E. DeVor, *Proceedings of ASME 2010 International Manufacturing Science and Engineering Conference*, 2, 447–457, 2010.)

penetration depth, D, starting at zero at the interface and showing how deep a particle traveled into the pore before becoming stuck. Negative penetration depth means that a particle has not entered a tortuous pore. Rather, its trapped location resides above the pores in the bulk flow region. The penetration depth will be used later in the interpretation of the fouling mechanisms.

The added bulk flow volume and the tortuous negative volume are meshed in the software GAMBIT. Compared to the standard cylindrical geometries typically used in this type of analysis [22–26,36], the tortuosity of this volume creates locations where features are small and difficult to mesh. Size functions, which alter volume element size as a function of distance from a location of interest, are implemented to alleviate this issue. The final mesh, seen in Figure 8.17, is imported into Fluent (ANSYS, Inc.) for fluid flow and particle simulation. The top areas represent velocity inlets and the bottom areas represent pressure outlets.

8.5.2 MODEL DEVELOPMENT

The fluid flow through a tortuous three-dimensional volume is modeled, as in the previous sections, by solving the Navier–Stokes equations with steady-state incompressible flow in the absence of gravity terms. Fouling of the tortuous geometry is modeled though injected particles becoming trapped in the membrane. The particle movement is calculated by numerical integration of Equation 8.2.

8.5.2.1 Wall Collision Model

Particles in fluids are subject to collisions with walls as they flow through the membrane pore structure. In previous models [26,28,30], post-collision movement was modeled as a mechanism isolating the particle at the wall, causing it to roll along the curvature of the wall. These models do not describe the complete influence of the collision physics on particle trajectory. Instead, it is likely that the particle will rebound off the wall and into the fluid stream before being recaptured by fluid forces. This is because (1) the momentum of the particle is conserved and its impulse after collision imparts a surface-departing velocity; and (2) varying boundary layer fluid velocities cause a larger force onto the noncollision surface than the collision surface thereby imparting both a surface-directed velocity and a moment on the particle. This combination of surface-departing velocity and surface-directing velocity isolates the particle at the membrane surface. The moment, from (2), causes particle rotation, thereby rolling the particle along the surface curvature. This section details the reflection mechanism prior to fluid force impartation.

A particle travels in a continuous trajectory until the trajectory is altered by a physical object such as a wall, also referred to in this section as a face. A wall can either be the microfilter membrane or a previously trapped particle surface.

When a face centroid is encountered to be less than a radius distance away from the particle centroid, the particle reflects in a direction according to specular reflection,

$$\Delta v = -(1+\alpha)(v_0 \cdot n)n \tag{8.15}$$

where Δv is the change in the velocity of the particle, α is the coefficient of restitution, v_0 is the incident velocity vector, and n is the vector normal of the collision surface. The change in velocity is then accounted for in the velocity in the succeeding time step, viz.,

$$v_f = v_0 + \Delta v \tag{8.16}$$

where v_f is the resultant velocity vector. The position of the particle in the succeeding time step is also calculated,

$$x_f = x_0 + v_f \Delta t \tag{8.17}$$

where x_f is the post-collision position, x_0 is the precollision position, and Δt is the time step.

The postcollision position and the resultant velocity, as calculated from Equations 8.16 and 8.17, are set as the positions and velocities for the succeeding time step. In that time step, the particle receives forces from hydrodynamics, electrostatics, and Brownian motion. Its position and velocity is then updated accordingly. This model operates by first selecting a face and determining the distance between the face centroid and the particle centroid. If the distance is less than the particle radius, a collision is marked and the incident velocity of the particle is considered so to determine if the particle is moving toward or away from the face. If the particle approaches the face then Equation 8.15 is used to determine a resultant velocity vector. The position and velocity for the succeeding time step are then calculated along with the values and effect of the hydrodynamic and electrostatic forces and Brownian motion.

In summary, this wall collision model determines that the particle collided with the closest collidable face F_{min} at a centroid-to-centroid distance of d_{min}. The resultant velocity of the particle is calculated from Equation 8.16 with influence from the incident velocity, the normalized face normal, and the coefficient of restitution.

8.5.2.2 Particle-Trapping Model

It can be visually recognized when a particle has become trapped in two dimensions [25,26,34]. However, the disposition of a particle is difficult to judge in three dimensions due to the tortuous nature of the pore path and the poor distinguishability between particle element lines and membrane mesh lines. Therefore, the wall collision model serves a dual purpose—not only used for the aforementioned wall collision scenarios, but also is paramount in determining when a particle has become trapped.

Being trapped is realized quantitatively when the location of a particle remains the same for consecutive time steps or suffers repeated wall collisions that isolate its position in the same location. The particle trapping model activates after it is determined that the particle will impact the wall but before the succeeding time step.

Particle trapping is determined by calculating the moving average of the particle position and finding the difference between the current position of the particle and the moving average. The positional moving average is used to determine how far or close the current position of a particle is in relation to its average position over time. If the ratio of the distance difference versus a characteristic length is consecutively less than 5% of the particle movement over multiple time steps, the particle is trapped. This is conceptualized as a particle remaining at the same position, due to any combination of hydrodynamic, electrostatic, Brownian motion, or wall-collision forces.

8.5.3 Fluid Flow and Initial Particle Simulation

The same methodology of determining the flow through the membrane and then sending particles one-at-a-time was used in this three-dimensional simulation as in those previously discussed for two dimensions. Particles were injected from the velocity inlet into the three-dimensional tortuous pore geometry one after the other. That is, a particle was not injected until the disposition of the previous particle was known and recorded. A particle can achieve trapped status if its position is the same after consecutive time steps; it achieves escaped status if it leaves through the pressure outlet. The three-step particle injection procedure is repeated as many times as necessary until the flux reaches steady state or the bulk flow region is completely full of particles.

Flow through the model was calculated by examining the effective flow area of numerous parallel cross-sections of a simulation and finding the minimum area. As the flow material is liquid and assumed incompressible, the minimum effective cross-sectional flow area regulates how much volumetric flow can pass. As particles become trapped, effective cross-sectional flow areas shrink, with some shrinking faster than others. The faster shrinking areas become the areas limiting the volumetric flow.

The flow rate at the start of the simulation was 0.325 nL/h. After 48 particles, the flow became steady state and 0.241 nL/h; a 27% decrease. The simulation ended at 48 particles as continued injection led to immediate blocking at the membrane surface by the congregation of particles filling the bulk flow region. Figure 8.18 shows the flux decline over the life of the simulation. This ability to model flux is the primary reason that three-dimensional CFD modeling is so powerful. The flux initially decreases sharply and then transitions to steady state. This is similar to experimental results from refs. [13,35]. This flow response is caused by one particle trapping in a significantly sized pore and initiating surface fouling. This behavior is

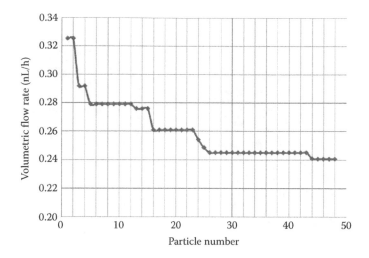

FIGURE 8.18 Volumetric flow rate decreases with number of particles. (Adapted from B. Yu, S. G. Kapoor, and R. E. DeVor, *Proceedings of ASME 2010 International Manufacturing Science and Engineering Conference, 2,* 447–457, 2010.)

realized toward the beginning of the simulation where a particle blocks a major pore and the flow drops substantially.

Figure 8.19 provides the visualization of the particles that became trapped. Although flow does have a small decrease just prior to the end of the simulation, additional injected particles past particle 48 did not have a significant impact on flux decline as a result of how the particles were stacked in the bulk flow region. Particles resting upon one another were not able to affect the established minimum cross-sectional area of the pores.

It is important to note that the flux decline profile of Figure 8.18 reaches a nonzero steady state similar to experimental results from refs. [13,35]. Also, this simulation shows that spherical particles do not allow a complete sealing of the inlet flow as in ref. [30]. However, prior to steady state, the injection of more particles into the flow causes blockage at the crevices of the existing trapped particles thereby reducing the flux as shown in the flux decline results.

One of the interesting aspects of the three-dimensional simulation is the revelation of the nature of particle depth penetration as the simulation progresses. Particle penetration depth is provided in Figure 8.20. Particles 1–10, at the start of the simulation, have positive penetration depth and show blocking within the pores. Injections after particle 4 are shown to be less deep than particle 4. This reveals that although particles can still enter other pores, once a major pore is partially blocked, particle stacking ensues nearby the already partially blocked pore leading to further restriction of that pore. Moreover, after particle 13, a majority of particles injected have zero or negative penetration depth. Here, particles stack at the pore inlets and are unable to travel deeper. Succeeding particles then accumulate at the inlets and grow into the bulk flow region. This further suggests that once particles are trapped, succeeding particles begin stacking. This resembles the fouling mechanism called caking.

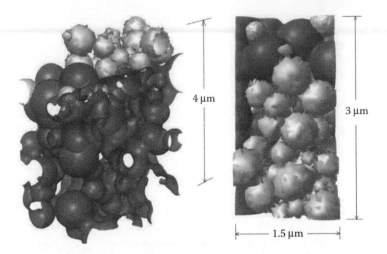

FIGURE 8.19 Isometric view and top–down view of the trapped particles (light) and the membrane (dark). (Adapted from B. Yu, S. G. Kapoor, and R. E. DeVor, *Proceedings of ASME 2010 International Manufacturing Science and Engineering Conference,* 2, 447–457, 2010.)

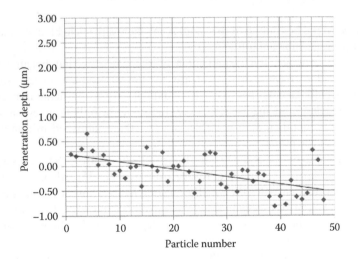

FIGURE 8.20 Relationship between penetration depth and injected number of particles. (Adapted from B. Yu, S. G. Kapoor, and R. E. DeVor, *Proceedings of ASME 2010 International Manufacturing Science and Engineering Conference,* 2, 447–457, 2010.)

Fouling by cake layer development was examined comprehensively by the microfiltration review done by Belfort et al. [9]. Their review highlighted the significant problem of cake layer development in many applications and feed fluids. This fouling mechanism was also discussed by Skerlos et al. [4] where the filtration of synthetic MWF defoamers in absence of lubricants formed a cake layer; and likewise by Wentz et al. [12] where the end result of filtering semisynthetic fluids resulted in cake layer formation as seen in the comparison of Figures 8.21a,b.

FIGURE 8.21 SEM images of a clean membrane (a) and a developed cake layer (b). (Adapted from J. E. Wentz et al., *Journal of Manufacturing Science and Engineering*, 130(4), 041014.1–9, 2008.)

Even though pores are on average larger than the diameters of particles injected, particles become trapped primarily due to partial pore blocking and particle–particle stacking internal to the pores and cake layer formation external to the pores thereafter. This is different than previous work showing particles being trapped at all depths of pore constrictions [30,36], instead of primarily on the upmost layer.

Caking behavior is attributed to the particle trapping model as it is through the wall-collision model that particles can become trapped in any location instead of just certain pore restrictions. Without the particle trapping model, the simulation would have been completed when all pores were blocked. This would not have shown any cake layer buildup.

8.5.4   Effect of Particle Distribution on Flux Decline

One of the advantages to using CFDs to model microfiltration fouling is that input parameters are easily changed. This allows the model to be used as a design tool. One of the parameters that can be changed is the particle size distribution. In the purification of MWFs particle size distribution changes as the fluid continues in its use cycle due to the shearing action of the system pump causing particles to overcome their electrostatic repulsion and becoming aggregated. A set of simulations were developed to investigate three size distributions, one representing the fluid when it was new and two states of progressive use. The particle size distributions are presented in Table 8.2.

New MWF is considered in this paper as fluid that has not been used for more than 200 min in system filtration. The majority of particles in this class can pass through a majority of pores due to their diameter being smaller than the average 500 nm pore diameter. For the two progressive stages of use, particles can pass through microfiltration material; however common, a fraction of particles do block pores. The fluid that has been used longer has a larger fraction of pore-blocking particles. Figure 8.22 shows the flux decline curves for the semiused (a) and new (b) particle size distributions. The distribution flux curve for the fluid that has been used the longest was provided in Figure 8.18.

TABLE 8.2

Governing Equations and Statistics for Particle sizes

MWF Class	Time of Use (min)	Particle Diameter Probability Density Function	Particle Diameter Range (nm)
New	180	$f(x) = 0.1191 - 4.760 \times 10^{-4}x$	50–250
Semiused	900	$f(x) = 0.0549 - 1.220 \times 10^{-4}x$	50–450
Used	2000	$f(x) = 0.0137 - 2.384 \times 10^{-5}x$	125–550

Source: Adapted from B. Yu, S. G. Kapoor, and R. E. DeVor, *Proceedings of ASME 2010 International Manufacturing Science and Engineering Conference,* 2, 447–457, 2010.

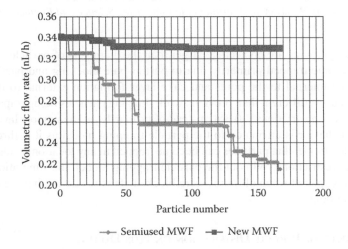

FIGURE 8.22 Flux decline for semiused (a), and new (b) MWF. (Adapted from B. Yu, S. G. Kapoor, and R. E. DeVor, *Proceedings of ASME 2010 International Manufacturing Science and Engineering Conference,* 2, 447–457, 2010.)

The curves in Figures 8.18 and 8.22 illustrate how particle size distributions can impact flux decline. The curves for used and semiused distributions both show continuous flux decline while the new distribution saw very little flux decline due to the much smaller particle sizes. The semiused distribution shows a 37% flux reduction while the new distribution shows only a 4.5% decrease in flux. The difference in flux decline between the size distributions can also be seen in trapped particle visualizations that show the larger particle aggregations within the membrane causing the partial blocking and cake formation leading to flux decline. Figure 8.23 shows the three-dimensional model visualizations of the fouled membranes from the simulations.

8.6 CONCLUSIONS

The evolution of pore-scale fluid dynamic modeling of microfiltration membranes has been presented starting with the earliest numerical work and proceeding through

FIGURE 8.23 Isometric/top views of the trapped particles (light) and the membrane (dark) in a semiused (left images) and new (right images) distribution. (Adapted from B. Yu, S. G. Kapoor, and R. E. DeVor, *Proceedings of ASME 2010 International Manufacturing Science and Engineering Conference,* 2, 447–457, 2010.)

three-dimensional models of realistic pore geometry that are able to predict flux decline. The following conclusions are derived from this chapter.

1. CFDs was used to investigate microfiltration membrane fouling. Simulations using realistic two-dimensional pore geometries showed for the first time that partial pore blocking due to bridging and aggregation can occur within tortuous membrane pores.
2. The incorporation of electrostatic and Brownian forces into two-dimensional models with realistic pore geometry showed that by adjusting the strength of the electrostatic repulsive force the aggregation and bridging can be eliminated, significantly reducing partial pore blocking.
3. A three-dimensional model including hydrodynamic, electrostatic, and Brownain motion forces was created based on FIB slices of an aluminum oxide membrane. This model was used to simulate the flux decline of microfiltration of semisynthetic MWF.
4. Simulations of membrane flux decline for three particle size distributions representing different MWF usage levels showed that particle size can significantly impact the severity of membrane fouling. In simulations using particle size distributions that were on the same level as pore diameters a cake layer was observed to form similar to experimental results.
5. The CFD simulations demonstrate the progression of membrane fouling over the fluid use cycle; in particular, how different fouling mechanisms influence flux at different points in time.
6. CFD-based models of fluid flow for the microfiltration process enable researchers to precisely recreate membrane geometries and the physical interactions that particles in the fluid media have with the membranes. Through changing input parameters like particle size distributions the use of CFDs to model membrane fouling allows membranes and particles to be designed to alleviate fouling.

REFERENCES

1. M. Cheryan, *Ultrafiltration and Microfiltration Handbook*, Technomic Publishing Company, Inc., Lancaster, PA, 1998.
2. W. S. W. Ho and K. K. Sirkar, *Membrane Handbook*, Van Nostrand Reinhold, New York, NY, 1992.
3. L. J. Zeman and A. L. Zydney, *Microfiltration and Ultrafiltration: Principles and Applications*, Marcel Dekker, Inc., New York, NY, 1996.
4. S. J. Skerlos, N. Rajagopalan, R. E. DeVor, S. G. Kapoor, and V. D. Angspatt, Ingredient-wise study of flux characteristics in the ceramic membrane filtration of uncontaminated synthetic metalworking fluids, Part 2: Analysis of underlying mechanisms, *Journal of Manufacturing Science and Engineering*, 122(4), 746–752, 2000.
5. N. Rajagopalan, T. Rusk, and N. Dianovsky, Purification of semi-synthetic metalworking fluids by microfiltration, *Tribology and Lubrication Technology*, 60(8), 38–44, 2004.
6. F. Zhao, M. Urbance, and S. Skerlos, Mechanistic model of coaxial microfiltration for semi-synthetic metalworking fluid emulsions, *Journal of Manufacturing Science and Engineering*, 126(3), 435–444, 2004.
7. R. R. Bhave, *Inorganic Membranes: Synthesis Characteristics and Applications*, Van Nostrand Reinhold, New York, NY, 1991.
8. A. Rushton, A. S. Ward, and R. G. Holdich, *Solid-Liquid Filtration and Separation Technology*, 2nd ed., Wiley-VCH, New York, NY, 2000.
9. G. Belfort, R. H. Davis, and A. L. Zydney, The behavior of suspensions and macromolecular solutions in crossflow microfiltration, *Journal of Membrane Science*, 96, 1–58, 1994.
10. S. J. Skerlos, N. Rajagopalan, R. E. DeVor, S. G. Kapoor, and V. D. Angspatt, Ingredient-wise study of flux characteristics in the ceramic membrane filtration of uncontaminated synthetic metalworking fluids, Part 1: Experimental investigation of flux decline, *Journal of Manufacturing Science and Engineering*, 122(4), 739–745, 2000.
11. S. J. Skerlos, N. Rajagopalan, R. E. DeVor, S. G. Kapoor, and V. D. Angspatt, Microfiltration of polyoxyalkylene metalworking fluid lubricant additives using aluminum oxide membranes, *Journal of Manufacturing Science and Engineering*, 123, 692–699, 2001.
12. J. E. Wentz, S. G. Kapoor, R. E. DeVor, and N. Rajagopalan, Partial pore blocking in microfiltration recycling of a semisynthetic metalworking fluid, *Journal of Manufacturing Science and Engineering*, 130(4), 041014.1–9, 2008.
13. K. L. Tung, Y. L. Chang, and C. J. Chuang, Effect of pore morphology on fluid flow through track-etched polycarbonate membrane, *Tamkang Journal of Science and Engineering*, 4(2), 127–132, 2001.
14. J. E. Wentz, S. G. Kapoor, R. E. DeVor, and N. Rajagopalan, Experimental investigation of membrane fouling due to microfiltration of semi-synthetic metalworking fluids, *Transactions of NAMRI/SME*, 33, 281–288, 2005.
15. J. Hermia, Constant pressure blocking filtration laws- application to power-law non-newtonian fluids, *Transactions of Institute of Chemical Engineers*, 60(2), 183–187, 1982.
16. A. B. Koltuniewicz, R. W. Field, and T. C. Arnot, Cross-flow and dead-end microfiltration of oily-water emulsion, Part 1: Experimental study and analysis of flux decline, *Journal of Membrane Science*, 102(1), 193–207, 1995.
17. L. Song, Flux decline in cross-flow microfiltration and ultrafiltration: Mechanisms and modeling of membrane fouling, *Journal of Membrane Science*, 139(2), 183–200, 1998.
18. J. A. Howell, O. Velicangil, M. S. Le, and A. L. Herrera-zepplin, ultrafiltration of protein solutions, *Annals of the New York Academy of Sciences*, 369(1), 355–366, 1981.

19. V. Starov, D. Lloyd, A. Filippov, and S. Glaser, Sieve mechanism of microfiltration separation, *Separation and Purification Technology*, 26, 51–59, 2002.
20. S. Kosvintsev, I. Cumming, R. Holdich, D. Lloyd, and V. Starov, Sieve mechanism of microfiltration separation, *Colloids and Surfaces A: Physiochemical Engineering Aspects*, 230, 167–182, 2004.
21. J. S. Chang, L. J. Tsai, and S. Vigneswaran, Experimental investigation of the effect of particle size distribution of suspended particles on microfiltration, *Water Science Technology*, 34(9), 133–140, 1996.
22. W. R. Bowen and A. O. Sharif, Hydrodynamic and colloidal interaction effects on the rejection of a particle larger than a pore in microfiltration and ultrafiltration membranes, *Chemical Engineering Science*, 53(5), 879–890, 1998.
23. W. R. Bowen, A. N. Filippov, A. O. Sharif, and V. M. Starov, A model of the interaction between a charged particle and a pore in a charged membrane surface, *Advances in Colloid and Interface Science*, 81(35–72), 1999.
24. W. R. Bowen and A. O. Sharif, Prediction of optimum membrane design: Pore entrance shape and surface potential, *Colloids and Surfaces A: Physiochemical and Engineering Aspects*, 201, 207–217, 2002.
25. M. M. Kim and A. L. Zydney, Effect of electrostatic, hydrodynamic, and Brownian forces on particle trajectories and sieving in normal flow filtration, *Journal of Colloid and Interface Science*, 269, 425–431, 2004.
26. M. M. Kim and A. L. Zydney, Particle-particle interactions during normal flow filtration: Model simulations, *Chemical Engineering Science*, 60, 4073–4082, 2005.
27. D. J. Jeffery, and Y. Onishi, Calculation of resistance and mobility functions for two unequal rigid spheres in low-reynolds number flow, *Journal of Fluid Mechanics*, 139, 261–290, 1984.
28. J. E. Wentz, S. G. Kapoor, R. E. DeVor, and N. Rajagopalan, Dynamic simulations of alumina membrane fouling from recycling of semi-synthetic metalworking fluids, *Journal of Manufacturing Science and Engineering*, 130(6), 061015.1–9, 2008.
29. Fluent 6.1 User's Guide, Fluent, Inc., Lebanon, NH, 2003.
30. S. Ham, J. E. Wentz, S. G. Kapoor, and R. E. DeVor, The impact of surface forces on particle flow and pore blocking in the microfiltration of metalworking fluids, *Journal of Manufacturing Science and Engineering*, 132(1), 011006.1–9, 2010.
31. M. Elimelech, J. Gregory, X. Jia, and R. A. Williams, *Particle Deposition and Aggregation: Measurement, Modeling, and Simulation*, Butterworth-Heinemann, Woburn, MA, 1995.
32. J. N. Israelachvili, *Intermolecular and Surface Forces*, Academic Press, Burlington, MA, 1991.
33. H. Ounis, G. Ahmadi, and J. B. McLaughlin, Brownian diffusion of submicrometer particles in the viscous sublayer, *Journal of Colloidal and Interface Science*, 143(1), 266–277, 1991.
34. B. Yu, S. G. Kapoor, and R. E. DeVor, Three-dimensional simulation of cross-flow microfilter fouling in tortuous pore profiles with semi-synthetic metalworking fluids, *Proceedings of ASME 2010 International Manufacturing Science and Engineering Conference*, 2, 447–457, 2010.
35. F. Zhao, A. Clarens, and S. J. Skerlos, Optimization of metalworking fluid microemulsion surfactant concentrations for microfiltration recycling, *Environmental Science & Technology*, 41(3), 1016–1023, 2006.
36. S. Ham, Fluid dynamic models to investigate pore blocking and flux decline in microfiltration using alumina membranes, MS Thesis, University of Illinois at Urbana-Champaign, 2009.

9 Computational Analysis of Enhanced Cooling Performance and Pressure Drop for Nanofluid Flow in Microchannels

Clement Kleinstreuer, Jie Li, and Yu Feng

CONTENTS

9.1 BACKGROUND INFORMATION

9.1.1 INTRODUCTION

High rates of heat transfer in mechanical, chemical, and biomedical *microsystems* require heat exchangers which are very small, light, and efficient. *Microchannels* made out of glass, silicon, or polymers form the basic elements of such microsystems. Improving the thermal performance of compact devices requires better coolants than conventional fluids such as oil, water, or ethylene glycol. One solution to microscale cooling problems is the addition of solid nanoparticles to the fluid. The resulting *nanofluids*, that is, dilute suspensions of nanoparticles in liquids, may significantly change the mixture's properties, most notably its thermal conductivity and viscosity.

Nanoparticles considered for microsystem cooling range from metals and metal oxides to carbon nanotubes with diameters of 1–100 nm. Indeed, prevailing experimental evidence indicates a greater enhancement of *nanofluid thermal conductivity*, k_{nf}, than predicted by the "effective medium" theory of Maxwell [1] or Hamilton and Crosser [2]. Such an increase of k_{nf} over $k_{base-fluid}$ varies with nanoparticle volume fraction and characteristics, for example, size, shape, material, surface charge/coating, and degree of particle aggregation, as well as with the type of base fluid, its temperature, conductivity, pH value, and additives. Nevertheless, although in the past-enhanced k_{nf} values have been reported when employing the transient hot-wire method, some data based on recent nonintrusive (optical) techniques could not confirm such high k_{nf} values, or even an increases over the values obtained with Maxwell's theory. Clearly, additional studies are warranted.

For a better understanding of the underlying physics of k_{nf} enhancement, six major sources should be considered: (i) micromixing because of Brownian motion of the nanoparticles affecting the surrounding fluid, (ii) higher pathway conduction of clustered nanoparticles or connected carbon nanotubes, (iii) liquid–molecule layering around nanoparticles causing lower heat resistance, (iv) larger heat conduction in the case of certain metallic nanoparticles, (v) thermal wave impact, and/or (vi) nanoparticle and wall–shear-layer interactions.

Experimental observations and theoretical models of nanofluid thermal conductivity enhancement have been critically reviewed by Kleinstreuer and Feng [3], Oezerinc et al. [4], Fan and Wang [5] as well as by Murshed et al. [6] and Timofeeva et al. [7], whereas heat-transfer applications with nanofluids as coolants are discussed in books by Das et al. [8], Li [9], and Kumar [10]. Kleinstreuer and Feng [3] focused on the review of k_{nf} measurement techniques and theoretical models. Comparing theoretical predictions and experimental findings of k_{nf} values, Oezerinc et al. [4] reiterated that significant discrepancies exist. Fan and Wang [5] provided an overview of recent contributions discussing thermal conductivity enhancement of nanofluids. They tabulated experimental observations for carbon nanotubes and nanospheres in different base liquids. Of the various theoretical models discussed, they favored the recently developed thermal wave theory with which experimental findings can be apparently modeled and explained. Murshed et al. [6] as well as Timofeeva et al. [7] took a more comprehensive view of nanofluids, evaluating nanoscale contributions to both mixture properties and engineering applications.

First, some representative experimental and numerical papers concerning possible k_{nf} enhancement are reviewed. Then, the computational development and cooling application of nanofluid flow in a microchannel is discussed to illustrate the impact of different thermal conductivity and effective viscosity models on convection heat transfer and pressure drop. Reduction of entropy generation for this fundamental test case is provided as well.

9.1.2 REVIEW OF EXPERIMENTAL EVIDENCE

Numerous experimental studies on nanofluid single-phase heat transfer have been reported in the literature. Most of these relied on the transient hot-wire method to measure thermal conductivity and/or indirectly Nusselt number values. For example, Pak and Cho [11] investigated alumina–water and titania–water nanofluids in turbulent convective heat transfer in tubes. Xuan and Roetzel [12] conveyed a heat–transfer correlation for nanofluids to capture the effect of energy transport by particle "dispersion." Wen and Ding [13] studied laminar nanofluid convective heat transfer and reported significant enhancement in the entry region. Heris et al. [14,15] analyzed the effects of alumina and copper oxide nanofluids on laminar heat transfer in a circular tube under constant wall temperature boundary condition. They reported an enhancement of the heat-transfer coefficient for both nanofluids with increasing nanoparticle concentrations as well as the Peclet number, and observed a larger enhancement for alumina than for copper oxide.

Buongiorno [16] suggested that a reduction of viscosity within the boundary layer and consequent thinning of the laminar sublayer lead to an abnormal increase in the convective heat-transfer coefficient in the turbulent flow regime. Jung et al. [17] measured the convective heat transfer and friction factor of nanofluids in rectangular microchannels. Nanofluids with 170-nm aluminum dioxide particles and with various particle volume fractions were used in their experiments. For a volume fraction of 1.8%, a 320% convective heat transfer increase in the laminar regime was measured compared with distilled water, without major frictional loss. They also found that the Nusselt number increases with the increasing Reynolds number in the laminar flow regime, which is contradictory to the result from conventional thermal Poiseuille-flow analysis.

Rea et al. [18] experimentally investigated laminar convective heat transfer and viscous pressure loss for alumina–water and zirconia–water nanofluids in a flow loop with a vertical heated tube. Their measured heat-transfer coefficient and pressure loss were in good agreement with the traditional model predictions for laminar flow; in other words, there was no abnormal heat-transfer enhancement or pressure loss observed within the measurement error.

Vajjha et al. [19] presented new correlations for the convective heat-transfer coefficient and friction factor developed from the experiments with different nanofluids assuming fully developed turbulent flow. Khiabani et al. [20] considered the impact of cylindrical particles on the heat transfer in a microchannel based on the solution of the lattice-Boltzmann equation for fluid flow, coupled with the energy equation for thermal transport and the particle dynamics equations for direct simulation of suspended particle transport. They found that each particle can locally enhance heat transfer, and hence the average heat-transfer performance can be improved. Hojjat et al. [21] experimentally investigated convective heat transfer of non-Newtonian nanofluids through a uniformly heated circular tube under turbulent flow conditions, noting again heat-transfer enhancement of such nanofluids. Considering Al_2O_3–ethylene glycol (EG) nanofluids, Hemalatha et al. [22] found experimentally that above a weight percentage of 0.6, particle–particle interactions may be important due to particle aggregation. Escher et al. [23] found that mixture properties of silica–water nanofluids, even at high concentrations (i.e., up to 31%), did not deviate more than 10% from the effective medium theory established by Maxwell. Furthermore, they demonstrated that any relative k_{nf} enhancement must be larger than the relative viscosity increase in order to gain effective cooling performance with nanofluid flow in microchannels. In a combined experimental and theoretical study, Gharagozloo and Goodson [24] analyzed temperature-dependent nanoparticle aggregation and diffusion. They concluded that aggregation produces an *unfavorable* nanofluid for heat transfer and suggested an optimal nanoparticle diameter of 130 nm for minimizing aggregation, sedimentation, and thermal diffusion. In order to illuminate potential causes of experimental uncertainties, such as nanoparticle aggregation effects or "time-dependent nanofluid characteristics," Xie et al. [25] summarized numerous experimental data sets for oxide and carbon–nanotube nanofluids in different liquids. They recommended that extra care should be taken in preparing homogeneous and stable mixtures, noting that additives such as acid, base, and/or surfactants are influential. They observed that k_{nf} enhancement increases monotonously with nanoparticle loading, whereas the temperature effect on nanofluids largely depends on the type of nanoparticle and base fluid pairing. A recent study by Lee et al. [26] introduced round-robin tests on thermal conductivity measurements of three samples of EG-based ZnO nanofluids. The experiments were conducted in five laboratories, where four of them used measurement apparatuses developed in house and one used a commercial device. On the basis of their results, the conventional thermal conductivity model underestimates the effective thermal conductivity of nanofluids. Thus, the effective thermal conductivity of nanofluids cannot be fully explained by the effective medium theory for well-dispersed nanoparticles.

Recently, several research groups attributed convective heat-transfer enhancement to the interactions of nanoparticles and the system/device walls (see [27–30]). Specifically, the random fluid-particle velocity field in the wall boundary layer is changed due to the interaction of nanoparticles with the channel walls as well as the surrounding fluid parcels. It may steepen the mixture temperature profile and hence lead to a large temperature gradient near the wall, resulting in a higher heat transfer rate. For example, Hwang et al. [27] measured the pressure drop and convective heat-transfer coefficient of water-based alumina nanofluids flowing through a uniformly heated circular tube in the fully developed laminar flow regime. They discussed the various parameter effects on the remarkable enhancement of the convective heat transfer coefficient and showed for the first time the flattened core velocity profile.

9.1.3 REVIEW OF COMPUTATIONAL ANALYSES

To investigate heat transfer enhancement by nanofluids computationally, two main approaches have been adopted in the literature. The first one is the two-phase model that takes into account the fluid- and solid-phase roles in the heat transfer process. The second one is the single-phase model where both the fluid phase and the solid particles are in a thermal equilibrium state and flow with the same local velocity. The second approach is simpler and requires less computational time. Also, if the main interest is focused on the heat-transfer process, the modified single-phase approach is more convenient than the two-phase model. For example, based on computational fluid dynamics (CFD) simulations, Fard et al. [32] found that the two-phase approach generates better predictions of nanofluid convective heat transfer compared with the single-phase model, Lofti et al. [33] numerically investigated convective heat transfer of nanofluids in horizontal tubes using a single-phase model and a two-phase mixture model as well as the two-phase Eulerian model. The comparison of calculated results with experimental values indicated that the two-phase mixture model was more precise.

Heyhat and Kowsary [34] investigated numerically the effect of nanoparticle migration on the flow pattern and convective heat transfer in a circular pipe. They claimed that the nonuniform nanoparticle distribution led to a higher heat-transfer coefficient while the wall shear stress decreased. Akbarina et al. [35] considered alumina–water nanofluid flow in two-dimensional rectangular microchannels to study heat-transfer enhancement due to the addition of nanoparticles to the base fluid, especially at low Reynolds numbers. They found that for a given Reynolds number, the major enhancement in the Nusselt number was not so much caused by higher nanoparticle concentrations but mainly due to an increase in flow rate to reach a set Reynolds number. Hence, constant Reynolds number studies of nanofluid flow may be insufficient when evaluating the heat-transfer characteristics. Kondaraju et al. [36] focused on deviations in experiments concerning the effective thermal conductivity of polydisperse nanoparticles. Both the initial particle distribution and any inhomogeneous coagulation were found to be major factors in the determination of effective thermal conductivity results for multisize particles in nanofluids. Concerning nanoparticle aggregation, Werth et al. [37] analyzed different particle forces leading to agglomeration of charged nano-powders. Jiang et al. [38] developed a model for the prediction of nanoparticle aggregation and sedimentation. Focusing on nonspherical nanoparticles, Evans et al. [39] employed Monte Carlo simulations and determined a positive impact of aggregation on the effective thermal conductivity.

For microchannel heat-sink applications, Li and Kleinstreuer [40] analyzed entropy generation of pure water and CuO–water nanofluid flow in trapezoidal microchannels. Similarly, Singh et al. [41] investigated the entropy generation of nanofluids due to flow friction and heat transfer for alumina–water nanofluids in tubes of three different diameters. An alternative approach to assuming that a nanofluid is a homogeneous mixture, Kalteh et al. [42] proposed Eulerian–Eulerian two-phase simulations to compute possible heat-transfer enhancement in microchannel flow.

9.1.4 MIXTURE VISCOSITY MODELS

Clearly, coarse and even fine particles added to liquid coolants (before the advent of nanofluids) may measurably increase the mixture's *effective viscosity*, which relates to high pumping power, sedimentation, and filter clogging. So, parallel to improved heat-transfer performance when using nanofluids, mixture viscosity data and models have to be considered. Focusing on viscosity models for nanofluids, Pak and Cho [11], for example, selected 13-nm alumina nanoparticles in water with volume fractions, φ, from 1.33 to 2.86% and measured μ_{eff} / μ_{bf} values ranging from 1.63 to 2.6. In contrast, Wang et al. [43] recorded significantly lower viscosity ratios (i.e., 1.12–1.88 when $1.02 < \varphi < 5\%$) for a 28-nm Al_2O_3–water mixture. More recent experimental studies (see [44–46] among others) confirmed the nonlinear dependence of the effective mixture viscosity on various mixture parameters. Thus, earlier correlations proposed by Einstein [47] and improved upon by Brinkman [48], Batchelor [49], and Graham [50], which included only the nanoparticle volume fraction, have to be revisited. For example, Koo and Kleinstreuer [51] suggested $\mu_{eff} = \mu_{static} + \mu_{Brownian}$, where the viscosity enhancement due to Brownian nanoparticle motion is a function of k_{nf}/k_{bf}, μ_{bf}, and the Prandtl number. Similarly, Masoumi et al. [52] postulated that μ_{eff} is the sum of μ_{bf} plus an apparent viscosity which is mainly caused by Brownian motion. Traditionally, the wall roughness of microchannels also contributes to an increase in friction factor and hence the required pumping power; however, more modern fabrication techniques may keep the (or result in) relative surface roughness typically below 0.5%.

9.2 THEORY

For the selected test case, it is assumed that the continuum hypothesis is valid, that is, the channel hydraulic diameter $D_h > 100\ \mu m$ (see Chapter 7 in ref. [53]). Thus, for steady three-dimensional laminar incompressible nanofluid flow in a microchannel, the continuity, momentum, energy, and species mass transfer equations have to be solved, considering temperature and volume-fraction-dependent mixture properties. In addition to the conservation laws and models for the mixture properties, the second law of thermodynamics has to be formulated for system optimization via reduction in entropy generation rate.

9.2.1 GOVERNING EQUATIONS

Continuity:

$$\frac{\partial}{\partial x_i}(\rho u_i) = 0 \tag{9.1}$$

Momentum:

$$u_j \frac{\partial}{\partial x_j}(\rho u_i) = -\frac{\partial p}{\partial x_i} + \frac{\partial}{\partial x_j}\left[\mu\left(\frac{\partial u_i}{\partial x_j} + \frac{\partial u_j}{\partial x_i}\right)\right] \tag{9.2}$$

Energy (within fluid):

$$u_i\left(\frac{\partial(\rho c_p T)}{\partial x_i}\right) = \frac{\partial}{\partial x_i}\left(k\frac{\partial T}{\partial x_i}\right) + \mu\Phi \tag{9.3}$$

where the dissipation function reads:

$$\Phi = \left(\frac{\partial u_i}{\partial x_j} + \frac{\partial u_j}{\partial x_i}\right)\frac{\partial u_i}{\partial x_j} \tag{9.4}$$

Energy (within solid):

$$k_s\frac{\partial^2 T}{\partial x_i^2} = 0 \tag{9.5}$$

For nanofluid flow and pure fluid flow, the corresponding physical properties are the density, thermal conductivities k_{nf} and k_{bf} and the viscosities μ_{nf} and μ_{bf}, respectively (see Section 9.2.2). The thermal conductivity of the silicon microchannel wall k_s was assumed to be constant. Uniform velocities were applied at the channel inlet, that is, $u = 0$, $v = 0$, $w = U_{in}$ to simulate any entrance effect. Exposed to the atmosphere, the outlet pressure was the static pressure, that is, $p_{gage} = 0$. The no-slip boundary condition was enforced at all solid walls. The thermal boundary condition at the bottom was a constant wall heat flux, whereas an adiabatic boundary condition is imposed on the top wall, symmetric boundary condition at the two side walls, and $T = T_0$ at the microchannel inlet. These thermal boundary conditions (see Figure 9.1) are standard assumptions, where the channel-cover functions as a perfect insulator while the sidewalls of the machined or edged microchannel are equally exposed to the heat source, expressed as a constant wall-heat flux.

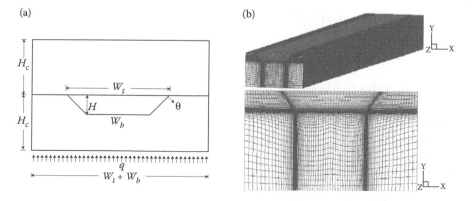

FIGURE 9.1 (a) Typical microchannel heat sink element and (b) finite volume mesh.

The models for the nanofluid properties, μ_{nf} and k_{nf}, are outlined in Section 9.2.2 and balance equations for entropy generation are discussed in Section 9.2.3, while the performance indicators are given in Section 9.2.4.

9.2.2 MIXTURE PROPERTIES

The basic nanofluid properties are a function of nanoparticle volume fraction φ and mixture temperature T. Such nanofluids are assumed to be dilute suspensions, that is, the homogeneous, noninteracting nanoparticles are well dispersed. Specifically, for dilute Al_2O_3–water nanofluids [54]:

$$\rho_{nf} = \varphi\rho_p + (1 - \varphi)\rho_{bf} \tag{9.6}$$

$$(\rho c_p)_{nf} = \varphi(\rho c_p)_p + (1 - \varphi)(\rho c_p)_{bf} \tag{9.7}$$

where the subscripts nf, bf, and p indicate nanofluid, base fluid, and particle, respectively, ρ is the density, c_p is the specific heat capacity, φ is the nanoparticle volume fraction.

The properties of the base fluid (water) are assumed to be temperature-dependent [54]:

$$\rho_{water} = 1000\left(1 - \frac{(\tilde{T} + 15.7914)}{508,929.2 \cdot (\tilde{T} - 205.0204)}(\tilde{T} - 277.1363)^2\right) \tag{9.8a}$$

$$c_{p,water} = 9616.873445 - 48.73648329 \cdot \tilde{T} \\ + 0.1444662 \cdot \tilde{T}^2 - 0.000141414 \cdot \tilde{T}^3 \tag{9.8b}$$

$$\mu_{water} = 0.02165 - 0.0001208 \cdot \tilde{T} + 1.7184e - 7 \cdot \tilde{T}^2 \tag{9.8c}$$

$$k_{water} = -1.1245 + 0.009734 \cdot \tilde{T} - 0.00001315 \cdot \tilde{T}^2 \tag{9.8d}$$

where $\tilde{T} = T/(1[K])$ is the nondimensional temperature.

9.2.2.1 Effective Dynamic Viscosity

Most of the reported data for nanofluid viscosities have been discussed in terms of formulations proposed by Einstein [47], Brinkman [48], Batchelor [49], and Graham [50], to name a few. The conventional viscosity models of nanofluids are summarized in Table 9.1. It turns out that none of the models mentioned can predict the viscosity of nanofluids very well for a wide range of nanoparticle volume fraction.

Nguyen et al. [55] investigated experimentally the influence of both the temperature and the particle size on the dynamic viscosity for two nanofluids, that is,

TABLE 9.1
Conventional Viscosity Models for Nanofluids

Model	Expression	Comments
Einstein [47]	$\mu_{nf} = \mu_{bf}(1 + 2.5\varphi)$	Spherical particles and low-volume fraction, that is, $\varphi < 2\%$
Brinkman [48]	$\mu_{nf} = \dfrac{\mu_{bf}}{(1-\varphi)^{2.5}}$	Extended Einstein expression
Batchelor [49]	$\mu_{nf} = \mu_{bf}(1 + 2.5\varphi + 6.5\varphi^2)$	Extended Einstein equation by considering the effect of Brownian motion on the bulk stress
Graham [50]	$\mu_{nf} = \mu_{bf}\left(1 + 2.5\varphi + 4.5\left[\dfrac{1}{(c/d_p)(2 + (c/d_p))(1 + (c/d_p))^2}\right]\right)$	d_p is the particle diameter and c is the inter-particle spacing

Al_2O_3–water and CuO–water combinations. More recently, employing a different measurement technique, Chandrasekar et al. [31] got similar viscosity data for Al_2O_3–water nanofluids.

Abu-Nada [56] performed a two-dimensional regression on experimental data of Nguyen et al. [55] and developed the following relation of viscosity in centi-poise as a function of the nondimensional temperature $\tilde{T}$ and volume fraction φ, which had a maximum error of 5%.

$$\mu_{nf\_Al_2O_3} = -0.155 - \frac{19.582}{\tilde{T}} + 0.794\varphi + \frac{2094.47}{\tilde{T}^2} - 0.192\varphi^2 - 8.11\frac{\varphi}{\tilde{T}}$$
$$- \frac{27,463.863}{\tilde{T}^3} + 0.0127\varphi^3 + 1.6044\frac{\varphi^2}{\tilde{T}} + 2.1754\frac{\varphi}{\tilde{T}^2} \qquad (9.9)$$

Masoumi et al. [52] proposed a new model by including the effect of Brownian motion of the nanoparticles on the viscosity of nanofluids in terms of $\mu_{nf} = \mu_{bf} + \mu_{app}$, where μ_{app} is the apparent viscosity defined by Masoumi et al. [52]. Temperature, nanoparticle diameter, volume fraction, nanoparticle density as well as the base fluid physical properties were all considered.

$$\mu_{nf} = \mu_{bf} + \frac{\rho_p v_B d_p^2}{72C\delta} \qquad (9.10)$$

where $v_B = (1/d_p)\sqrt{(18\kappa_b T/\pi\rho_p d_p)}$ is the Brownian velocity, κ_b is the Boltzmann constant, $\delta = \sqrt[3]{(\pi/6\varphi)}d_p$ is the distance between nanoparticles, $C = \text{fct}(\mu_{bf}, d_p, \varphi)$ is the correction factor.

The correlation factor C was calculated by using experimental data for water-based nanofluids consisting of 13 and 28 nm Al$_2$O$_3$ nanoparticles as well as the 36 nm Al$_2$O$_3$–water nanofluid by Nguyen et al. [55].

For a better understanding the performance of nanofluids in microchannels, the extended Einstein's equation by Brinkman [48] and the effective nanofluid viscosity postulated by Masoumi et al. [52] were analyzed. Figure 9.2 compares the two models as a function of volume fraction and three different nanoparticle sizes for the model by Masoumi et al. [52]. As already noticed by Ngugen et al. [55], the conventional Brinkman model (see Table 9.1) underpredicts the nanofluid viscosity. Furthermore, the functional dependence μ_{nf} (φ) is highly nonlinear in the model by Masoumi et al. [52], especially for $\varphi > 3\%$ and $d_p < 40$ nm (see Figure 9.2a). Such a viscosity enhancement of nanofluids may increase the pressure drop or the requirement of pumping power. As expected, the temperature influence on μ_{nf} (see Figure 9.2b) is much less dramatic, being expressed as μ_{bf} (T).

9.2.2.2 Effective Thermal Conductivity

For the thermal performance analysis of nanofluid flow in microconduits, several thermal conductivity models have been employed (see [57] among others). In this chapter, three different models were applied and compared, that is, the conventional Maxwell model, the correlation by Patel et al. [58], as well as the newly developed Feng–Kleinstreuer (F–K) model [57]. Maxwell [1] derived a "static" thermal conductivity model for conventional fluids containing at that time micrometer/millimeter particles. It was assumed that the effective thermal conductivity of the mixture is a function of the thermal conductivity of the suspensions and the base fluid as well as the volume fraction of the suspensions:

$$k_{static} = \left(1 + \frac{3\left(\dfrac{k_p}{k_{bf}} - 1\right)\varphi}{\left(\dfrac{k_p}{k_{bf}} + 2\right) - \left(\dfrac{k_p}{k_{bf}} - 1\right)\varphi} \right) k_{bf} \qquad (9.11)$$

The emergence of nanofluids as a new field of nanoscale heat transfer, with applications to microsystem cooling, is directly related to miniaturization trends and nanotechnology. The unexpected high thermal conductivity of nanofluids documented in many experiments showed that the effective thermal conductivity of nanofluids depends not only on the nanostructures of the suspensions but also on the dynamics of nanoparticles in liquid. Kleinstreuer and Feng [57] postulated that the thermal conductivity of nanofluids consists of a static part (k_{static}) after Maxwell [1] and a micromixing part (k_{mm}), that is, enhancement due to Brownian motion of nanoparticles. The F–K model can be expressed as

$$k_{nf} = k_{static} + k_{mm} \qquad (9.12)$$

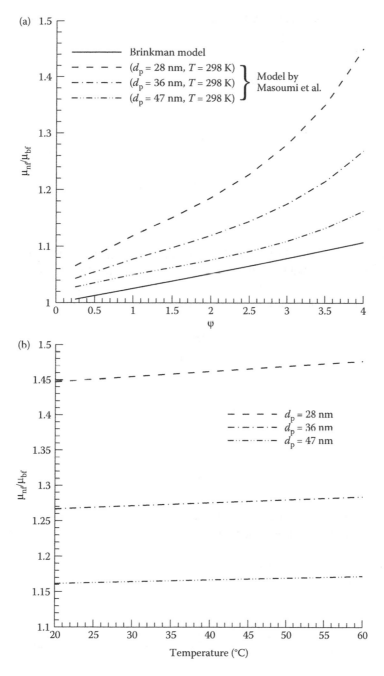

FIGURE 9.2 Dynamic viscosity models for nanofluids: (a) viscosity versus volume fraction and (b) viscosity change with temperature ($\varphi = 4\%$, model by Masoumi et al. [52]).

The static part is Equation 9.11, whereas the micromixing part, based on sound physics, is given by (see [57])

$$k_{mm} = 49,500 \cdot \frac{\kappa_B \tau_p}{2m_p} \cdot C_c \cdot \left(\rho c_p\right)_{nf} \cdot \varphi^2$$

$$\cdot \left(T \ln T - T\right) \cdot \frac{\exp(-\zeta \omega_n \tau_p) \sinh\left(\sqrt{\dfrac{\left(3\pi\mu_{bf}d_p\right)^2}{4m_p^2} - \dfrac{K_{P-P}}{m_p}} \dfrac{m_p}{3\pi\mu_{bf}d_p}\right)}{\tau_p \sqrt{\dfrac{\left(3\pi\mu_{bf}d_p\right)^2}{4m_p^2} - \dfrac{K_{P-P}}{m_p}}} \qquad (9.13)$$

Here, C_c is equal to 38 for metal oxide nanofluids which can be derived theoretically (which also holds for the number 49,500), instead of being obtained via a curve-fitting technique [59].The damping coefficient ζ, natural frequency ω_n, and characteristic time interval τ_p can be expressed as

$$\zeta = \frac{3\pi d_p \mu_{bf}}{2m_p \omega_n} \qquad (9.14)$$

$$\omega_n = \sqrt{\frac{K_{P-P}}{m_p}} \qquad (9.15)$$

$$\tau_p = \frac{m_p}{3\pi\mu_{bf}d_p} \qquad (9.16)$$

Specifically, for metal oxide nanofluids, the magnitude of particle–particle interaction intensity K_{P-P} is determined for different particle diameters as

$$K_{P-P} = \rho_p\sqrt{d_p \cdot 10^{-9}}\left(\frac{32.1724 \cdot 273K}{T} - 19.4849\right) \quad \text{for } 20\,\text{nm} < d_p \leq 50\,\text{nm}$$

$$(9.17)$$

$$K_{P-P} = \rho_p\sqrt{d_p \cdot 10^{-9}}\left(\frac{24.6402 \cdot 273K}{T} - 18.7592\right) \quad \text{for } d_p > 50\,\text{nm} \quad (9.18)$$

In light of experimental evidence, the F–K model is suitable for several types of metal oxide nanoparticles in water with volume fractions up to 5% and mixture temperatures below 350 K.

In contrast, Patel et al. [58] provided a correlation for the effective thermal conductivity of nanofluids, based on a regression analysis of several experimental data sets:

$$k_{nf} = k_{bf}(1 + 0.135 \times (k_p/k_{bf})^{0.273} \times \varphi^{0.467} \times [(T - 273)/20]^{0.547} \times (100/d_p)^{0.234})$$

(9.19)

where T is the temperature of nanofluids in degree Kelvin; d_p is the average nanoparticle diameter in nanometers. Apparently, the correlation is valid for suspensions of spherical nanoparticles of 10–150 nm diameter, a thermal conductivity range of 20–400 W/mK; base fluids having thermal conductivities of 0.1–0.7 W/mK, particle volume fractions of 0.1–3%, and suspension temperatures from 20°C to 50°C. The model by Kleinstreuer and Feng [57] is based on physical principles without the use of empirical matching factors. In contrast, the k_{nf} correlation by Patel et al. [58], having a much simpler form, is easier to use but lacks physical insight and a broad range of applications.

Figures 9.3 and 9.4 compare the F–K model and Patel's correlation with some recent benchmark experimental data sets. Overall, the F–K model generates a better matching in trend and precision for different volume fractions and temperatures; although, the measured k_{nf} increase with nanoparticle volume fraction above 3% by Mintsa et al. [60] is surprisingly low (see Maxwell model prediction).

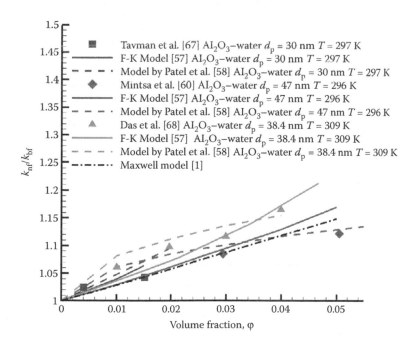

FIGURE 9.3 Comparison of thermal conductivity models for nanofluids at different volume fractions.

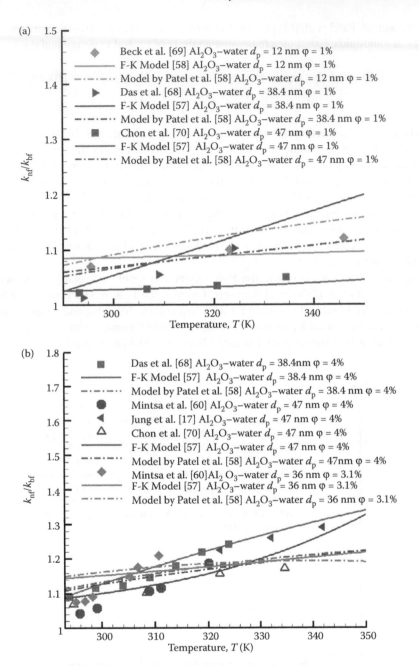

FIGURE 9.4 Comparison of thermal conductivity models for nanofluids at different temperature: (a) small volume fraction and (b) large volume fraction.

9.2.3 ENTROPY GENERATION

If the mixture is Newtonian and it obeys Fourier's law of heat conduction, the total entropy generation rate per unit volume ($S_{gen} \equiv S_G$ in W/K $\times$ m^3) can be expressed as [40]

$$S_{gen} \equiv S_G = \frac{k}{T^2}\left[\left(\frac{\partial T}{\partial x}\right)^2 + \left(\frac{\partial T}{\partial y}\right)^2 + \left(\frac{\partial T}{\partial z}\right)^2\right]$$

$$+\frac{\mu}{T}\left\{ \begin{array}{l} 2\left[\left(\frac{\partial u}{\partial x}\right)^2 + \left(\frac{\partial v}{\partial y}\right)^2 + \left(\frac{\partial w}{\partial z}\right)^2\right] + \left(\frac{\partial u}{\partial y} + \frac{\partial v}{\partial x}\right)^2 + \\ \left(\frac{\partial u}{\partial z} + \frac{\partial w}{\partial x}\right)^2 + \left(\frac{\partial v}{\partial z} + \frac{\partial w}{\partial y}\right)^2 \end{array} \right\} \tag{9.20}$$

Equation 9.20 encapsulates the irreversibilities due to heat transfer and frictional effects, that is,

$$S_{gen} = S_{gen}(\text{thermal}) + S_{gen}(\text{frictional}) \tag{9.21}$$

Specifically, the dimensionless entropy generation rate induced by fluid friction can be defined as follows:

$$S_{G,F} = S_{gen}(\text{frictional})\frac{kT_0^2}{q^2} \tag{9.22a}$$

where

$$S_{gen}(\text{frictional}) = \frac{\mu}{T}\left\{2\left[\left(\frac{\partial u}{\partial x}\right)^2 + \left(\frac{\partial v}{\partial y}\right)^2 + \left(\frac{\partial w}{\partial z}\right)^2\right]\right.$$

$$\left. + \left(\frac{\partial u}{\partial y} + \frac{\partial v}{\partial x}\right)^2 + \left(\frac{\partial u}{\partial z} + \frac{\partial w}{\partial x}\right)^2 + \left(\frac{\partial v}{\partial z} + \frac{\partial w}{\partial y}\right)^2\right\} \tag{9.22b}$$

while for the thermal entropy source, we have

$$S_{G,T} = S_{gen}(\text{thermal})\frac{kT_0^2}{q^2} \tag{9.23a}$$

where

$$S_{gen}(\text{thermal}) = \frac{k}{T^2}\left[\left(\frac{\partial T}{\partial x}\right)^2 + \left(\frac{\partial T}{\partial y}\right)^2 + \left(\frac{\partial T}{\partial z}\right)^2\right] \tag{9.23b}$$

Finally,

$$S_{G,total} = S_{gen} \frac{kT_0^2}{q^2} = S_{G,F} + S_{G,T} \qquad (9.24)$$

where T_0 is the fluid inlet temperature, and q is the wall heat flux.

In order to assess the overall entropy generated in the entire flow field for different scenarios, the integral form is used:

$$\hat{S}_{G,total} = \frac{1}{\dot{m}c_p} \iiint_V S_{gen} \, dV \qquad (9.25)$$

where the fractions of entropy generation due to friction and heat transfer are, respectively:

$$\xi_F = \frac{\hat{S}_{G,F}}{\hat{S}_{G,total}} \quad \text{and} \quad \xi_T = \frac{\hat{S}_{G,}}{\hat{S}_{G,tc}} \qquad (9.26a,b)$$

9.2.4 THERMAL PERFORMANCE COMPARISONS

Pumping power is needed to drive the working fluid in microchannels, which is defined as the product of the pressure drop across the channel and the volumetric flow rate:

$$P = \Delta p \cdot Q \qquad (9.27)$$

In order to compare the thermal performance of nanofluids with pure liquids, the thermal resistance is employed:

$$\theta = \frac{T_{w,ave} - T_{in}}{q} \qquad (9.28)$$

where $T_{w,ave}$ is the average wall temperature, T_{in} is the fluid inlet temperature, and q is the heat added to the microchannel (or the heat removed by the moving fluid).

9.2.5 NUMERICAL METHOD

The numerical solution of the Eulerian transport equations were carried out with a user-enhanced, unstructured finite-volume-based program, that is, CFX 12.1 from ANSYS, Inc. (Canonsburg, PA). The computations were performed on an IBM Linux Cluster at North Carolina State University's High Performance Computing Center (Raleigh, NC) and on a local dual Xeon Intel 3.2G Dell desktop (C M-P Laboratory,

MAE Department, NC State University). Mesh independence was examined and verified by increasing the nodal number by 50% which produced a maximum result change of just 1.33% for the velocity and 0.8% for the temperature field. The unstructured mesh for a typical case contained 671,224 hexahedral elements with 631,050 nodes for the fluid domain and 269,486 elements with 247,266 nodes for the solid domain. Figure 9.1 shows the representative geometry and the finite volume mesh we generated. Mesh independence was examined and verified by increasing the nodal number by 50% which produced a maximum result change of just 1.23%. Furthermore, the solutions of the flow field were assumed to be converged when the dimensionless mass and momentum as well as the thermal energy residual ratios were below 10^{-6}. Improving the convergence criteria to $<10^{-7}$ had a negligible effect on the simulation results. A typical simulation run took about 24 h. Additional model validations were achieved by comparing numerical results of velocity and temperature fields with an analytical solution as well as existing numerical and experimental data sets, as given in Section 9.3.

9.3   RESULTS AND DISCUSSION

9.3.1   FLOW FRICTION VALIDATION STUDIES

The Fanning friction factor was used to evaluate viscous effects of flow through microchannels. The Fanning friction factor f, representing the ratio of fluid shearing strength at the wall to the average kinetic energy of the fluid per unit volume, is defined as

$$f = \frac{\Delta p D_h}{2\rho U^2 L} \tag{9.29}$$

where Δp is the pressure drop between the inlet and outlet of the microchannel, ρ is the density of working fluid, D_h and L are the hydraulic diameter and the length of the microchannel, respectively, and U is the average fluid velocity.

For fully developed channel flow, the friction factor has an inverse correlation with the Reynolds number, that is, fRe should be a constant for a particular geometry. The fully developed velocity profile was first computed using an extended domain upstream, and the outlet velocity profile from this upstream domain was applied as the inlet velocity profile to the primary domain. Adopting the same geometry/aspect ratios and operational parameters, we compared our simulation results with the analytical result for a mini tube (fRe = 16) with diameter of 1.812 mm and the analytical result of Shah [61] for a trapezoidal microchannel, that is, according to Shah [61], fRe = 14.637 for base angle $\alpha = 60°$, $H/W_b = 0.75$. $D_h = 187$ μm has been used for this validation. As shown in Figure 9.5, the simulation results match the experimental correlations for both mini-tube and trapezoidal microchannel. It is also apparent that for fully developed flow employing 4% alumina–water nanofluid, no measurable increase in the friction factor occurred. Numerically, for the Reynolds number range of 100–1000, fRe = 16.28 ~ 16.29 in the mini-tube and fRe = 14.58 ~ 14.62 for flow in the microchannel; thus, the maximum change is <2%.

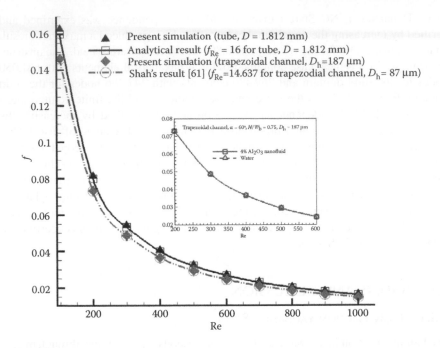

FIGURE 9.5 Model validation (friction factor vs. the Reynolds number).

9.3.2 Convective Heat-Transfer Validation Results

To compare convective heat transfer characteristics, pure water in a mini-tube was investigated. As shown in Figure 9.6a, the present simulation shows a very good match with the experimental data of Hwang et al. [27] as well as the correlation by Shah [62,63]. For an additional validation of the F–K model (see Equations 9.11 through 9.18, and Figures 9.3 and 9.4), its thermal performance of water-based alumina nanofluid flow in a 1.02-mm tube was compared with the experimental data of Lai et al. [64]. Employing the same geometry and operational conditions (i.e., a volume flow rate of 5 mL/min), the use of the F–K model generates acceptable results for the local heat transfer coefficient (see Figure 9.6b).

9.3.3 Friction Factor and Pressure Drop Results

As indicated in Section 9.2, the friction factor and pressure drop as well as the thermal performance of alumina nanofluids were computed for a particular trapezoidal microchannel with hydraulic diameter $D_h = 194.5$ μm, base angle $\alpha = 54.7°$, and length $L = 3$ cm. The conventional dynamic viscosity model by Brinkman [48] and the newly developed model by Masoumi et al. [52] were compared, the new F–K model for heat transfer enhancement was applied, and entropy generation of thermal nanofluid flow in the microchannel was computed as well.

  Figure 9.7 compares the pressure gradients at different Reynolds numbers for water and different nanofluid parameters. Although the conventional Brinkman

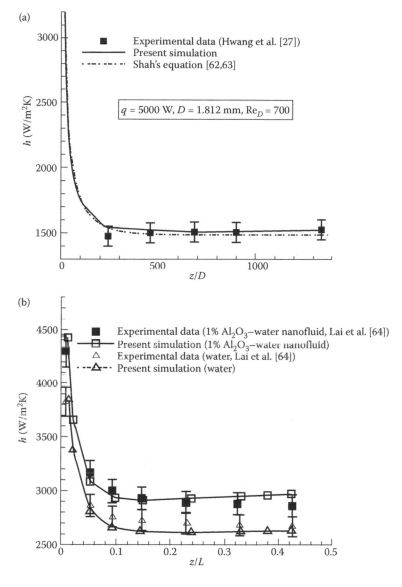

FIGURE 9.6 (a) and (b) F–K model validation by comparing the heat transfer coefficient in a mini-tube with existing experimental data and theoretical prediction.

viscosity model with a 1% nanoparticle volume fraction generates almost the same results as for pure water, the pressure gradient increases when employing the more realistic model by Masoumi et al. [52], especially when the particle size is small and/or the volume fraction is large. Of practical interest is the power requirement necessary to generate different pressure drops across the microchannel length (see Figure 9.8). Fortunately, there is not much of a difference in pressure drops when using nanofluids with relatively small volume fractions, that is, about a 4% increase

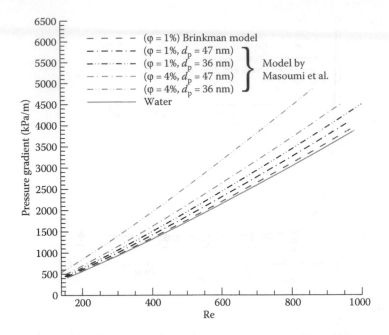

FIGURE 9.7 Pressure gradient versus the Reynolds number for nanofluid flow using two different viscosity models.

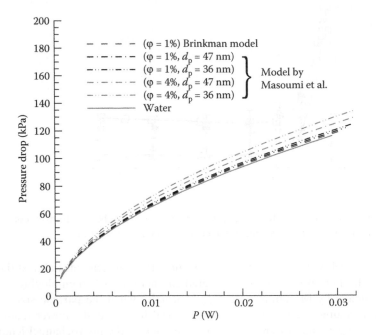

FIGURE 9.8 Pressure drop versus pumping power for nanofluid flow using two different viscosity models.

for Al_2O_3–water with 1% volume fraction and 47 nm particles and approximately 7% for $\varphi = 4\%$ and $d_p = 36$ nm, when employing the viscosity model by Masoumi et al. [52]. Thus, low-concentration nanofluids do not require much more additional pumping power for mixture flow in microchannels. Chein and Chuang [65] and Li and Kleinstreuer [66] provided very similar results based on experiments and numerical simulations, respectively.

9.3.4 Convective Heat Transfer

In order to compare the thermal performance of nanofluid flow in the microchannel, the local heat-transfer coefficient as well as the thermal resistance for different pumping powers were compared for pure water flow and nanofluid flow with different volume fractions, employing the viscosity model by Masoumi et al. [52] and the k_{nf} model by Kleinstreuer and Feng [57].

The local heat-transfer coefficient developing along the microchannel at the same inlet Reynolds number and heat flux for different fluids is shown in Figure 9.9a. As expected, the local heat transfer coefficient increases when using nanofluids, where smaller nanoparticles yield elevated heat transfer coefficients. It should be noted that the Reynolds number of nanofluids depends on the kinematic viscosity of the nanofluids, which increases with an increase in nanoparticle volume fraction and smaller nanoparticles, implying that the inlet velocity should be increased to keep the Reynolds number constant. Thus, to eliminate the dependence of h_z on a varying U_{in}, Figure 9.9b depicts the heat-transfer enhancement for $U_{in} = $ constant, demonstrating that the heat-transfer coefficient increases due to higher nanoparticle volume fractions and lower nanoparticle diameters. Interestingly, the thermal resistance (see Equation 9.28) decreases when employing nanofluids, especially for nanofluids with larger volume fractions and smaller nanoparticles (see Figure 9.10). The reason is that the average wall temperature is lower due to the higher thermal performance of nanofluids without a significant increase in pumping power. Overall, the average enhancement of thermal performance for Al_2O_3–water with a volume fraction of 4% is about 10% for 36 nm nanoparticles and 7% for 47 nm nanoparticles.

9.3.5 Entropy Generation

Focusing on the impact of inlet temperature and the Reynolds number, entropy generation in trapezoidal microchannels is compared for flow of pure water and the 47-nm alumina–water nanofluid. As shown in Figure 9.11a, for a constant inlet velocity total entropy generation decreases slightly with increasing inlet temperature. Assuming a constant inlet Reynolds number produces quite a different outcome, that is, $S_{gen}(T_{in})$ increases measurably (Figure 9.11b). For both cases nanofluids generate lower entropy rates than pure water. The thermal conductivity of the nanofluid is enhanced with an increase in bulk temperature (T_{in}), yielding lower temperature gradients (see Equation 9.20). As a result, total entropy generation is decreased with the increase of inlet temperature. In contrast, to maintain a constant inlet Reynolds number, a lower inlet velocity is computed to balance the effects of viscosity and density changes resulting from the increasing inlet temperature. The lower bulk velocity

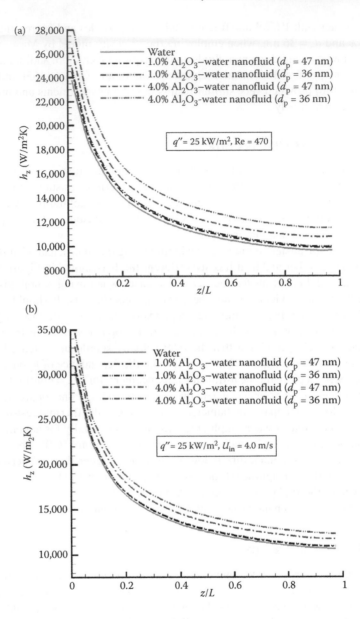

FIGURE 9.9 Local heat-transfer coefficient developing along the microchannel using the new viscosity model by Masoumi et al. [52] and F–K thermal conductivity model: (a) at constant Reynolds number (b) at constant inlet velocity.

caused larger temperature gradients which in turn increased the total entropy generation. Clearly, nanofluids generate less entropy than pure water (see Figures 9.11 and 9.12) because of the milder velocity and temperature gradients due to flatter velocity profiles and better heat transfer in the microchannel. The total entropy generation decreases with higher Reynolds numbers (Figure 9.12). This is quite beneficial when

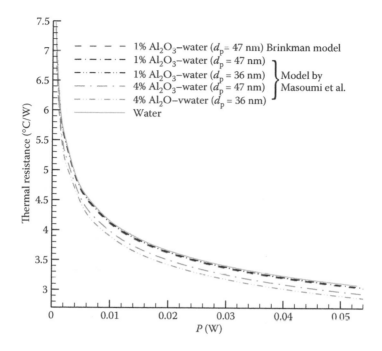

FIGURE 9.10 Thermal resistance versus pumping power considering different viscosity models for different nanofluids.

dealing with high heat flux conditions, for example, heat sinks, where heat-transfer-induced entropy generation is dominant, especially at low Reynolds numbers. However, at higher Reynolds numbers the frictional part of entropy generation (see Equation 9.22) may be dominant. Hence, it may be desirable to operate a microsystem in a Reynolds number range associated with relatively low S_{gen} values. In addition to the Reynolds number, the channel geometry (i.e., aspect ratio) is also an important parameter for the minimization of a system's total entropy generation, as pointed out by Li and Kleinstreuer [40].

9.4 CONCLUSION

Actual thermal conductivity enhancement of nanofluids k_{nf} over the "effective medium" theory of Maxwell is still subject to debate; although, most experimentalists reported measurably elevated k_{nf} values. Associated theoretical models addressing all the important physical phenomena explaining possible k_{nf} enhancement are needed as well. After a brief review of the most recent papers concerning k_{nf} experimental observations and theoretical modeling, steady laminar thermal nanofluid flow and entropy generation in a trapezoidal microchannel were numerically analyzed. Specifically, two different viscosity models and thermal conductivity models for water flow and alumina–water nanofluid flow were compared after extensive computer model validations. The results show that nanofluids do measurably enhance the

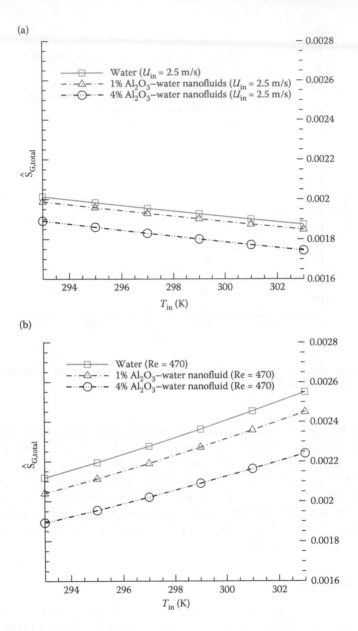

FIGURE 9.11 Entropy generation: (a) varying inlet temperature for the same inlet velocity and (b) varying inlet temperature for the same Reynolds number.

thermal performance of microchannel mixture flow with a small increase in needed pumping power. Nanofluids with smaller nanoparticles at the same volume fraction exhibit a better convective heat-transfer performance, but they require more pumping power. The entropy-generation analysis indicates that S_{gen} reduces when using nanofluids due to their improved thermal transport mechanism.

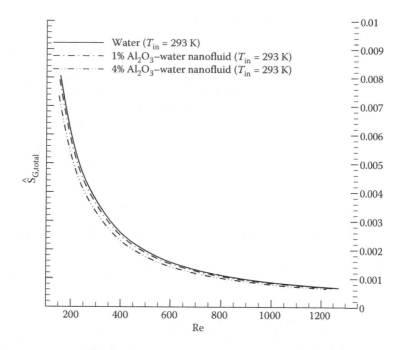

FIGURE 9.12 Entropy generation versus Reynolds number for different fluids.

ACKNOWLEDGMENT

The use of ANSYS-CFX12.1 (Ansys, Inc., Canonsburg, PA) is gratefully acknowledged.

REFERENCES

1. J.C. Maxwell, *A Treatise on Electricity and Magnetism*, 3rd Edition, Clarendon Press, Oxford, UK, 1891.
2. R.L. Hamilton, O. K. Crosser, Thermal conductivity of heterogeneous two component systems, *Industrial and Engineering Chemistry Fundamentals*, 1, 187–191, 1962.
3. C. Kleinstreuer, Y. Feng, Experimental and theoretical studies of nanofluid thermal conductivity enhancement: A review, *Nanoscale Research Letters*, 6, 229, 2011.
4. S. Ozerinc, S. Kakac, A. Guvenc, Y. Yazicioglu, Enhanced thermal conductivity of nanofluids: A state-of-art review, *Microfluidics and Nanofluidics*, 8, 145–170, 2010.
5. S.M.S. Murshed, K.C. Leong, C. Yang, Thermophysical and electrokinetic properties of nanofluids: A critical review, *Applied Thermal Engineering*, 28, 2109–2125, 2008.
6. E.V. Timofeeva, W. Yu, D.M. France, D. Singh, J.L. Routbort, Nanofluids for heat transfer: An engineering approach, *Nanoscale Research Letters*, 6, 182, 2011.
7. J. Fan, L. Wang, Review of heat conduction in nanofluids, *ASME Journal of Heat Transfer*, 133, 040801-1–14, 2011.
8. S.K. Das, S. U.S. Choi, W. Yu, T. Pradeep, *Nanofluids: Science and Technology*, John Wiley & Sons, Inc., Publication, New Jersey, USA, 2008.
9. D. Li (Editor), *Encyclopedia of Microfluidics and Nanofluidics*, Springer, Amsterdam, Berlin, 2008.

10. C.S. Kumar (Editor), *Microfluidic Devices in Nanotechnology: Applications*, John Wiley & Sons, Inc., Publication, Hoboken, New Jersey, 2010.

11. B.C. Pak, Y.I. Cho, Hydrodynamic and heat transfer study of dispersed fluids with submicron metallic oxide particles, *Experimental Heat Transfer*, 11, 151, 1998.

12. Y. Xuan, W. Roetzel, Conceptions for heat transfer correlation of nanofluids, *International Journal of Heat and Mass Transfer*, 43, 3701–3707, 2000.

13. D. Wen, Y. Ding, Experimental investigation into convective heat transfer of nanofluids at the entrance region under laminar flow conditions, *International Journal of Heat Mass Transfer*, 47, 5181–5188, 2004.

14. S.Z. Heris, M.N. Esfahany, G. Etemad, Investigation of CuO/water nanofluid laminar convective heat transfer through a circular tube, *Journal of Enhanced Heat Transfer*, 13, 279–289, 2006.

15. S.Z. Heris, M.N. Esfahany, S. Gh. Etemad, Experimental investigation of convective heat transfer of Al$_2$O$_3$/water nanofluid in circular tube, *International Journal of Heat Fluid Flow*, 28, 203–210, 2007.

16. J. Buongiorno, Convective transport in nanofluids, *ASME Journal of Heat Transfer*, 128, 240–250, 2006.

17. J.-Y. Jung, H.-S. Oh, H.-Y. Kwak, Forced convective heat transfer of nanofluids in microchannels, *International Journal of Heat and Mass Transfer*, 52, 466–472, 2009.

18. U. Rea, T. McKrell, L. Hu, J. Buongiorno, Laminar convective heat transfer and viscous pressure loss of aluminar-water and zirconia-water nanofluids, *International Journal of Heat and Mass Transfer*, 52, 2042–2048, 2009.

19. P.S. Vajjha, D.K. Das, D.P. Kulkarni, Development of new correlations for convective heat transfer and friction factor in turbulent regime for nanofluids, *International Journal of Heat and Mass Transfer*, 53, 4607–4618, 2010.

20. R. H. Khiabani, Y. Joshi, C.K. Aidun, Heat transfer in microchannels with suspended solid particles: Lattice-Boltzmann based computations, *ASME Journal of Heat Transfer*, 132, 041003–1–9, 2010.

21. M. Hojjat, S.Gh. Etemad, R. Bagheri, J. Thibault, Convective heat transfer of non-Newtonian nanofluids through a uniformly heated circular tube, *International Journal of Thermal Science*, 50, 525–531, 2011.

22. J. Hemalatha, T. Prabhakaran, R. P. Nalini, A comparative study on particle-fluid interactions in micro and nanofluids of aluminium oxide, *Microfluidics and Nanofluidics*, 10, 263–270, 2011.

23. W. Escher, T. Brunschwiler, N. Shalkevick, A. Shalkevich, T. Burgi, B. Michel, D. Poulikakos, On the cooling of electronics with nanofluids, *ASME Journal of Heat Transfer*, 133, 051401–1–11, 2011.

24. P.E. Gharagozloo and K.E. Goodson, Temperature-dependent aggregation and diffusion in nanofluids, *International Journal of Heat and Mass Transfer*, 54, 797–806, 2011.

25. H. Xie, W. Yu, Y. Li, L. Chen, Discussion on the thermal conductivity enhancement of nanofluids, *Nanoscale Research Letters*, 6, 124, 2011.

26. W.-H. Lee, C.-K. Rhee, J. Koo et al., Round-robin test on thermal conductivity measurement of ZnO nanofluids and comparison of experimental results with theoretical bounds, *Nanoscale Research Letters*, 6, 258–282, 2011.

27. K.S. Hwang, S.P. Jang, S.U.S. Choi, Flow and convective heat transfer characteristics of water based Al$_2$O$_3$ nanofluids in fully developed laminar flow regime, *International Journal of Heat and Mass Transfer*, 52, 193–199, 2009.

28. X. Wu, H. Wu, P. Cheng, Pressure drop and heat transfer of Al$_2$O$_3$-H$_2$O nanofluids through silicon microchannels, *Journal of Micromechanics and Microengineering*, 19, 105020-1-11, 2009.

29. D. Kim, Y. Kwon, Y. Cho et al., Convective heat transfer characteristics of nanofluids under laminar and turbulent flow conditions, *Current Applied Physics*, 9, e119123, 2009.

30. H. Xie, Y. Li, W. Yu, Intriguingly high convective heat transfer enhancement of nano-fluid coolants in laminar flows, *Physics Letters A*, 374, 2566–2568, 2010.
31. M. Chandrasekar, S. Suresh, A.C. Bose, Experimental studies of heat transfer and friction factor characteristics of Al_2O_3/water nanofluid in a circular pipe under laminar flow with wire coil inserts, *Experimental Thermal and Fluid Science*, 34, 122–130, 2010.
32. M. H. Fard, M. N. Esfahany, M.R. Talaie, Numerical study of convective heat transfer of nanofluids in a circular tube two-phase model versus single-phase model, *International Communications in Heat and Mass Transfer*, 37, 91–97, 2009.
33. R. Lotfi, Y. Saboohi, A.M. Rashidi, Numerical study of forced convective heat transfer for nanofluids: Comparison of different approaches, *International Communications in Heat and Mass Transfer*, 37, 74–78, 2010.
34. M.M. Heyhat, F. Kowsary, Effect of particle migration on flow and convective heat transfer of nanofluids flowing through a circular pipe, *ASME Journal of Heat Transfer*, 132, 062401-1-9, 2010.
35. A. Akbarinia, M. Abdolzadeh, R. Laur, Critical investigation of heat transfer enhancement using nanofluids in microchannels with slip and non-slip flow regiomes, *Applied Thermal Engineering*, 31, 556–565, 2011.
36. S. Kondaraju, E.K. Jin, J.S. Lee, Effect of the multi-sized nanoparticle distribution on the thermal conductivity of nanofluids, *Microfluidics and Nanofluidics*, 10, 133–144, 2011.
37. J. H. Werth, M. Linsenuhler, S.M. Dammer, Z. Farkas, H. Hinichsen, K.E. Wirth, D.E. Wolf, Agglomeration of charged nanopowders in suspensions, *Powder Technology*, 133, 106–112, 2011.
38. W. Jiang, G. Ding, H. Peng, H. Hu, Modeling of nanoparticles aggregation and sedimentation in nanofluid, *Current Applied Physics*, 10, 934–941, 2011.
39. W. Evans, R. Prasher, J. Fish, P. Mcakin, P. Phelan, P. Keblinski, Effect of aggregation and interfacial thermal resistance on thermal conductivity of nanocompsites and colloidal nanofluids, *International Journal of Heat and Mass Transfer*, 51, 1431–1438, 2008.
40. J. Li, C. Kleinstreuer, Entropy generation analysis for nanofluid flow in microchannels, *ASME Journal of Heat Transfer*, 132, 122401-1-8, 2010.
41. P.K. Singh, K.B. Anoop, T. Sundararajan, S.K. Das, Entropy generation due to flow and heat transfer in nanofluids, *International Journal of Heat and Mass Transfer*, 53, 4757–4767, 2010.
42. M. Kalteh, A. Abbassi, M. Saffar-Avval, J. Harting, Eulerian-Eulerian two-phase numerical simulation of nanofluid laminar forced convection in a microchannel, *International Journal of Heat and Fluid Flow*, 32, 107–116, 2011.
43. X. W. Wang, X. F. Xu, S. U. S. Choi, Thermal conductivity of nanoparticle-fluid mixture, *Journal of Thermal Physics and Heat Transfer*, 13, 474–480, 1999.
44. H. Chen, Y. Ding, C. Tan, Rheological behavior of nanofluids, *New Journal of Physics*, 9, 367-1–24, 2007.
45. C.T. Nguyen, F. Desgranges, N. Galanis et al., Viscosity data for Al_2O_3/water nano-fluids-hysteresis: is heat transfer enhancement using nanofluids reliable, *International Journal of Thermal Science*, 47, 103–111, 2008.
46. S.M.S. Murshed, K.C., Leong, C. Yang, Investigations of thermal conductivity and viscosity of nanofluids, *International Journal of Thermal Science*, 47, 560–568, 2008.
47. A. Einstein, A new determination of the molecular dimensions, *Annals of Physics*, 19, 289–306, 1906.
48. H.C. Brinkman, The viscosity of concentrated suspensions and solutions, *Journal of Chemistry Physics*, 20, 571–581, 1952.
49. G.K. Batchelor, The effect of Brownian motion on the bulk stress in the suspension of spherical particles, *Journal of Fluid Mechanics*, 128, 240, 1977.
50. A.L., Graham, On the viscosity of suspensions of solid spheres, *Applied Scientific Research* 37, 275, 1981.

51. J. Koo, C. Kleinstreuer, Laminar nanofluid flow in microheat-sinks, *International Journal of Heat and Mass Transfer*, 48 (13), 2652–2661, 2005.
52. N. Masoumi, N. Sohrabi and A. Behzadmehr, A new model for calculating the effective viscosity of nanofluids, *Journal of Physics D: Applied Physics*, 42, 055501-1–6, 2009.
53. C. Kleinstreuer, *Modern Fluid Dynamics*, Springer, New York, NY, USA, 2010.
54. Y. Feng, C. Kleinstreuer, Nanofluid convective heat transfer in a parallel-disk system, *International Journal of Heat and Mass Transfer*, 53, 4619–4628, 2010.
55. C.T. Nguyen, F. Desgranges, G. Roy et al., Temperature and particle size dependent viscosity data for water-based nanofluids-Hysteresis phenomenon, *International Journal of Heat and Fluid Flow*, 28, 1492–1506, 2007.
56. E. Abu-Nada, Effects of variable viscosity and thermal conductivity of Al_2O_3-water nanofluid on heat transfer enhancement in natural convection, *International Journal of Heat and Fluid Flow*, 30, 679–690, 2009.
57. C. Kleinstreuer, Y. Feng, Thermal nanofluid property model with application to nanofluid flow in a parallel-disk system Part I: A new thermal conductivity model for nanofluid flow, *Journal of Heat Transfer*, 134(5), 051002, 2012.
58. H. E. Patel, T. Sundararajan, S.K. Das, An experimental investigation into the thermal conductivity enhancement in oxide and metallic nanofluids, *Journal of Nanoparticle Research*, 12, 1015–1031, 2010.
59. Y. Feng, *A New Thermal Conductivity Model for Nanofluids with Convection Heat Transfer Application*, MS Thesis, NC State University, Raleigh, NC, USA, 2009.
60. H. A. Mintsa, G. Roy, C. T. Nguyen, D. Doucet, New temperature dependent thermal conductivity data for water-based nanofluids, *International Journal of Thermal Sciences*, 48, 363–371, 2009.
61. R.K. Shah, Laminar flow friction and forced convection heat transfer in ducts of arbitrary geometry, *International Journal of Heat and Mass Transfer*, 18, 849–862, 1975.
62. R.K. Shah, A.L. London, Laminar flow forced convection in ducts, *Supplement 1 to Advances in Heat Transfer*, Academic Press, New York, 1978.
63. R.K. Shah, M.S. Bhatti, Laminar convective heat transfer in ducts, in: S. Kakac, R.K. Shah, W. Aung (Eds.), *Handbook of Single-Phase Convective Heat Transfer*, Wiley, New York, 1987 (Chapter 3).
64. W.Y. Lai, S. Vinod, P.E. Phelan, and R. Prasher, Convective heat transfer for water-based Alumina nanofluids in a single 1.02-mm tube, *ASME Journal of Heat Transfer*, 131, 112401, 2009.
65. R. Chein, J. Chuang, Analysis of microchannel heat sink performance using nanofluids, *Applied Thermal Engineering*, 25, 3104–3114, 2005.
66. J. Li, C. Kleinstreuer, Thermal performance of nanofluid flow in microchannels, *International Journal of Heat and Fluid Flow*, 29, 1221–1232, 2008.
67. I. Tavman, A. Turgut, M. Chirtoc, K. Hadjov, O. Fudym, S. Tavman, Experimental study on thermal conductivity and viscosity of water-based nanofluids, *Heat Transfer Research*, 41, 339–351, 2010.
68. S.K. Das, N. Putra, P. Thiesen, W. Roetzel, Temperature dependence of thermal conductivity enhancement for nanofluids, *Journal of Heat Transfer,* 125, 567–574, 2003.
69. M. P. Beck, Y. Yuan, P. Warrier, A.S. Teja, The thermal conductivity of alumina nanofluids in water, ethylene glycol, and ethylene glycol + water mixture, *Journal of Nanoparticle Research*, 12, 1469–1477, 2010.
70. C. H. Chon, K. D. Kihm, S. P. Lee, S. U. S. Choi, Empirical correlation finding the role of temperature and particle size for nanofluid (Al_2O_3) thermal conductivity enhancement, *Applied Physics Letters*, 87, 153107-1–3, 2005.

10 Natural Convection in Nanofluids

Massimo Corcione

CONTENTS

10.1 INTRODUCTION

Buoyancy-induced convection is the heat-removal strategy preferred by many thermal engineering designers when small power consumption, negligible operating noise, and high reliability of the system are main concerns. However, the inherently poor energy efficiency of natural convection, in comparison with equivalent or similar forced convection cases, and the intrinsic low thermal conductivity of conventional coolants such as water, ethylene glycol, and mineral oils, limit the amount of heat that can be dissipated via buoyancy-driven cooling.

In this context, in the past decades, a considerable research effort has been dedicated to the development of new techniques for heat-transfer enhancement, such as those based on the use of extended surfaces and/or turbulators as well as to the study of new geometries and configurations. Yet, these remedies are not able to satisfy completely the severe cooling requirements of modern devices.

A possible solution to mitigate the problem is the replacement of traditional heat-transfer fluids with nanofluids, that is, liquid suspensions of solid nanoparticles, whose effective thermal conductivity is known to be higher than that of the corresponding pure base liquid.

The majority of the studies available in the literature on convective heat transfer in nanofluids are related to forced convection flows, proving that nanoparticle suspensions have undoubtedly a great potential for heat-transfer enhancement. These findings are discussed in the review articles recently compiled by Daungthongsuk and Wongwises [1], Murshed et al. [2], and Kakaç and Pramuanjaroenkij [3]. Conversely, the relatively few works performed on buoyancy-induced heat transfer in nanofluids, most of which are numerical studies dealing with enclosed flows, lead to contradictory conclusions. These results leave unanswered the question if the use of nanoparticle suspensions for natural convection applications is actually advantageous with respect to pure liquids. In fact, according to some authors, the addition of nanoparticles to a base liquid implies a more or less remarkable enhancement of the heat-transfer rate, while, according to others, a deterioration may occur.

The reason for such conflicting results can be explained by considering that the heat-transfer performance of nanofluids in natural convection flows is a direct consequence of two opposite effects arising, respectively, from the increase of the effective thermal conductivity and the increase of the effective dynamic viscosity that occur as the nanoparticle volume fraction is augmented. In other words, the dispersion of a certain concentration of nanoparticles into a base liquid can bring about either an enhancement or a degradation of the heat-transfer performance in buoyancy-induced flows, depending on whether the increased thermal conductivity effect is larger or smaller than the increased viscosity effect.

Besides the experimental investigation, the typical approach used to study the main heat-transfer features of nanoparticle suspensions is based on the assumption that nanofluids behave like single-phase fluids. This means that the mass, momentum, and energy transfer governing equations for pure fluids, as well as any heat-transfer correlation available in the literature, can be directly extended to nanoparticle suspensions, provided that the thermophysical properties appearing in them are the nanofluid effective properties (the numerical approach based on the two-phase model is not usually performed, possibly due to higher computational difficulties). Thus, the use of robust theoretical models or empirical equations, capable to predict the nanofluid effective thermal conductivity and dynamic viscosity as accurately as possible, is crucial for obtaining realistic results. Unfortunately, most of the numerical studies of buoyancy-driven nanofluids based on the single-phase model miss this requirement, for one reason or another, thus leading to unreliable results.

Actually, one of the commonest causes of erroneous outcomes is the use of the Einstein equation [4,5] or the Brinkman equation [6] for predicting the nanofluid effective dynamic viscosity, whose values are notoriously underestimated by these models, as, for example, demonstrated by the experiments executed by Chen et al. [7,8] and Chevalier et al. [9] on the rheological behavior of nanofluids. Moreover, the effective thermal conductivity is often calculated by the Maxwell–Garnett model [10] or other traditional mean-field theories, such as the Hamilton–Crosser model [11] and the Bruggemann model [12], that appear to be suitable to this end when the nanofluid is at "room" temperature—see, for example, Refs. [13,14]. On the contrary, these models fail dramatically when the temperature of the suspension is one or some tens of degrees higher than 20–25°C, as shown in Das et al. [15], Li and Peterson [16], and Yu et al. [17]. Misleading conclusions may also derive from the calculation of the nanofluid effective physical properties by either partly inconsistent theoretical models or correlations based on experimental data that are inexplicably in contrast with the main body of the literature results.

Framed in this general background, the aim of this chapter is to provide an extensive, updated review of the open literature dealing with natural convection heat transfer in nanofluids, with the main aim to point out and discuss thoroughly any possible flaws of the results reported on this topic. A further scope of this work is to introduce a pair of easy-to-apply empirical correlations for predicting the effective thermal conductivity and dynamic viscosity of nanofluids that, matching sufficiently well a high number of experimental data available in the literature, may be usefully employed for numerical simulation purposes and thermal design tasks. Finally, a collection of new findings on the thermal performance of buoyancy-driven nanofluids is presented, and new data relevant to the existence of an optimal particle loading are discussed, taking into account the effects of the geometry of the system, the operating conditions, the nanoparticle diameter, and the solid–liquid combination.

10.2 LITERATURE REVIEW

10.2.1 SIDE-HEATED RECTANGULAR ENCLOSURES

The first, well-documented studies of natural convection of nanofluids in enclosures differentially heated at their sides, which were published in 2003 by Khanafer et al. [18] and Putra et al. [19], are a typical example of conflicting results most frequently cited in the literature. In detail, in their two-dimensional numerical study, Khanafer and co-workers showed that, owing to the dispersion of copper nano-sized particles having a diameter of 10 nm into pure water, the amount of heat transferred across a square cavity increased remarkably with increases of the nanoparticle volume fraction at any investigated Grashof number. Such enhanced performance was ascribed to the irregular and random motion of the suspended nanoparticles, whose effects were simulated by assuming that the nanofluid effective conductivity could be expressed as the sum of a static term evaluated by the Maxwell-Garnett model [10] and a dynamic term calculated following the model developed by Amiri and Vafai [20] to account for the thermal dispersion in porous media.

In contrast, according to the experimental results obtained by Putra and colleagues for a horizontal cylindrical vessel differentially heated at its ends and containing Al_2O_3 (d_p = 131.2 nm) + H_2O or CuO (d_p = 87.3 nm) + H_2O both having volume fractions of 1% and 4%, the average Nusselt number for the enclosure decreased with increasing the nanoparticle concentration, with a degree of deterioration depending on both the nanoparticle material and the aspect ratio of the enclosure. The authors were not able to explain these findings. However, the disagreement between these results lacks substance, as Khanafer and co-workers defined the Nusselt number using the thermal conductivity of the base fluid k_f, whereas Putra and colleagues based the Nusselt number on the effective thermal conductivity of the nanofluid k, which may lead to ambiguous interpretations of the data. In fact, a Nusselt number that would describe the heat-transfer performance of the enclosure with fidelity should increase with increasing the average coefficient of convection h, and vice versa. This is what happens when the Nusselt number is defined according to the point of view of Khanafer and coworkers, that is, $Nu = hL/k_f$, since the ratio L/k_f is independent of the nanoparticle volume fraction. On the contrary, when the Nusselt number is defined as $Nu = hL/k$, Nu may either increase or decrease with increasing h, depending if $\partial h/\partial k$ is positive or negative. In particular, it may occur that for a given nanoparticle volume fraction φ, the coefficient of convection of the nanofluid is higher than that of the pure base liquid, but if $\partial h/\partial k < 0$ the average Nusselt number of the nanofluid is smaller than that of the pure base liquid. This is the case of the experimental data of Putra and colleagues for $\varphi = 0.01$. Indeed, if the Nusselt numbers reported by Putra and colleagues for $\varphi = 0.01$ are rearranged in the same form used by Khanafer and co-workers, that is, properly multiplied by the ratio k/k_f, a heat transfer enhancement may be clearly observed. At $\varphi = 0.04$ a slight heat-transfer degradation occurred for both nanofluids, thus implying the existence of an optimal particle loading.

Enhanced heat transfer in side-heated cavities, consequent to the addition of nanoparticles to water, was also reported in the numerical studies executed later by Jou and Tzeng [21], Tiwari and Das [22], Oztop and Abu-Nada [23], Abu-Nada and Oztop [24], Öğüt [25], and Kahveci [26]. The nanoparticle materials were Cu in Refs. [21,22, 24], Cu, Al_2O_3, or TiO_2 in Ref. [23], and Cu, Ag, CuO, Al_2O_3, or TiO_2 in Refs. [25,26].

An optimal particle loading for maximum heat transfer across a cavity filled with Al_2O_3 ($d_p = 27$ nm) + H_2O was discovered experimentally by Nnanna [27], who explained its existence as the consequence of an excessive increase in viscosity that occurred above a certain optimal nanoparticle concentration. The existence of such an optimal nanoparticle volume fraction can also be inferred from a careful analysis of the experimental data recently published by Ho et al. [28] for a square enclosure filled with Al_2O_3 ($d_p = 33$ nm) + H_2O. Actually, it must be acknowledged that the effective dynamic viscosity plays a role as crucial as the role played by the effective thermal conductivity in determining the heat transfer performance of nanofluids in natural convection flows. This means that the results obtained through any numerical simulation procedure that is not based on a reliable viscosity model are more or less unrealistic. By chance, all the numerical studies cited above in Refs. [18,21–26] are based on the Brinkman equation [6] that, as said earlier, is known to significantly underpredict the effective dynamic viscosity of nanofluids.

The relevance of the nanofluid effective dynamic viscosity in enclosed natural convection flows was stressed also by Santra et al. [29] and Ho et al. [30], who showed that the effect of the nanoparticle volume fraction on the cavity Nusselt number depended significantly on the viscosity model adopted. In both studies, the use of the Brinkman model resulted in an increase of the average Nusselt number. Conversely, a more or less pronounced decrease of the average Nusselt number was observed when the nanofluid effective dynamic viscosity was calculated through empirical correlations. In detail, Santra and colleagues used a correlation constructed by fitting the values of zero-shear viscosity obtained experimentally by Kwak and Kim [31] for ethylene glycol suspensions of CuO nanoparticles having the shape of prolate spheroids with a mean diameter of 12 nm and an aspect ratio of approximately three. For their part, Ho and co-workers used the correlation derived by Maïga et al. [32] on the basis of the measurements executed by Wang et al. [33] for water-based nanofluids containing Al_2O_3 nanoparticles with a diameter of 28 nm. However, no optimal particle loading was detected either by Santra and colleagues or by Ho and coworkers, whose results deserve to be further discussed.

First of all, it must be said that the study performed by Santra et al. [29] was based on the assumption that the enclosure was filled with a mixture of water and spherical copper nanoparticles having a diameter of 100 nm, whose effective thermal conductivity was evaluated through the model proposed by Patel et al. [34]. In contrast, as cited earlier, Santra and colleagues calculated the effective dynamic viscosity by a correlation based on the Kwak–Kim experimental data relative to a liquid suspension of tiny rod-like particles, which gave rise to a remarkable overestimation of the viscosity effects. In fact, for the same nanoparticle concentration, the overall contact surface area between small rod-like nanoparticles and base fluid is significantly greater than that existing between relatively large spherical nanoparticles and base fluid. This implies that the nanofluid containing rod-like nanoparticles is characterized by a larger amount of friction occurring at the solid/liquid interface and, correspondingly, a higher effective dynamic viscosity. Note that the increase of the nanofluid effective dynamic viscosity with decreasing the diameter of the suspended nanoparticles at a fixed concentration (which implies an increase of the area of the solid/liquid contact surface), is largely demonstrated by a number of experimental studies readily available in the literature, such as those performed by Prasher et al. [35] and Chevalier et al. [9].

Regarding the work executed by Ho et al. [30], the nanofluid effective thermal conductivity was evaluated from a correlation originally developed for slurries by Charuyakorn et al. [36] that ignores any nanoscale phenomenon. This approach necessarily gives more or less underestimated values, with the consequence that the viscosity effects prevail over the thermal conductivity effects.

Circumstantially, another central point relative to natural convection flows in nanofluids has been raised, that is, the fact that the trustworthiness of the solutions of any numerical work on this subject relies significantly also on the accuracy of the model adopted for predicting the nanofluid effective thermal conductivity. In this connection, it is worth noting that in numerous studies, the effective thermal conductivity is calculated by the Maxwell–Garnett model [10] that, as mentioned earlier, is known to fail dramatically in predicting the increased thermal conductivity of nanofluids, at least when the temperature of the suspension is higher than "room" temperature.

Abu-Nada et al. [37] performed a numerical study on side-heated cavities filled with Al_2O_3 ($d_p = 47$ nm) + H_2O or CuO ($d_p = 29$ nm) + H_2O. The effective thermal conductivity was evaluated through the empirical correlation proposed by Chon et al. [38], whereas the effective dynamic viscosity was calculated by a pair of correlations based on the experimental data of Nguyen et al. [39]. It was found that, for the convection-dominated regime, the average Nusselt number decreased with increasing the nanoparticle volume fraction, with the degree of deterioration depending on both the nanoparticle material and the cavity aspect ratio. However, also in this case, the results were meaningfully affected by an overestimation of the effective dynamic viscosity. In fact, the effective dynamic viscosities measured by Nguyen and coworkers for Al_2O_3 ($d_p = 47$ nm) + H_2O were higher than those measured for Al_2O_3 ($d_p = 36$ nm) + H_2O. This is in contrast with most results available in the literature, according to which the nanofluid effective dynamic viscosity is inversely proportional to the size of the suspended nanoparticles. Moreover, as the data relative to $d_p = 36$ nm are in substantial good agreement with the results obtained by Chevalier et al. [9] for $d_p = 35$ nm, it follows that the data reported for Al_2O_3 ($d_p = 47$ nm) + H_2O tend to be overstimated. In addition, the values of viscosity of CuO ($d_p = 29$ nm) + H_2O detected by Nguyen and coworkers are larger than those available in the literature for nanofluids containing nanoparticles having similar size, which is the case of the data reported by Masuda et al. [40] for $d_p = 27$ nm, Pak and Cho [41] for $d_p = 27$ nm, and Wang et al. [33] for $d_p = 28$ nm. The reasons behind such overestimated values are difficult to understand, although a possible explanation may be related to the use of an unknown surfactant which could have unusually affected the mechanical behavior of the suspensions prepared for experiments.

Lin and Violi [42] carried out a numerical investigation on natural convection in a differentially heated square cavity filled with Al_2O_3 + H_2O, under the assumption of a nonuniform nanoparticle diameter having a fractal distribution between the assigned minimum and maximum values. The effective thermal conductivity of the nanofluid was calculated through the model developed by Xu et al. [43], while the effective dynamic viscosity was evaluated by an equation proposed by Jang et al. [44]. It was found that, for a fixed volume fraction, the heat-transfer rate increased with decreasing the average diameter of the suspended nanoparticles and increasing the uniformity of their size. In addition, unless the nanoparticle diameter was very small, that is, 5 nm, the average Nusselt numbers computed for the nanofluid were lower than those of the pure base liquid at any Grashof number investigated. Again, some considerations of the role played by the choice of the viscosity equation on the reliability of the results are regarded as necessary. In fact, the viscosity equation proposed by Jang and colleagues includes two empirical constants whose values were determined on the basis of a number of capillary viscosity measurements performed using Al_2O_3 ($d_p = 30$ nm) + H_2O with low nanoparticle concentrations in the range between 0.02% and 0.3%, flowing through microtubes having diameters of 310–1773 μm. Surprisingly, the nanofluid effective dynamic viscosity was found to decrease with increasing the microtube diameter, which was expressed in terms of a direct proportionality to the ratio between the fixed size of the suspended nanoparticles and the variable tube diameter. This persuaded Lin and Violi to believe that the nanofluid viscosity could be assumed to be proportional to the ratio between the nanoparticle diameter and the side of the cavity. Two inconsistencies may be detected in this model. First, the use of an equation derived for a fixed nanoparticle diameter to evaluate the effects of the nanoparticle size on the viscosity of the nanofluid does not make much sense, especially as the direct proportionality of the effective viscosity to the nanoparticle size has no experimental evidence. Second, the hypothesis that the value of the nanofluid viscosity can be related to the characteristic length of the cavity has little physical support. Furthermore, the extrapolation of the validity of the equation proposed by Jang and colleagues to nanoparticle volume fractions well outside the 0.02–0.3% experimental range seems rather drastic.

Quite recently, Abu-Nada and Chamkha [45] executed a numerical study of differentially heated enclosures filled with CuO nanoparticles having an average diameter of 29 nm dispersed in a 60:40 (in weight) mixture of ethylene glycol and water. The investigative approach was the same as that carried out 1 year earlier by Abu-Nada et al. [37], except that now the nanofluid effective thermal conductivity was predicted by the Jang–Choi model [46], and the effective dynamic viscosity was evaluated using an experimental correlation developed by Namburu et al. [47]. The data reported in the study are very similar to those obtained in Ref. [37]. In particular, for the convection-dominated regime, the average Nusselt number was found to decrease with increasing the nanoparticle volume fraction, with a degree of deterioration that increased as the cavity aspect ratio was increased. Indeed, as in the previous work, the results obtained were affected by the use of decidedly high values of the dynamic viscosity. To give a measure of such overestimation, note that, according to the experiments performed by Namburu and coworkers, the viscosity of the nanofluid with 3% and 6% solid-phase concentrations at ambient temperature was roughly equal to 1.7 and 3.5 times the viscosity of the base fluid at same temperature. This seems excessive when compared with other experimental data relative to nanofluids with suspended nanoparticles of similar size. Moreover, it is worth pointing out that in some situations, the Jang–Choi model tends to underpredict numerous experimental data sets available in the literature, which is further reason to explain why the increased viscosity effect prevailed more or less remarkably on the increased thermal conductivity effect, determining the decrease of the Nusselt number with increases of the concentration of the suspended nanoparticles.

10.2.2 BOTTOM-HEATED RECTANGULAR ENCLOSURES

The first two studies performed on natural convection in enclosures heated from below, published by Kim et al. [48] and Wen and Ding [49] in 2004 and 2005, respectively, are another example of conflicting results. Specifically, Kim and coworkers implemented a theoretical investigation of both the convective instability driven by buoyancy and the heat-transfer characteristics of nanofluid layers heated from below for different values of the density ratio $\delta_1 = \rho_s/\rho_f$, the heat capacity ratio $\delta_2 = (\rho c)_s/(\rho c)_f$, the thermal conductivity ratio $\gamma = k_s/k_f$, and the nanoparticle shape factor n, where the subscripts "s" and "f" denote the solid and fluid phases, respectively. The nanofluid effective thermal conductivity was calculated by either the Hamilton–Crosser model [11] (which for spherical particles, i.e., for $n = 3$, reduces to the Maxwell–Garnett model [10]) or the Bruggemann model [12], while the effective dynamic viscosity was evaluated by either the Einstein equation [4,5] or the Brinkman equation [6], whose predictions are practically the same. According to the results obtained, δ_1 and δ_2 played a destabilizing role, whereas γ and n acted as stabilizers of the nanofluid. Even more important was that the amount of heat transferred across the nanofluid layer increased remarkably with increasing the nanoparticle volume fraction, and that all four parameters cited above contributed positively to the enhancement of the heat-transfer performance of the nanofluid.

Conversely, the experimental data reported by Wen and Ding for a horizontal enclosure consisting of a pair of differentially heated discs separated by a small gap filled with $TiO_2 + H_2O$ revealed that the average Nusselt number of the enclosure decreased with increasing the nanoparticle volume fraction, particularly at low Rayleigh numbers.

The disagreement between these results can be explained by considering that the data obtained by Kim and colleagues are somewhat unrealistic, owing to the use of the Einstein equation or the Brinkman equation for the calculation of the effective dynamic viscosity, and to the use of the Hamilton–Crosser model or the Bruggemann model for the evaluation of the effective thermal conductivity of the nanofluid. As far as the Wen–Ding experimental data are concerned, a possible reason for the observed heat-transfer degradation may be sought in the effectiveness of the nanofluid preparation method used in experiments. In fact, the primary TiO_2 nanoparticles having a nominal diameter of 34 nm were supplied by the manufacturer in the form of agglomerates that had to be processed in order to obtain a stable nanofluid. However, at the end of the setup procedure, which included an ultrasonification and a prolonged shear mixing of the nanoparticle dispersion, the measured average size of the solid-phase suspended in water was ~170 nm, owing to the presence of several aggregates that could have possibly caused an anomalously high increase of the dynamic viscosity.

Enhanced heat transfer in a cavity heated from below, consequent to the addition of nanoparticles to water, was reported in the numerical study performed by Wang et al. [50]. In this work, the effective thermal conductivity was calculated through a pair of empirical correlations based on the experimental data obtained by Eastman et al. [51] and Wen and Ding [52] for $Al_2O_3 + H_2O$, and by Xuan and Li [53] for $Cu + H_2O$, while the effective dynamic viscosity was predicted by the Brinkman equation [6], whose use is known to affect significantly the reliability of the results.

The importance of the model used to predict the rheological behavior of the nanofluid was emphasized by Hwang et al. [54], who carried out a theoretical study of turbulent natural convection in a bottom-heated rectangular cavity filled with $Al_2O_3 + H_2O$. The effective thermal conductivity was evaluated by the model proposed by Jang and Choi [46], whereas the effective dynamic viscosity was calculated by either the Einstein equation [4,5] or an empirical correlation based on the experimental data by Pak and Cho [41]. As expected, the use of the Einstein equation resulted in an enhanced heat-transfer performance. In contrast, a decrease in the amount of heat convected across the cavity was detected when the effective dynamic viscosity of the nanofluid was calculated by the Pak–Cho correlation. However, this result is not surprising. Indeed, Hwang and coworkers assumed that the nanoparticle diameter could be as large as 50 nm, but the alumina nanoparticles used by Pak and Cho in their experiments had an average size of 13 nm, which has brought a noteworthy overestimation of the viscosity effects. In fact, as discussed earlier, for a fixed nanoparticle volume fraction, the overall contact surface area between smaller nanoparticles and the base fluid is significantly greater than that existing between larger nanoparticles and the base fluid. This implies that the nanofluid that contains smaller nanoparticles is characterized by a larger amount of friction occurring at the solid/liquid interface. Correspondingly, there is higher effective dynamic viscosity, as shown in Prasher et al. [35] and Chevalier et al. [9]. Moreover, as already pointed out, in some cases, especially at temperatures higher than "room" temperature, the Jang–Choi model tends to underpredict the effective thermal conductivity, which is a further justification of the results derived by Hwang and coworkers. However, whatever was the viscosity model adopted, the dispersion of solid nanoparticles in the base liquid was found to increase the stability of the fluid layer.

Finally, with regard to the thermal instability of nanofluid layers heated from below, a different result was derived analytically by Tzou [55], who, temporarily leaving out the dependence of the nanofluid thermophysical properties on the volume fraction, concluded that the combination of the Brownian motion and thermophoresis of the suspended nanoparticles provided a strong destabilization of the nanofluid layer, whose effect was to decrease the critical Rayleigh number by as much as two orders of magnitude compared with that of the pure base liquid.

10.2.3 Differentially-Heated Horizontal Annuli

Natural convection in horizontal annuli heated at the inner cylinder and cooled at the outer cylinder using nanofluids was studied first in 2008 by Abu-Nada et al. [56], who performed a numerical investigation reporting enhanced heat transfer with respect to the case of pure base liquid at a rate depending on both the Rayleigh number of the base fluid and the ratio between the thickness of the annulus and the diameter of the inner cylinder. However, like in many other works, the nanofluid thermal conductivity and dynamic viscosity were predicted by the Maxwell–Garnett model [10] and the Brinkman equation [6], respectively, which limits considerably the reliability of the results obtained.

Successively, two more numerical investigations were executed by Abu-Nada [57,58] for water-based nanofluids containing either Al_2O_3 or CuO suspended spherical nanoparticles, with diameters of 47 and 29 nm, respectively. In both studies, the nanofluid thermal conductivity was evaluated by the empirical correlation proposed by Chon et al. [38], while the effective dynamic viscosity was calculated by a correlation derived using the raw experimental data of Nguyen et al. [39], following the same approach previously used in Ref. [37]. It was found that, for the convection-dominated regime, the average Nusselt number decreased with increasing the nanoparticle volume fraction. These results are seriously affected by an overestimation of the effective dynamic viscosity of the nanofluid. In fact, as already discussed in relation with Ref. [37], the dynamic viscosities measured by Nguyen and colleagues for Al_2O_3 ($d_p = 47$ nm) + H_2O were higher than those measured for Al_2O_3 ($d_p = 36$ nm) + H_2O, which would imply a direct proportionality between the effective dynamic viscosity of the nanofluid and the nanoparticle size. This is in sharp contrast with the majority of the literature results. In addition, the viscosity values detected for CuO ($d_p = 29$ nm) + H_2O are larger than those reported by other authors for nanofluids containing suspended nanoparticles having a similar diameter (see, e.g., [40,41,33]).

10.2.4 External Flow Configurations

The very few papers available in the literature on external natural convection flows in nanoparticle suspensions deal with the basic geometry of a vertical flat plate.

In 2007, Polidori et al. [59] executed a theoretical study based on the boundary layer approach, for both conditions of uniform heat flux (UHF) and uniform wall temperature (UWT) at the plate surface. The nanofluid investigated was $Al_2O_3 + H_2O$, whose effective thermal conductivity was calculated by the Maxwell–Garnett model [10]. The effective dynamic viscosity was evaluated using either the Brinkman equation [6] or the equation derived by Maïga et al. [32] by way of regression analysis of the experimental data reported by Wang et al. [33] for Al_2O_3 ($d_p = 28$ nm) + H_2O, with the main aim to emphasize the key role of viscosity in determining the heat-transfer performance of nanofluids in natural convection laminar flows. The results showed that the use of the Brinkman equation yielded a heat-transfer enhancement for both UHF and UWT conditions. In contrast, the use of the empirical correlation developed by Maïga and coworkers led to a very slight heat transfer enhancement, around 0.6%, for a 2.5% concentration of the suspended nanoparticles, followed by a decreasing trend, which was ascribed to the dominant effect of the kinematic viscosity. However, owing to the smoothness of the maximum, the authors did not give much importance to this result. Actually, if the thermal conductivity had been calculated by a model closer to reality than the Maxwell–Garnett model, the maximum for heat-transfer enhancement would have been much more accentuated.

Kuznetsov and Nield [60] found the similarity solutions of the boundary layer flow on the basis of the two-phase, four-equation, nonhomogeneous equilibrium model developed by Buongiorno [61] that incorporates the effects of Brownian motion and thermophoresis. The results were presented in the form of dimensionless correlations expressing the Nusselt number ratioed with the Rayleigh number raised to the one-fourth power as a function of a buoyancy-ratio parameter, a Brownian motion parameter, and a thermophoresis parameter, for different values of the Prandtl and Lewis numbers. In particular, for any investigated pair of the Prandtl and Lewis numbers, the aforementioned ratio was found to be a decreasing function of each of the other three independent dimensionless parameters.

Finally, in a work based on the same theoretical approach of Ref. [59], Popa et al. [62] extended the investigation to the turbulent regime and to $CuO + H_2O$. In this case, the effective thermal conductivity of the nanofluids was calculated by a pair of correlations proposed by Mintsa et al. [63] for water-based nanofluids containing either Al_2O_3 or CuO nanoparticles, having a diameter of 36–47 or 29 nm, respectively. With respect to the prediction of the effective dynamic viscosity, the cited equation derived by Maïga et al. [32] was used for $Al_2O_3 + H_2O$, whereas the correlation developed by Nguyen et al. [39] on the basis of their own experimental data was adopted for $CuO + H_2O$. It was found that the heat-transfer performance of the nanofluid decreased with increasing the nanoparticle volume fraction, in both laminar and turbulent flows, much more for $CuO + H_2O$ than for $Al_2O_3 + H_2O$, which can basically be imputed to an overestimation of the dynamic viscosity effects. In fact, the experimental correlation proposed by Mintsa and colleagues for predicting the thermal conductivity of $Al_2O_3 + H_2O$ was obtained using nanoparticles with an average size of 36–47 nm, while the viscosity equation developed by Maïga and coworkers is relative to a water-based nanofluid containing Al_2O_3 nanoparticles with a mean diameter of 28 nm. The increase of the dynamic viscosity with decreasing the nanoparticle diameter was demonstrated by Prasher et al. [35] and Chevalier et al. [9]. Moreover, the values of dynamic viscosity of CuO ($d_p = 29$ nm) + H_2O measured by Nguyen and colleagues are larger than those available in the literature for nanofluids containing nanoparticles having similar size (see, e.g., Refs. [40,41,33]). In addition, a further contribution to the observed decrease of the heat-transfer performance with increases of φ may originate from an underestimation of the thermal conductivity effects. In fact, it must be noticed that the thermal conductivities measured by Mintsa and collaborators are slightly lower than those reported by other authors for liquid suspensions of nano-sized particles having a similar diameter, see, for example, Das et al. [15] and Lee et al. [64] for Al_2O_3 ($d_p = 38.4$ nm) + H_2O, and Das et al. [15] for CuO ($d_p = 28.6$ nm) + H_2O.

10.3 EVALUATION OF THE EFFECTIVE PROPERTIES

According to the foregoing review, most of the inconsistencies or contradictions of the results available for natural convection in nanofluids can be ascribed to unproper evaluations of the effective physical properties. In this connection, two empirical correlating equations for predicting the nanofluid effective thermal conductivity and dynamic viscosity, useful for numerical simulation purposes and thermal design tasks, are here presented and discussed. In addition, the expressions for the nanofluid effective mass density, specific heat at constant pressure, and coefficient of thermal expansion, derived by the mixing theory, are also reviewed.

10.3.1 THERMAL CONDUCTIVITY

The inadequacy of the traditional mean-field theory in predicting the nanofluid effective thermal conductivity with a sufficiently good approximation, unless the temperature is about 20–25°C, has motivated the development of several new models. A number of these models assign a key role to the effect of the interfacial nanolayer, whose existence was suggested by Choi et al. [65] on the basis of the work of Yu et al. [66,67], who reported the observation of molecular fluid layering in a liquid at the solid/liquid interface using x-ray reflectivity—see Yu and Choi [68], Xue [69], Xie et al. [70], and Leong et al. [71]. A second group of models incorporate two different contributions: one static and one dynamic. The former contribution depends on the composition of the nanofluid, while the latter contribution accounts for the effect of the micromixing convection caused by the Brownian motion of the nanoparticles that is assumed to be a decisive mechanism of energy transfer—see Kumar et al. [72], Koo and Kleinstreuer [73], Jang and Choi [46], Patel et al. [34], Ren et al. [74], Prasher et al. [75,76], Xuan et al. [77], Xu et al. [43], Prakash and Giannelis [78], and Murshed et al. [79]. Note that the models discussed in Refs. [46] and [75–77] consider also the role of the interfacial Kapitza resistance [80] whose temperature-discontinuity effect could degrade significantly the nanofluid heat-transfer performance. In contrast, the combined effects of the Brownian motion and the interfacial nanolayer are taken into account in the models proposed in Refs. [74] and [78], as well as in Ref. [79] wherein the additional contributions of the nanoparticle surface chemistry and the interaction potential are also considered. Finally, some models take into account the nanoparticle aggregation that causes percolation effects (see Refs. [81–83]). Other models combine the microconvection due to the nanoparticle Brownian motion and the aggregation occurring among individual nanoparticles and/or nanoparticle clusters (see Refs. [84,85]). However, all these models show large discrepancies among each other, which clearly represents a restriction to their applicability. Moreover, many of them include empirical constants whose values were often determined on the basis of a limited number of experimental data or were not clearly defined.

For these reasons, an empirical correlating equation based on a wide variety of experimental data available in the literature has been developed for the nanofluid effective thermal conductivity k normalized by the thermal conductivity of the base fluid k_f [86]. In this regard, it is worth pointing out that a certain dispersion of the experimental data reported by different authors for the same type of nanofluid is unavoidable. Indeed, in some cases the discrepancies among the data may also reach the order of 50%, which could be due to the different measurement techniques used in experiments, as well as to the different degrees of dispersion/agglomeration obtained for the suspended nanoparticles and to the accuracy of evaluation of their shape and size. Therefore, in deriving the correlation, some data sets found in the literature have been discarded, because either the data were in a sharp contrast with the main body of the literature results without any convincing physical evidence, or the investigation procedure was not properly described in detail, or specific chemical surfactants were used in experiments which could have meaningfully altered the thermo-mechanical behavior of the suspension.

The empirical correlation, derived by way of regression analysis of the data extracted from the sources enumerated in Table 10.1 [15,38,40,63,64,87–90] with a 1.86% standard deviation of error and a ±4% range of error, is

$$\frac{k}{k_f} = 1 + 4.4 \, \mathrm{Re}^{0.4} \, \mathrm{Pr}_f^{0.66} \left(\frac{T}{T_{fr}}\right)^{10} \left(\frac{k_s}{k_f}\right)^{0.03} \varphi^{0.66}, \tag{10.1}$$

where Re is the nanoparticle Reynolds number, Pr_f is the Prandtl number of the base liquid, T is the nanofluid temperature in Kelvin degrees, T_{fr} is the freezing point of the base liquid, k_s is the thermal conductivity of the solid phase, and φ is the nanoparticle volume fraction.

The nanoparticle Reynolds number is defined as $\mathrm{Re} = (\rho_f \, u_B \, d_p)/\mu_f$, where ρ_f and μ_f are the mass density and dynamic viscosity of the base fluid, respectively, and d_p and u_B are the nanoparticle diameter and the nanoparticle Brownian velocity, respectively. Once u_B is calculated as the ratio between d_p and the time τ_D required to cover such a distance $\tau_D = (d_p)^2/6D$—see Ref. [91]—in which D stands for the Einstein diffusion coefficient, the nanoparticle Reynolds number is given by

$$\mathrm{Re} = \frac{2\rho_f k_b T}{\pi \mu_f^2 d_p}, \tag{10.2}$$

TABLE 10.1
Thermal Conductivity Experimental Data Used for Deriving Equation 10.1

Literature Source	Nanofluid Type	Nanoparticle Size (nm)	Measuring Method
Masuda et al. [40]	TiO_2 + water	27	Transient hot-wire
Lee et al. [64]	CuO + water	23.6	Transient hot-wire
	Al_2O_3 + water	38.4	
	CuO + ethylene glycol	23.6	
	Al_2O_3 + ethylene glycol	38.4	
Eastman et al. [87]	Cu + ethylene glycol	10	Transient hot-wire
Das et al. [15]	CuO + water	28.6	Temperature oscillation
	Al_2O_3 + water	38.4	
Chon et al. [38]	Al_2O_3 + water	47	Transient hot-wire
Chon and Kihm [88]	Al_2O_3 + water	47	Transient hot-wire
	Al_2O_3 + water	150	
Murshed et al. [89]	Al_2O_3 + water	80	Transient hot-wire
	Al_2O_3 + ethylene glycol	80	
Mintsa et al. [63]	CuO + water	29	Transient hot-wire
Duangthongsuk and Wongwises [90]	TiO_2 + water	21	Transient hot-wire

where $k_b = 1.38 \times 10^{-23}$ J/K is the Boltzmann's constant. Note that in Equations 10.1 and 10.2, all the physical properties are calculated at the nanofluid temperature T.

It may be seen that the thermal conductivity ratio k/k_f increases as φ and T increase and as d_p decreases. Moreover, it may be noticed that k/k_f depends marginally on the solid nanoparticle material, as denoted by the extremely small exponent of k_s/k_f. The distributions of k/k_f versus φ that emerge from Equation 10.1 for $Al_2O_3 + H_2O$ with d_p and T as parameters are displayed in Figure 10.1, where the prediction of the Maxwell–Garnett model [10] is also reported for comparison, showing that the degree of failure of this model applied to nanofluids increases as T increases and d_p decreases.

Although the mechanisms behind the thermal behavior of nanofluids are still in debate, a tentative conclusion may be reached on the basis of the observations reported in the foregoing. In fact, as the data used to derive Equation 10.1 are relative to combinations of solid and liquid phases having values of k_s/k_f that span over two orders of magnitude (from nearly 15 for $TiO_2 + H_2O$ to more than 1500 for Cu + ethylene glycol), the substantial independence of k/k_f from k_s/k_f gives strength to the possibility that the Brownian motion of the suspended nanoparticles plays a nonnegligible role. Of course, when the nanoparticle concentration increases, the chance that the suspended nanoparticles aggregate, thus forming complex massive clusters, gets progressively higher, with the consequence that the contribution of the Brownian motion to the enhanced thermal conductivity tends to decrease. However, at same time, the percolation contribution increases. In this regard, the fact that the growth of k/k_f with φ occurs with a decreasing slope, as reflected by the 0.66

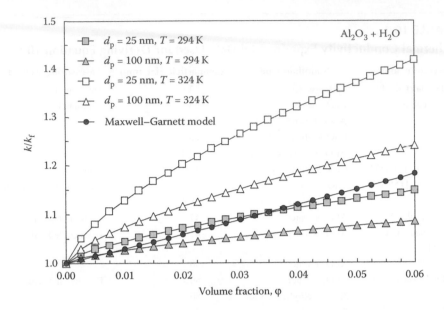

FIGURE 10.1   Distributions of k/k_f vs. φ for $Al_2O_3 + H_2O$ with d_p and T as parameters.

exponent of φ in Equation 10.1, may lead to the belief that percolation is a heat-transport mechanism proportionally less effective than micro-convection.

10.3.2   Dynamic Viscosity

Although the traditional theories fail dramatically in predicting the increased dynamic viscosity of nanofluids, only few models have recently been proposed for describing their rheological behavior. This is exemplified by the models developed by Koo [92] and Masoumi et al. [93] that account for the effects of the Brownian motion of the suspended nanoparticles, and the model proposed by Ganguly and Chakraborty [94] that is based on the kinetics of the agglomeration–deagglomeration phenomena due to interparticle interactions. However, as these models contain empirical correction factors based on an extremely small number of experimental data, their regions of validity are someway limited.

An empirical correlating equation based on a large number of experimental data selected from literature has therefore been developed for the nanofluid effective dynamic viscosity μ normalized by the dynamic viscosity of the base fluid μ_f [86]. The correlation, derived by way of regression analysis of the data extracted from the sources listed in Table 10.2 [7–9,19,33,35,40,41,95–98] with a 1.84% standard deviation of error and a ±4.5% range of error, is

$$\frac{\mu}{\mu_f} = \frac{1}{1 - 34.87(d_p/d_f)^{-0.3}\varphi^{1.03}}, \tag{10.3}$$

TABLE 10.2

Dynamic Viscosity Experimental Data Used for Deriving Equation 10.3

Literature Source	Nanofluid Type	Nanoparticle Size (nm)	Viscometer/Rheometer
Masuda et al. [40]	TiO_2 + water	27	—
Pak and Cho [41]	TiO_2 + water	27	Cone/plate (Brookfield)
Wang et al. [33]	Al_2O_3 + water	28	—
Putra and co-workers [19,95]	Al_2O_3 + water	38	Rotating disk-type
Prasher et al. [35]	Al_2O_3 + propylene glycol	27	Controlled stress-type
	Al_2O_3 + propylene glycol	40	
	Al_2O_3 + propylene glycol	50	
He et al. [96]	TiO_2 + water	95	Bohlin CVO (Malvern)
Chen et al. [7,8]	TiO_2 + ethylene glycol	25	Bohlin CVO (Malvern)
Chevalier et al. [9]	SiO_2 + ethanol	35	Capillary-type
	SiO_2 + ethanol	94	
	SiO_2 + ethanol	190	
Lee et al. [97]	Al_2O_3 + water	30	VM-10A (CBC Co.)
Garg et al. [98]	Cu + ethylene glycol	200	AR-G2 (TA Instruments)

where d_f is the equivalent diameter of a base fluid molecule given by

$$d_f = 0.1 \left[\frac{6M}{N\pi\rho_{f0}} \right]^{1/3},$$
(10.4)

in which M is the molecular weight of the base fluid, N is the Avogadro number, and ρ_{f0} is the mass density of the base fluid calculated at temperature $T_0 = 293$ K.

It may be seen that according to Equation 10.3, the dynamic viscosity ratio μ/μ_f is independent of both the solid nanoparticle material and the temperature, at least for particle volume fractions not >0.1 and temperatures not too far from "room" temperature. Another interesting feature is that μ/μ_f increases as d_p decreases and φ increases. This can tentatively be explained if we reasonably assume that the dynamic viscosity of a nanofluid is related to the amount of friction occurring at the contact surface between nanoparticles and base liquid and to the strength of the colloidal interactions and the degree of aggregation existing among the suspended particles. The amount of friction between nanoparticles and base fluid depends on the extent of the overall contact surface, whose area, for an assigned volume fraction, is inversely proportional to the nanoparticle diameter. In contrast, for a fixed diameter of the suspended nanoparticles, both the strength of the colloidal interactions and the degree of aggregation are expected to increase with increases of the nanoparticle concentration.

The distributions of μ/μ_f versus φ that emerge from Equation 10.3 for water-based nanofluids with d_p as a parameter are displayed in Figure 10.2, wherein the predictions of the Brinkman equation [6] (that are practically the same as those of the Einstein

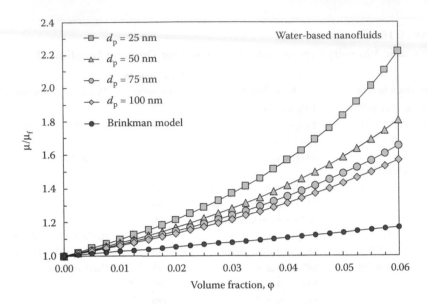

FIGURE 10.2 Distributions of μ/μ_f vs. φ for water-based nanofluids with d_p as a parameter.

equation [4,5]) are additionally delineated, pointing out that the error deriving from its application to nanofluids increases remarkably with decreasing the nanoparticle size.

10.3.3 Other Physical Properties

The other effective properties of the nanofluid can easily be calculated through the mixing theory, as is typically done in the majority of the studies performed in this field.

The mass density of the nanofluid ρ is given by

$$\rho = (1 - \varphi)\rho_f + \varphi\rho_s, \tag{10.5}$$

where ρ_f and ρ_s are the mass densities of the base fluid and the solid nanoparticles, respectively.

The heat capacity at constant pressure per unit volume of the nanofluid ρc is

$$\rho c = (1 - \varphi)(\rho c)_f + \varphi(\rho c)_s, \tag{10.6}$$

where $(\rho c)_f$ and $(\rho c)_s$ are the heat capacities at constant pressure per unit volume of the base fluid and the solid nanoparticles, respectively. Accordingly, the specific heat at constant pressure of the nanofluid, c, is calculated as

$$c = \frac{(1 - \varphi)(\rho c)_f + \varphi(\rho c)_s}{(1 - \varphi)\rho_f + \varphi\rho_s}. \tag{10.7}$$

Finally, the coefficient of thermal expansion of the nanofluid β is defined by

$$\rho\beta = -\frac{d\rho}{dT}. \tag{10.8}$$

If we substitute Equation 10.5 into Equation 10.8, and replace the temperature derivatives of ρ_f and ρ_s with $(\rho\beta)_f$ and $(\rho\beta)_s$, respectively, we have

$$\rho\beta = (1 - \varphi)(\rho\beta)_f + \varphi(\rho\beta)_s, \tag{10.9}$$

thus obtaining

$$\beta = \frac{(1 - \varphi)(\rho\beta)_f + \varphi(\rho\beta)_s}{(1 - \varphi)\rho_f + \varphi\rho_s}. \tag{10.10}$$

The distributions of the mass density ratio ρ/ρ_f, the ratio between the specific heats at constant pressure c/c_f, and the ratio between the coefficients of thermal expansion β/β_f, plotted against the nanoparticle volume fraction for $Al_2O_3 + H_2O$ at $T = 309$ K, are shown in Figure 10.3, where the distributions of the ratio between the heat capacities at constant pressure per unit volume $\rho c/(\rho c)_f$ and the ratio between the temperature derivatives of the mass densities $\rho\beta/(\rho\beta)_f$ are also represented.

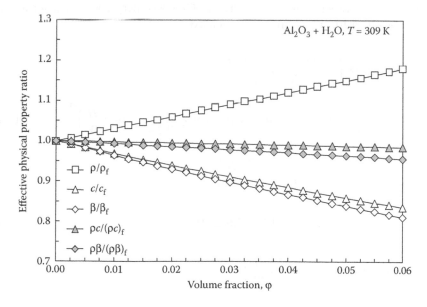

FIGURE 10.3 Distributions of the other property ratios vs. φ for $Al_2O_3 + H_2O$ at $T = 309$ K.

10.4 OPTIMAL PARTICLE LOADING

The existence of an optimal particle loading for buoyancy-driven nanofluids and its dependence on the geometry of the system, the operating conditions, the nanoparticle diameter, and the solid–liquid combination is demonstrated by a theoretical approach based on the common assumption that nanofluids behave like single-phase fluids. In fact, as the suspended nanoparticles have usually small size and concentration, the hypothesis of a solid–liquid mixture statistically homogeneous and isotropic seems to be absolutely reasonable. Thus, if we additionally assume that the nanoparticles and base fluid are in local thermal equilibrium and no-slip motion occurs between the solid and liquid phases, to all intents and purposes the nanofluid can be treated as a pure fluid. Accordingly, all the convective heat-transfer correlations available in the literature for single-phase flows can be easily extended to the corresponding nanofluid applications, once the thermophysical properties appearing in them are substituted by the effective properties of the nanofluid calculated at the reference temperature.

Such a formulation finds experimental confirmation in the study performed by Chang et al. [99], who demonstrated that the same Nusselt–Rayleigh correlation valid for layers of pure water is applicable with good approximation to nanofluid layers consisting of Al_2O_3 ($d_p = 250$ nm) + H_2O if both Nusselt and Rayleigh numbers are calculated using the thermophysical properties of the suspension.

The pair of enclosed-flow configurations consisting of a vertical rectangular cavity differentially heated at sides with perfectly insulated bottom and top walls, and a horizontal rectangular cavity heated from below and cooled from above with adiabatic sidewalls, are considered [100,101].

For side-heated configurations, the Berkovski–Polevikov correlations for aspect ratios ranging between 1 and 10, or the MacGregor–Emery correlation for aspect ratios larger than 10, are usually recommended:

$$Nu = 0.18 \left(\frac{Pr}{0.2 + Pr} Ra \right)^{0.29} \left(\frac{H}{W} \right)^{0.13}$$

$$1 \leq \frac{H}{W} \leq 2, \quad 10^{-3} \leq Pr \leq 10^{5}, \quad 10^{3} \leq \frac{Pr}{0.2 + Pr} Ra \left(\frac{H}{W} \right)^{-3}; \qquad (10.11)$$

$$Nu = 0.22 \left(\frac{Pr}{0.2 + Pr} Ra \right)^{0.28} \left(\frac{H}{W} \right)^{-0.09}$$

$$2 \leq \frac{H}{W} \leq 10, \quad Pr \leq 10^{5}, \quad Ra \leq 10^{13}; \qquad (10.12)$$

$$Nu = 0.42 \, Pr^{0.012} \, Ra^{0.25} \left(\frac{H}{W} \right)^{-0.05}$$

$$10 \leq \frac{H}{W} \leq 40, \quad 1 \leq Pr \leq 2 \times 10^{4}, \quad 10^{4} \leq Ra \left(\frac{H}{W} \right)^{-3} \leq 10^{7}, \qquad (10.13)$$

where H and W are the cavity height and width, respectively, Pr is the Prandtl number, and Nu and Ra are the Nusselt and Rayleigh numbers, respectively, both based on the height of the enclosure.

For bottom-heated configurations, the Churchill–Ozoe correlation can be used:

$$\mathrm{Nu} = \left\{ \left[1 + 1.446\left(1 - \frac{\mathrm{Ra_{cr}}}{\mathrm{Ra}} \right) \right]^{15} + \left[\frac{\mathrm{Ra} \times f(\mathrm{Pr})}{1420} \right]^{5} \right\}^{1/15}$$

$$f(\mathrm{Pr}) = \left[1 + \left(\frac{0.5}{\mathrm{Pr}} \right)^{9/16} \right]^{-16/9}, \mathrm{Ra} \geq \mathrm{Ra_{cr}}, \tag{10.14}$$

where $\mathrm{Ra_{cr}} = 1708$ is the critical Rayleigh number for the onset of convection. The correlation holds when the enclosure is sufficiently long and wide in the horizontal direction, so that the effect of the short vertical sides is negligible. Again, both Nu and Ra are defined using the height of the enclosure as characteristic length.

The heat-transfer correlations listed above are employed to assess the effect of the nanoparticle concentration on the heat transfer enhancement, E, defined as

$$E = \frac{h}{h_f} - 1 = \frac{\mathrm{Nu}}{\mathrm{Nu_f}} \times \frac{k}{k_f} - 1, \tag{10.15}$$

where h_f, $\mathrm{Nu_f}$, and k_f are the coefficient of convection, the Nusselt number and the thermal conductivity of the base fluid, and h, Nu, and k are the corresponding effective quantities of the nanofluid. The effective Nusselt number $\mathrm{Nu} = hH/k$ is calculated through Equations 10.11 through 10.14 in which the Rayleigh and Prandtl numbers of the pure base fluid are replaced by the effective Rayleigh and Prandtl numbers of the nanofluid, while the thermal conductivity ratio k/k_f is calculated by Equation 10.1. Note that in the preceding equations the physical properties of both the fluid and the nanoparticles are evaluated at the reference average temperature $T_{av} = (T_h + T_c)/2$, where T_h and T_c are the temperatures of the heated and cooled walls of the enclosure.

10.4.1 Vertical Enclosures

The effect of the nanoparticle concentration on the heat-transfer enhancement for a side-heated vertical enclosure filled with $Al_2O_3 + H_2O$ is displayed in Figures 10.4 through 10.6, for different nanoparticle diameters, average temperatures of the nanofluid, and aspect ratios of the enclosure, respectively.

It is apparent that owing to the dispersion of a progressively larger amount of solid nanoparticles into the base liquid, the heat-transfer enhancement increases up to a point, which is due to the increased effective thermal conductivity of the nanofluid. Notice that the impact of the increased effective thermal conductivity is higher when the nanoparticle diameter is smaller, the nanofluid average temperature is higher, and the aspect ratio of the enclosure is higher. The value of φ corresponding to the peak of E is defined as the optimal particle loading φ_{opt}. As the nanoparticle concentration is

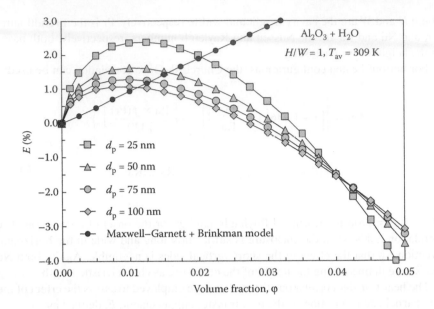

FIGURE 10.4 Distributions of E (%) vs. φ for a side-heated square enclosure filled with $Al_2O_3 + H_2O$ at $T_{av} = 309$ K with d_p as a parameter.

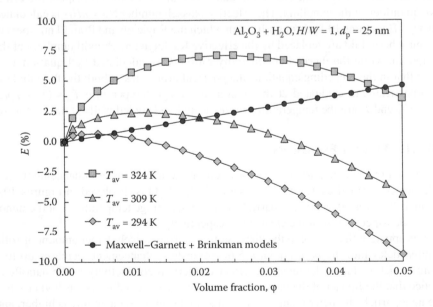

FIGURE 10.5 Distributions of E (%) vs. φ for a side-heated square enclosure filled with $Al_2O_3 + H_2O$ ($d_p = 25$ nm) with T_{av} as a parameter.

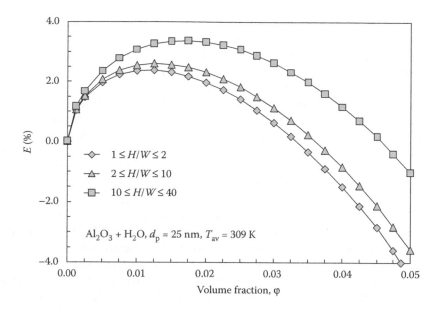

FIGURE 10.6 Distributions of E (%) vs. φ for side heated rectangular enclosures filled with $Al_2O_3 + H_2O$ ($d_p = 25$ nm) at $T_{av} = 309$ K with H/W as a parameter.

further increased above ψ_{opt}, the heat transfer enhancement decreases, which is due to the excessive growth of the nanofluid effective viscosity. In fact, as discussed earlier, the nanofluid behavior in natural convection flows is a direct consequence of the two opposite effects that originate from the increase of the effective thermal conductivity and dynamic viscosity occurring as the nanoparticle volume fraction is increased.

According to Figures 10.1 and 10.2, the first effect, which tends to enhance the heat-transfer performance, prevails at small volume fractions, whereas the second effect, which tends to degrade the heat-transfer performance, prevails at large concentrations. Obviously, when the effect of the increased dynamic viscosity outweighs the effect of the increased thermal conductivity, the heat-transfer enhancement becomes negative, which means that the convective thermal performance of the nanofluid is lower than that of the pure base liquid. In Figures 10.4 and 10.5, the distributions of E versus φ obtained by using the Maxwell–Garnett and Brinkman models for calculating the effective thermal conductivity and dynamic viscosity are also delineated, confirming the weakness of these models in capturing the actual thermo-mechanical features of nanofluids.

It is worth pointing out that the optimal particle loading increases slightly with decreasing the size of the suspended nanoparticles, while increases more remarkably when the nanofluid temperature and/or the aspect ratio of the enclosure are increased. In fact, both k/k_f and μ/μ_f increase as d_p is reduced, which implies that the effect of the nanoparticle size on φ_{opt} is almost imperceptible. Conversely, as k/k_f enhances significantly when T_{av} is increased, while μ/μ_f keeps constant, the nanoparticle volume fraction at which the increase in viscosity becomes excessive increases drastically with increasing the nanofluid average temperature.

With regard to the effect of the aspect ratio of the cavity on φ_{opt}, it must be noticed that when H/W is increased, the resistance encountered by the fluid to flow across the enclosure reduces proportionally, which implies that the growth of the effective dynamic viscosity starts becoming excessive with respect to the growth of the effective thermal conductivity at a larger volume fraction.

To conclude, the effects of the solid–liquid combination on the heat-transfer enhancement are pointed out in Figure 10.7 for a square geometry, showing that the effect of the base fluid is much more pronounced than that of the nanoparticle material. This can be justified by considering that, based on Equation 10.15, E depends on both Nu/Nu_f and k/k_f. From Equation 10.11, if we consider that for many liquids the Prandtl number is generally much larger than 0.2, we derive that Nu/Nu_f is a primary function of the ratio between the Rayleigh numbers of the nanofluid and the base fluid, which is equal to the ratio of the product $[\rho\beta/(\rho\beta)_f] \times [\rho c/(\rho c)_f]$ to the product $(k/k_f) \times (\mu/\mu_f)$. In contrast, according to Figure 10.3, both $\rho\beta/(\rho\beta)_f$ and $\rho c/(\rho c)_f$ remain practically constant with increasing φ. This means that the heat-transfer enhancement E depends mostly on the thermal conductivity ratio k/k_f and the dynamic viscosity ratio μ/μ_f. Hence, taking into account that k/k_f depends very little on the nanoparticle material, whereas μ/μ_f is completely independent of the nanoparticle material, we can conclude that E is affected much more by the liquid phase than by the solid phase. Obviously, as the thermal conductivity of water is more than the double of the thermal conductivity of ethylene glycol, the heat-transfer enhancement produced by dispersing solid particles into to the base liquid is less marked for water than for ethylene glycol.

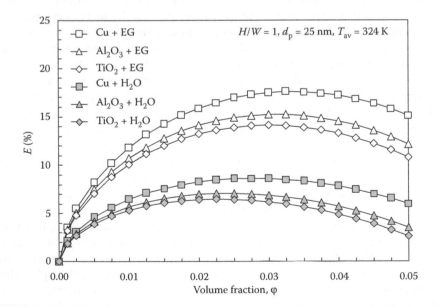

FIGURE 10.7 Distributions of E (%) vs. φ for a side-heated square enclosure filled with different nanofluids, assumed $d_p = 25$ nm and $T_{av} = 324$ K.

For the specific case of a square enclosure filled with $Al_2O_3 + H_2O$, the percent optimal particle loading can be calculated by the following empirical dimensional algebraic equation:

$$\varphi_{opt}(\%) = (5 \times 10^{-4})\,[t_{av}(°C)]^{2.335}\,[d_p(nm)]^{-0.19}, \qquad (10.16)$$

where t_{av} (°C) is the nanofluid average temperature in Celsius degrees, and d_p(nm) is the nanoparticle diameter in nanometers.

10.4.2 HORIZONTAL ENCLOSURES

Similarly to the vertical enclosures, also for the horizontal enclosures the heat transfer enhancement is higher when the diameter of the suspended nanoparticles is smaller and the nanofluid average temperature is higher, showing a peak at a certain optimal particle loading. Moreover, E depends mostly on the base liquid and much less on the nanoparticle materials.

However, a number of differences exist between the horizontal and vertical configurations. The first of them is the dependence of E on the Rayleigh number of the base fluid, as displayed for $Al_2O_3 + H_2O$ in Figure 10.8, where a nonmonotonic variation of E with Ra_f may be clearly noticed. This means that once the values of d_p and T_{av} as well as the solid–liquid combination are assigned, the heat-transfer enhancement has a peak at a certain Rayleigh number of the base fluid, as shown in Figure 10.9. In particular, such Rayleigh number increases slightly as the concentration of the suspended nanoparticles is increased.

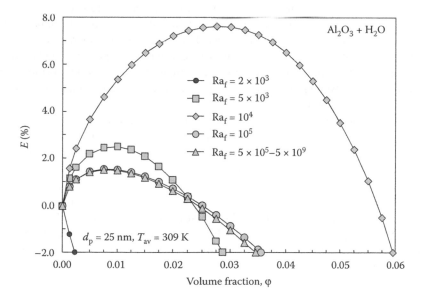

FIGURE 10.8 Distributions of E (%) vs. φ for a bottom-heated enclosure filled with $Al_2O_3 + H_2O$ ($d_p = 25$ nm) at $T_{av} = 309$ K with Ra_f as a parameter.

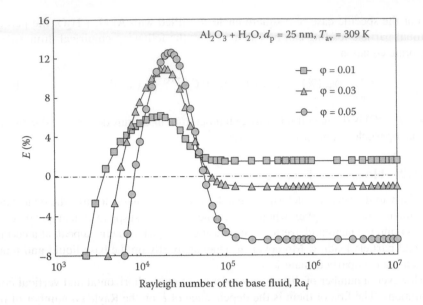

FIGURE 10.9 Distributions of E (%) vs. Ra_f for a bottom-heated enclosure filled with $Al_2O_3 + H_2O$ ($d_p = 25$ nm) at $T_{av} = 309$ K with φ as a parameter.

Moreover, it is worth observing that the addition of nanoparticles to the base liquid produces a significant increase in the thermal performance of the cavity when the flow regime is laminar.

Vice versa, the heat-transfer enhancement may be rather low or even negative for turbulent flows. Actually, due to the low heat and momentum transfer performance typical of the laminar regime, the addition of nanoparticles to the base fluid results in an increased thermal conductivity effect that prevails on the increased viscosity effect. On the contrary, as the turbulent flow is featured by a higher heat and momentum transfer performance, in this regime the increased viscosity effect is more pronounced and may even predominate over the increased thermal conductivity effect, particularly at high volume fractions.

With regard to the optimal particle loading, two sets of distibutions of φ_{opt} versus Ra_f are represented in Figures 10.10 and 10.11, for different values of T_{av} and d_p, respectively. It may be seen that φ_{opt} increases as the nanofluid average temperature increases, while it either increases or decreases with increasing the nanoparticle size, according as the flow regime is laminar or turbulent. In addition, φ_{opt} has a peak at the base-fluid Rayleigh number whose corresponding effective Rayleigh number of the nanofluid marks the beginning of the transition from laminar to turbulent flow.

The percent optimal particle loading for $Al_2O_3 + H_2O$ contained in a horizontal enclosure heated from below can be obtained by the following empirical dimensional algebraic equations:

$$\varphi_{opt}(\%) = \left\{ (2.725 \times 10^{-5})[t_{av}(°C)]^{0.795} \right\} Ra_f - 0.2475 [t_{av}(°C)]^{0.48}$$

$$\text{for } Ra_f \leq (2 \times 10^4) [t_{av}(°C)]^{-0.35} \tag{10.17}$$

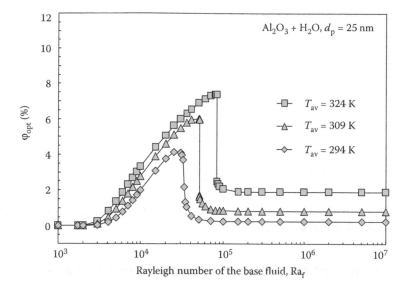

FIGURE 10.10 Distributions of φ_{opt} (%) vs. Ra_f for a bottom-heated enclosure filled with $Al_2O_3 + H_2O$ ($d_p = 25$ nm) with T_{av} as a parameter.

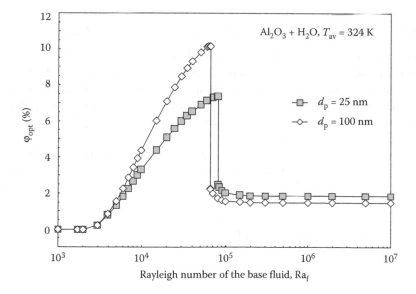

FIGURE 10.11 Distributions of φ_{opt} (%) vs. Ra_f for a bottom-heated enclosure filled with $Al_2O_3 + H_2O$ at $T_{av} = 324$ K with d_p as a parameter.

$$\varphi_{opt}(\%) = \left\{ 0.7147 \, [t_{av}(°C)]^{0.0503} \, [d_p(nm)]^{0.2969} \right\} Ln(Ra_f)$$

$$+ -7.013 \, [t_{av}(°C)]^{-0.02596} \, [d_p(nm)]^{0.3172} \quad \text{for } (2 \times 10^4) \, [t_{av}(°C)]^{-0.35}$$

$$\leq Ra_f \leq (0.85 \times 10^4) \, [t_{av}(°C)]^{1.26} \, [d_p(nm)]^{-0.135} \tag{10.18}$$

$$\varphi_{opt}(\%) = (3.3 \times 10^{-4}) \, [t_{av}(°C)]^{2.46} \, [d_p(nm)]^{-0.3}$$

$$\text{for } Ra_f \geq (2.5 \times 10^4)[t_{av}(°C)]^{0.44} \tag{10.19}$$

where $t_{av}(°C)$ is the nanofluid average temperature in Celsius degrees, and d_p (nm) is the nanoparticle diameter in nanometers.

10.5 FLOW STRUCTURES AND TEMPERATURE PATTERNS

Numerical simulations have been performed in the hypothesis of variable effective properties, in order to investigate the main heat and momentum transfer features of a nanofluid enclosed in a square cavity differentially heated at its sides. The resulting flow structures and temperature patterns are reported and discussed. A comparison is also made with the velocity and temperature fields obtained by using the Maxwell–Garnett and Brinkman models for predicting the nanofluid effective thermal conductivity and dynamic viscosity, under the more usual assumption of constant physical properties.

10.5.1 FORMULATION OF THE PROBLEM AND COMPUTATIONAL PROCEDURE

A square enclosure of width W filled with $Al_2O_3 + H_2O$ is differentially heated at the vertical sidewalls that are kept at uniform temperatures T_h and T_c, while the top and bottom walls are adiabatic. A zero emissivity is assumed for the confining walls, which physically corresponds to perfectly polished surfaces, thus implying that the situation investigated here involves pure natural convection, owing to the absence of any contribution by radiation.

The flow is considered steady, two-dimensional, laminar, and incompressible, with negligible viscous dissipation and pressure work. In addition, the nanoparticle suspension is assumed to behave like a single-phase, Newtonian fluid with variable effective physical properties.

Once the above assumptions are incorporated into the mass, momentum, and energy transfer equations, and the following dimensionless variables are introduced

$$X = \frac{x}{W}, \quad Y = \frac{y}{W} \tag{10.20}$$

$$U = \frac{u}{\mu_r/(\rho_r W)}, \quad V = \frac{v}{\mu_r/(\rho_r W)}, \quad P = \frac{p - p_r}{\mu_r^2/(\rho_r W^2)}, \quad \theta = \frac{T - T_r}{T_h - T_c} \tag{10.21}$$

$$\rho^* = \frac{\rho}{\rho_r}, \quad \mu^* = \frac{\mu}{\mu_r}, \quad c^* = \frac{c}{c_r}, \quad k^* = \frac{k}{k_r}, \tag{10.22}$$

the governing equations reduce to

$$\frac{\partial(\rho^* U)}{\partial X} + \frac{\partial(\rho^* V)}{\partial Y} = 0 \tag{10.23}$$

$$\rho^* \left(U \frac{\partial U}{\partial X} + V \frac{\partial U}{\partial Y} \right) = -\frac{\partial P}{\partial X} + \frac{\partial}{\partial X} \left[2\mu^* \frac{\partial U}{\partial X} - \frac{2}{3}\mu^* \left(\frac{\partial U}{\partial X} + \frac{\partial V}{\partial Y} \right) \right]$$
$$+ \frac{\partial}{\partial Y} \left[\mu^* \left(\frac{\partial U}{\partial Y} + \frac{\partial V}{\partial X} \right) \right] \tag{10.24}$$

$$\rho^* \left(U \frac{\partial V}{\partial X} + V \frac{\partial V}{\partial Y} \right) = -\frac{\partial P}{\partial Y} + \frac{\partial}{\partial Y} \left[2\mu^* \frac{\partial V}{\partial Y} - \frac{2}{3}\mu^* \left(\frac{\partial U}{\partial X} + \frac{\partial V}{\partial Y} \right) \right]$$
$$+ \frac{\partial}{\partial X} \left[\mu^* \left(\frac{\partial U}{\partial Y} + \frac{\partial V}{\partial X} \right) \right] + \frac{Ra}{Pr} \cdot \frac{\rho_r - \rho}{\rho_c - \rho_h} \tag{10.25}$$

$$\frac{\partial(\rho^* c^* U\theta)}{\partial X} + \frac{\partial(\rho^* c^* V\theta)}{\partial Y} = \frac{1}{Pr} \left[\frac{\partial}{\partial X} \left(k^* \frac{\partial \theta}{\partial X} \right) + \frac{\partial}{\partial Y} \left(k^* \frac{\partial \theta}{\partial Y} \right) \right], \tag{10.26}$$

where Ra and Pr are the effective Rayleigh and Prandtl numbers defined as

$$Ra = \frac{\rho_r c_r g (\rho_c - \rho_h) W^3}{k_r \mu_r}, \quad Pr = \frac{c_r \mu_r}{k_r}. \tag{10.27}$$

In the preceding equations x and y are the horizontal and vertical Cartesian coordinates, u and v are the x-wise and y-wise velocity components, p is the pressure, T is the temperature, g is the acceleration of gravity, ρ is the effective mass density, μ is the effective dynamic viscosity, c is the effective specific heat at constant pressure, and k is the effective thermal conductivity, where ρ, μ, c, and k are given by Equations 10.5, 10.3, 10.7, and 10.1, respectively. Note that the effective properties with subscript "r" are calculated at the reference temperature T_r, set equal to the temperature T_c of the cooled sidewall of the enclosure, whereas ρ_h and ρ_c are the effective mass densities calculated at temperatures T_h and T_c, respectively.

Fixed $T_r = T_c$, the thermal boundary conditions expressed in dimensionless form are (a) $\theta = 1$ at the heated sidewall; (b) $\theta = 0$ at the cooled sidewall; and (c) $\partial\theta/\partial Y = 0$ at the adiabatic top and bottom walls. With regard to the velocity boundary conditions, the no-slip condition $U = V = 0$ is assumed along the four confining walls.

The system of Equations 10.23 through 10.26 along with the boundary conditions stated above is solved through a control-volume formulation of the finite-diffrence method. The pressure–velocity coupling is handled through the SIMPLE-C algorithm described by Van Doormaal and Raithby [102]. The advection fluxes are evaluated by the QUICK discretization scheme proposed by Leonard [103]. The computational spatial domain is filled with a nonuniform grid, having a higher concentration of grid lines near the boundary walls, and a coarser uniform spacing throughout the remainder interior of the cavity. Starting from the first-approximation fields of the dependent variables across the cavity, that is, uniform dimensionless temperature set to 0 and nanofluid at rest, the discretized system of algebraic governing equations is solved iteratively by way of a line-by-line application of the Thomas algorithm. Under-relaxation is enforced in all steps of the computational procedure to ensure adequate convergence. The solution of the velocity and temperature fields is considered to be converged when the maximum absolute values of the mass source as well as the percentage changes of the dependent variables at any grid-node between two consecutive iterations are smaller than the prespecified values of 10^{-4} and 10^{-6}, respectively. Furthermore, the condition that the relative difference between the incoming and outgoing heat-transfer rates at the heated and cooled sides is smaller than the preassigned value of 10^{-4} must be verified.

After convergence is attained, the effective average Nusselt numbers at the heated and cooled sidewalls, Nu_h and Nu_c, are calculated with the expressions

$$Nu_h = \frac{h_h W}{(k_f)_h} = \frac{Q_h}{(k_f)_h (T_h - T_c)} = -\left(\frac{k}{k_f}\right)_h \cdot \int_0^1 \frac{\partial \theta}{\partial X}\bigg|_{X=0} dY \qquad (10.28)$$

$$Nu_c = \frac{h_c W}{(k_f)_c} = \frac{Q_c}{(k_f)_c (T_c - T_h)} = -\left(\frac{k}{k_f}\right)_c \cdot \int_0^1 \frac{\partial \theta}{\partial X}\bigg|_{X=1} dY, \qquad (10.29)$$

where h_h and h_c are the average coefficients of convection at the heated and cooled sidewalls, respectively, $(k_f)_h$ and $(k_f)_c$ are the values of the thermal conductivity of the base fluid calculated at temperatures T_h and T_c, respectively, Q_h and Q_c are the heat-transfer rates per unit length added to the nanofluid by the heated vertical wall and withdrawn from the nanofluid by the cooled vertical wall, respectively, and $(k/k_f)_h$ and $(k/k_f)_c$ are the values of the thermal conductivity ratio calculated at temperatures T_h and T_c, respectively. The temperature gradients in Equations 10.28 and 10.29 are evaluated by a second-order temperature profile embracing the wall-node and the two adjacent fluid-nodes. The integrals are computed numerically by means of the trapezoidal rule. Note that since the Nusselt numbers Nu_h and Nu_c defined above are based on the thermal conductivity of the pure liquid k_f, an increase or a decrease of their values with increasing the nanoparticle volume fraction unequivocally corresponds to an increase or a decrease of the thermal performance, which is due to the fact that the ratio W/k_f is independent of φ. Of course, because at steady-state the heat-transfer rates per unit length Q_h and Q_c are the same, the following relationship between Nu_h and Nu_c holds:

$$\mathrm{Nu}_h (k_f)_h = \mathrm{Nu}_c (k_f)_c \tag{10.30}$$

The heat-transfer enhancement $E = h/h_f - 1$, defined in Equation 10.15, is calculated with reference to the heated sidewall of the enclosure through the expression

$$E = \frac{\mathrm{Nu}_h}{(\mathrm{Nu}_f)_h} - 1 \tag{10.31}$$

Obviously, according to Equation 10.30, the same value of E can be obtained by replacing subscript "h" with "c" in Equation 10.31, that corresponds to make reference to the cooled sidewall rather than to the heated sidewall of the enclosure for the computation of the heat-transfer enhancement.

Numerical tests related to the dependence of the results on the mesh-spacing have been performed for several combinations of the five controlling parameters, that is, Ra_f, d_p, φ, T_c, and T_h. The optimal grid-size values, that is, those used for computations, are such that further refinements do not produce noticeable modifications either in the heat-transfer rates or in the flow field. Specifically, the percentage changes of Nu_h and Nu_c, and those of the maximum velocity components U_{max} and V_{max} on the vertical and horizontal midplanes of the cavity are smaller than the pre-established accuracy value, that is, 1%. The typical number of nodal points used for computations lies in the range between 40×40 and 120×120. Some test runs have also been executed with the first-approximation uniform value of θ set to 0.5 or 1, rather than 0, with the aim to determine what effect these starting conditions could have on the flow structures and temperature patterns. Solutions practically identical to those obtained assuming $\theta = 0$ across the cavity were obtained for all the configurations examined. Finally, with the scope to validate the numerical code used for this study, the solutions obtained for a differentially heated square enclosure filled with air, whose physical properties were assumed constant, have been compared with the benchmark results of de Vahl Davis [104] and other authors, that is, Mahdi and Kinney [105], Hortman et al. [106], and Wan et al. [107], showing a very good degree of agreement.

10.5.2 Discussion of the Results

A selection of local results is presented in Figures 10.12 through 10.16, in which the isotherm and streamline contours are plotted for different sets of values of Ra_f, d_p, φ, T_c, and T_h, in order to highlight the effects of these independent variables on the temperature and velocity fields. The contour lines of the isotherm plots correspond to equally spaced values of the dimensionless temperature θ in the range between 0 and 1. The streamline plots correspond to equally spaced absolute values of the normalized dimensionless stream function $\Psi/|\Psi|_{max}$ ranging from 0 to 1, where Ψ is defined by the relations:

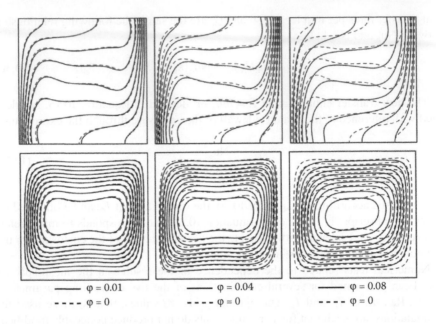

$\varphi = 0.01$
$\varphi = 0$

$\varphi = 0.04$
$\varphi = 0$

$\varphi = 0.08$
$\varphi = 0$

FIGURE 10.12 Effect of the nanoparticle volume fraction on the isotherm and streamline contour plots for $\mathrm{Ra_f} = 10^5$, $T_c = 303$ K, $\Delta T = 10$ K, and $d_p = 25$ nm, at $\varphi = 0.01$ (left), $\varphi = 0.04$ (middle), and $\varphi = 0.08$ (right); dashed line plots refer to pure fluid.

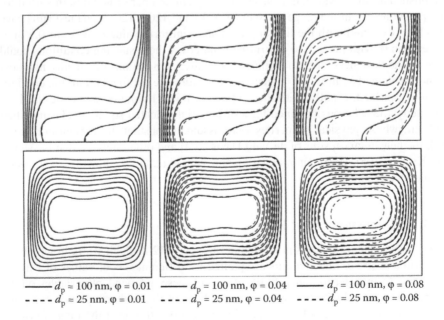

$d_p = 100$ nm, $\varphi = 0.01$
$d_p = 25$ nm, $\varphi = 0.01$

$d_p = 100$ nm, $\varphi = 0.04$
$d_p = 25$ nm, $\varphi = 0.04$

$d_p = 100$ nm, $\varphi = 0.08$
$d_p = 25$ nm, $\varphi = 0.08$

FIGURE 10.13 Effect of the nanoparticle diameter on the isotherm and streamline contour plots for $\mathrm{Ra_f} = 10^5$, $T_c = 303$ K, $\Delta T = 10$ K, and $d_p = 25$ nm (dashed line plots) and 100 nm (continuous line plots), at $\varphi = 0.01$ (left), $\varphi = 0.04$ (middle), and $\varphi = 0.08$ (right).

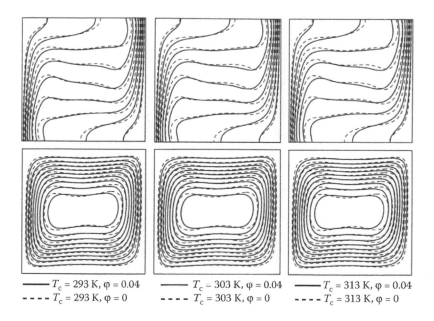

FIGURE 10.14 Effect of the temperature of the cooled wall on the isotherm and streamline contour plots for $Ra_f = 10^5$, $\varphi = 0.04$, $\Delta T = 10$ K, and $d_p = 25$ nm, with $T_c = 293$ K (left), $T_c = 303$ K (middle), and $T_c = 313$ K (right); dashed lines plots refer to pure fluid.

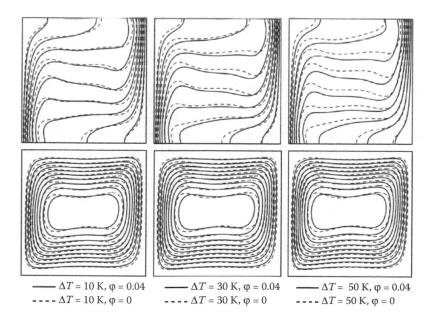

FIGURE 10.15 Effect of the sidewall temperature difference on the isotherm and streamline contour plots for $Ra_f = 10^5$, $\varphi = 0.04$, $T_c = 293$ K, and $d_p = 25$ nm, with $\Delta T = 10$ K (left), $\Delta T = 30$ K (middle), and $\Delta T = 50$ K (right); dashed line plots refer to pure fluid.

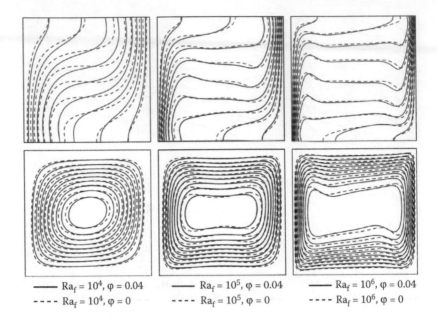

——— $Ra_f = 10^4$, $\varphi = 0.04$ ——— $Ra_f = 10^5$, $\varphi = 0.04$ ——— $Ra_f = 10^6$, $\varphi = 0.04$

- - - - $Ra_f = 10^4$, $\varphi = 0$ - - - - $Ra_f = 10^5$, $\varphi = 0$ - - - - $Ra_f = 10^6$, $\varphi = 0$

FIGURE 10.16 Effect of the base-fluid Rayleigh number on the isotherm and stream-line contour plots for $\varphi = 0.04$, $T_c = 303$ K, $\Delta T = 10$ K, and $d_p = 25$ nm, at $Ra_f = 10^4$ (left), $Ra_f = 10^5$ (middle), and $Ra_f = 10^6$ (right); dashed line plots refer to pure fluid.

$$\rho^* U = \frac{\partial \Psi}{\partial Y}, \quad \rho^* V = -\frac{\partial \Psi}{\partial X}. \tag{10.32}$$

As expected, for all the configurations examined the flow field consists of a single cell that derives from the rising of the hot fluid adjacent to the heated sidewall and its descent along the opposite cooled sidewall. It may be observed that when the volume fraction of the suspended nanoparticles increases and their average size decreases (see Figures 10.12 and 10.13), the consequent growth of the effective dynamic viscosity of the nanofluid entails a decrease in the motion intensity, as reflected by the expansion of the streamlines toward the core of the enclosure. Correspondingly, the isotherm lines tend to be less compressed toward the heated and cooled sidewalls of the cavity, but, since at same time also the effective thermal conductivity increases, the decrease of the local temperature gradients at both sidewalls does not necessarily mean that a local heat-transfer degradation occurs.

As far as the effects of temperatures T_c and T_h are concerned (see Figures 10.14 and 10.15), the thermal field is affected much more by the increase of the temperature difference ΔT between the sidewalls of the cavity for a fixed value of T_c, than by the increase of T_c (or T_h) for a fixed value of ΔT, as denoted by the progressively more pronounced deformation of the isotherms that occurs as ΔT is magnified. In fact, for a fixed T_c, the degree of compression of the isotherms toward the hot sidewall must decrease with increasing ΔT so as to ensure that the heat-transfer rates at both cavity sidewalls are the same, owing to the meaningful growth of the effective thermal conductivity with temperature. Regarding the effects of temperatures T_c and T_h on the velocity field, the value of $|\Psi|_{max}$ increases nonnegligibly as T_c and/or T_h are increased, due to the decrease of the effective dynamic viscosity with increases of the nanofluid temperature (following the same law of the pure base liquid).

Moreover, the degree of distortion of both the velocity and temperature fields with respect to the case of pure base liquid is practically insensitive to the Rayleigh number of the base fluid, at least for relatively high Rayleigh numbers, that is, $Ra_f > 10^5$ (see Figure 10.16). Conversely, for $Ra_f \leq 10^5$ a decrease in the Rayleigh number leads to a reduction of the local temperature gradients and a less marked fluid stratification in the core of the cavity, which can be imputed to the fact that the velocity diminution produced by the addition of solid nanoparticles to the base liquid is percentually more remarkable when the momentum transfer is low, that is, at small Rayleigh numbers.

Finally, a comparison between the solutions obtained in this study and the local fields that would have been derived by adopting the conventional Maxwell–Garnett and Brinkman models for the calculation of the effective thermal conductivity and dynamic viscosity of the nanofluid, in conjunction with the common assumption of constant physical properties, is presented in Figure 10.17, demonstrating the high degree of failure of the simulation procedures based on the traditional mean-field theories. In addition, it is believed that, owing to the significant increase of the

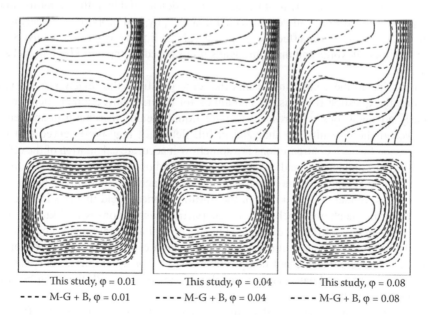

—— This study, $\varphi = 0.01$ —— This study, $\varphi = 0.04$ —— This study, $\varphi = 0.08$

- - - - M-G + B, $\varphi = 0.01$ - - - - M-G + B, $\varphi = 0.04$ - - - - M-G + B, $\varphi = 0.08$

FIGURE 10.17 Comparison between the results of this study (continuous lines) and a constant-properties study using the Maxwell–Garnett model and Brinkman equation (dashed lines) for $Ra_f = 10^5$, $T_c = 293$ K, $\Delta T = 50$ K, and $d_p = 25$ nm, at $\varphi = 0.01$–0.08.

nanofluid thermal conductivity with temperature, the hypothesis of constant properties should be avoided, at least when the temperature difference across the enclosure is higher than 10 K (on the other hand, when the temperature difference becomes significant in relation to the average temperature, the Boussinesq approximation ceases to be valid).

As far as the overall results are concerned (not shown for the sake of brevity), both the heat-transfer enhancement E and the optimal particle loading φ_{opt} increase as the diameter of the suspended nanoparticles decreases and the average temperature of the nanofluid $T_{av} = (T_h + T_c)/2$ increases, being nearly independent of the base-fluid Rayleigh number, at least for values of Ra_f not too low. Regarding the dependence on T_{av}, it must be said that for any assigned value of the average temperature of the nanofluid both E and φ_{opt} are much more sensitive to increases in ΔT for a fixed value of T_c, rather than increases in T_c for a fixed value of ΔT (note that also the local fields are affected much more by ΔT than by T_c, as discussed earlier).

10.6   CONCLUSIONS

A detailed overview of the research papers dealing with buoyancy-induced convection in nanofluids has been conducted, and the main results have been thoroughly discussed. It has been found that most investigations on this topic have been executed numerically under the assumption that nanofluids behave like single-phase fluids, which implies that the mass, momentum, and energy transfer governing equations for pure fluids, as well as any heat transfer correlation available in the literature, may easily be extended to nanofluids, provided that the physical properties appearing in them are replaced by the effective properties of the liquid suspension. Conversely, the experimental works and the numerical studies based on the two-phase model are very few.

According to the conflicting results that emerged from the review process carried out, the question if the use of nanofluids for natural convection applications is actually advantageous with respect to pure liquids seems to remain unanswered. However, if these results are analyzed in full details, it may be observed that the main reason for such contradictions is a somehow inaccurate calculation of the effective thermal conductivity and dynamic viscosity, whose increased values with no doubt affect meaningfully the heat-transfer performance of nanofluids in buoyancy-driven flows. In fact, the weakness of the traditional mean-field theories in describing the thermo-mechanical behavior of nanoparticle suspensions is well known. Yet, many investigators still rely on them. On the other hand, unrealistic conclusions may also derive from the evaluation of the nanofluid effective properties either by theoretical models that are not completely consistent or by empirical correlations based on experimental data that are inexplicably in contrast with the main body of the literature results.

Therefore, a pair of empirical correlating equations for predicting the effective thermal conductivity and dynamic viscosity of nanofluids, useful both for numerical simulation purposes and for thermal engineering design tasks, have been introduced and commented. In particular, their use has permitted to confirm the existence of an optimal nanoparticle loading for maximum heat transfer, which is a function of the geometry of the system, the operating conditions, the average size of the suspended nanoparticles, and the solid–liquid combination.

To conclude, a number of numerical simulations have been executed for a water suspension of alumina nano-sized particles, assuming that the nanofluid behaves like a single-phase, Newtonian fluid, in the further hypothesis of variable effective physical properties, with the main aim to determine the flow structures and temperature patterns inside a square cavity differentially heated at its sides. The comparison of these results with the velocity and temperature fields obtained using the Maxwell–Garnett model and the Brinkmann equation for the calculation of the effective thermal conductivity and dynamic viscosity, respectively, under the usual hypothesis of constant physical properties, has demonstrated the high degree of failure of the simulation procedures based on the traditional mean-field theories.

NOMENCLATURE

c	Effective specific heat at constant pressure of the nanofluid
c_f	Specific heat at constant pressure of the base fluid
c_s	Specific heat at constant pressure of the solid nanoparticles
c^*	Dimensionless effective specific heat at constant pressure of the nanofluid
D	Einstein diffusion coefficient
d_f	Equivalent diameter of a base fluid molecule
d_p	Diameter of the nanoparticle
E	Heat-transfer enhancement
g	Gravitational acceleration
H	Height of the enclosure
h	Coefficient of convection of the nanofluid
h_f	Coefficient of convection of the base fluid
k	Effective thermal conductivity of the nanofluid
k_b	Boltzmann's constant = 1.38066×10^{-23} J/K
k_f	Thermal conductivity of the base fluid
k_s	Thermal conductivity of the solid nanoparticles
k^*	Dimensionless effective thermal conductivity of the nanofluid
M	Molecular weight of the base fluid
N	Avogadro number = 6.022×10^{23} mol^{-1}
Nu	Effective Nusselt number of the nanofluid
Nu$_f$	Nusselt number of the base fluid
P	Dimensionless pressure
p	Pressure
Pr	Effective Prandtl number of the nanofluid
Pr$_f$	Prandtl number of the base fluid
Q_c	Dimensionless heat-transfer rate per unit length added to the nanofluid by the heated sidewall of the enclosure

Q_h Dimensionless heat-transfer rate per unit length withdrawn from the nanofluid
          by the cooled sidewall of the enclosure
Ra   Effective Rayleigh number of the nanofluid
Ra_{cr} Critical Rayleigh number
Ra_f Rayleigh number of the base fluid
Re   Nanoparticle Reynolds number
T Temperature of the nanofluid
T_{av} Average temperature of the enclosed fluid
T_c Temperature of the cooled sidewall of the enclosure
T_{fr} Freezing point of the base liquid
T_h Temperature of the heated sidewall of the enclosure
U Dimensionless horizontal velocity component
u Horizontal velocity component
u_B Mean Brownian velocity of the nanoparticle
V Dimensionless vertical velocity component
v Vertical velocity component
W Width of the enclosure
X Dimensionless horizontal Cartesian coordinate
x Horizontal Cartesian coordinate
Y Dimensionless vertical Cartesian coordinate
y Vertical Cartesian coordinate

Greek Symbols

β Effective coefficient of thermal expansion of the nanofluid
β_f Coefficient of thermal expansion of the base fluid
β_s Coefficient of thermal expansion of the solid nanoparticles
φ Nanoparticle volume fraction
φ_{opt} Optimal particle loading
μ Effective dynamic viscosity of the nanofluid
μ_f Dynamic viscosity of the base fluid
μ^* Dimensionless effective dynamic viscosity of the nanofluid
θ Dimensionless temperature of the nanofluid
ρ Effective mass density of the nanofluid
ρ_f Mass density of the base fluid
ρ_s Mass density of the solid nanoparticles
ρ^* Dimensionless effective mass density of the nanofluid
τ_D Time required to cover a distance d_p moving at velocity u_B
ψ Dimensionless stream function

Subscripts

c   At the temperature of the cooled sidewall of the enclosure
h   At the temperature of the heated sidewall of the enclosure
r   At the reference temperature

REFERENCES

1. W. Daungthongsuk and S. Wongwises, A critical review of convective heat transfer in nanofluids, *Renew. Sust. Energy Rev.*, 11, 797–817, 2007.
2. S.M.S. Murshed, K.C. Leong and C. Yang, Thermophysical and electrokinetic properties of nanofluids—A critical review, *Appl. Therm. Eng.*, 28, 2109–2125, 2008.
3. S. Kakaç and A. Pramuanjaroenkij, Review of convective heat transfer enhancement with nanofluid, *Int. J. Heat Mass Transf.*, 52, 3187–3196, 2009.
4. A. Einstein, Eine neue Bestimmung der Molekuldimension, *Ann. Phys.*, 19, 289–306, 1906.
5. A. Einstein, Berichtigung zu meiner Arbeit: Eine neue Bestimmung der Molekuldimension, *Ann. Phys.*, 34, 591–592, 1911.
6. H.C. Brinkman, The viscosity of concentrated suspensions and solutions, *J. Chem. Phys.*, 20, 571, 1952.
7. H. Chen, Y. Ding, and C. Tan, Rheological behaviour of nanofluids, *New J. Phys.*, 9, 367, 2007.
8. H. Chen, Y. Ding, Y. He, and C. Tan, Rheological behaviour of ethylene glycol based titania nanofluids, *Chem. Phys. Lett.*, 444, 333–337, 2007.
9. J. Chevalier, O. Tillement and F. Ayela, Rheological properties of nanofluids flowing through microchannels, *Appl. Phys. Lett.*, 91, 233103, 2007.
10. J.C. Maxwell, *A Treatise on Electricity and Magnetism*, 3rd edn., Dover, New York, 1954.
11. R.L. Hamilton and O.K. Crosser, Thermal conductivity of heterogeneous two component systems, *Ind. Eng. Chem. Fundam.*, 1, 187–191, 1962.
12. D.A.G. Bruggemann, Berechnung Verschiedener Physikalischer Konstanten von Heterogenen Substanzen, I. Dielektrizitatskonstanten und leitfahigkeiten der mischkorper aus isotropen substanzen, *Ann. Phys.*, 24, 636–679, 1935.
13. J. Eapen, W.C. Williams, J. Buongiorno, L.-W. Hu, S. Yip, R. Rusconi and R. Piazza, Mean-field versus microconvection effects in nanofluid thermal conduction, *Phys. Rev. Lett.*, 99, 095901, 2007.
14. J. Buongiorno et al., A benchmark study on the thermal conductivity of nanofluids, *J. Appl. Phys.*, 106, 094312, 2009.
15. S.K. Das, N. Putra, P. Thiesen, and W. Roetzel, Temperature dependence of thermal conductivity enhancement for nanofluids, *J. Heat Transf.*, 125, 567–574, 2003.
16. C.H. Li and G.P. Peterson, Experimental investigation of temperature and volume fraction variations on the effective thermal conductivity of nanoparticle suspensions (nanofluids), *J. Appl. Phys.*, 99, 084314, 2006.
17. W. Yu, H. Xie, L. Chen, and Y. Li, Investigation on the thermal transport properties of ethylene glycol-based nanofluids containing copper nanoparticles, *Powder Technol.*, 197, 218–221, 2010.
18. K. Khanafer, K. Vafai, and M. Lightstone, Buoyancy-driven heat transfer enhancement in a two-dimensional enclosure utilizing nanofluids, *Int. J. Heat Mass Transf.*, 46, 3639–3653, 2003.
19. N. Putra, W. Roetzel, and S.K. Das, Natural convection of nano-fluids, *Heat Mass Transf.*, 39, 775–784, 2003.
20. A. Amiri and K. Vafai, Analysis of dispersion effects and nonthermal equilibrium, non-Darcian, variable porosity, incompressible flow through porous media, *Int. J. Heat Mass Transf.*, 37, 939–954, 1994.
21. R.-Y. Jou and S.-C. Tzeng, Numerical research of nature convective heat transfer enhancement filled with nanofluids in rectangular enclosures, *Int. Comm. Heat Mass Transf.*, 33, 727–736, 2006.

22. R.K. Tiwari and M.K. Das, Heat transfer augmentation in a two-sided lid-driven differentially heated square cavity utilizing nanofluids, *Int. J. Heat Mass Transf.*, 50, 2002–2018, 2007.
23. H.F. Oztop and E. Abu-Nada, Numerical study of natural convection in partially heated rectangular enclosures filled with nanofluids, *Int. J. Heat Fluid Flow*, 29, 1326–1336, 2008.
24. E. Abu-Nada and H.F. Oztop, Effects of inclination angle on natural convection in enclosures filled with Cu-water nanofluid, *Int. J. Heat Fluid Flow*, 30, 669–678, 2009.
25. E.B. Öğüt, Natural convection of water-based nanofluids in an inclined enclosure with a heat source, *Int. J. Therm. Sci.*, 48, 2063–2073, 2010.
26. K. Kahveci, Buoyancy driven heat transfer of nanofluids in a tilted enclosure. *J. Heat Transf.*, 132, 062501, 2010.
27. A.G.A. Nnanna, Experimental model of temperature-driven nanofluid, *J. Heat Transf.*, 129, 697–704, 2007.
28. C.J. Ho, W.K. Liu, Y.S. Chang, and C.C. Lin, Natural convection heat transfer of alumina-water nanofluid in vertical square enclosures: An experimental study, *Int. J. Therm.Sci.*, 49, 1345–1353, 2010.
29. A.K. Santra, S. Sen, and N. Chakraborty, Study of heast transfer characteristics of copper-water nanofluid in a differentially heated square cavity with different viscosity models, *J. Enhanc. Heat Transf.*, 15, 273–287, 2008.
30. C.J. Ho, M.W. Chen, and Z.W. Li, Numerical simulation of natural convection of nanofluid in a square enclosure: Effects due to uncertainties of viscosity and thermal conductivity, *Int. J. Heat Mass Transf.*, 51, 4506–4516, 2008.
31. K. Kwak and C. Kim, Viscosity, and thermal conductivity of copper oxide nanofluid dispersed in ethylene glycol, *Korea-Aust. Rheol. J*, 17, 35–40, 2005.
32. S.E.B. Maïga, C.T. Nguyen, N. Galanis, and G. Roy, Heat transfer behaviours of nanofluids in a uniformly heated tube, *Superlattices Microstruct.*, 35, 543–557, 2004.
33. X. Wang, X. Xu and S.U.S. Choi, Thermal conductivity of nanoparticle-fluid mixture, *J. Thermophys. Heat Transf.*, 13, 474–480, 1999.
34. H.E. Patel, T. Sundararajan, T. Pradeep, A. Dasgupta, N. Dasgupta, and S.K. Das, A micro-convection model for the thermal conductivity of nanofluids, *Pramana—J. Phys.*, 65, 863–869, 2005.
35. R. Prasher, D. Song, J. Wang, and P. Phelan, Measurements of nanofluid viscosity and its implications for thermal applications, *Appl. Phys. Lett.*, 89, 133108, 2006.
36. P. Charuyakorn, S. Sengupta, and S.K. Roy, Forced convection heat transfer in microencapsulated phase change material slurries, *Int. J. Heat Mass Transf.*, 34, 819–833, 1991.
37. E. Abu-Nada, Z. Masoud, H.F. Oztop, and A. Campo, Effects of nanofluid variable properties on natural convection in enclosures, *Int. J. Therm. Sci.*, 49, 479–491, 2010.
38. C.H. Chon, K.D. Kihm, S.P. Lee, and S.U.S. Choi, Empirical correlation finding the role of temperature and particle size for nanofluid (Al_2O_3) thermal conductivity enhancement, *Appl. Phys. Lett.*, 87, 153107, 2005.
39. C.T. Nguyen, F. Desgranges, G. Roy, N. Galanis, T. Maré, S. Boucher, and H. Angue Mintsa, Temperature and particle-size dependent viscosity data for water-based nanofluids—Hysteresis phenomenon, *Int. J. Heat Fluid Flow*, 28, 1492–1506, 2007.
40. H. Masuda, A. Ebata, K. Teramae, and N. Hishinuma, Alteration of thermal conductivity and viscosity of liquid by dispersing ultra-fine particles (dispersion of γ-Al_2O_3, SiO_2, and TiO_2 ultra-fine particles), *Netsu Bussei*, 4, 227–233, 1993.
41. B.C. Pak and Y.I. Cho, Hydrodynamic and heat transfer study of dispersed fluids with submicron metallic oxide particles, *Exp. Heat Transf.*, 11, 151–170, 1998.
42. K.C. Lin and A. Violi Natural convection heat transfer of nanofluids in a vertical cavity: Effects of non-uniform particle diameter and temperature on thermal conductivity, *Int. J. Heat Fluid Flow*, 31, 236–245, 2010.

43. J. Xu, B. Yu, M. Zou, and P. Xu, A new model for heat conduction of nanofluids based on fractal distributions of nanoparticles, *J. Phys. D: Appl. Phys.*, 39, 4486–4490, 2006.

44. S.P. Jang, J.-H. Lee, K.S. Hwang, and S.U.S. Choi, Particle concentration and tube size dependence of viscosities of Al_2O_3-water nanofluids flowing through micro- and mini-tubes, *Appl. Phys. Lett.*, 91, 243112, 2007.

45. E. Abu-Nada and A.J. Chamkha, Effect of nanofluid variable properties on natural convection in enclosures filled with a CuO-EG-Water nanofluid, *Int. J. Therm. Sci.*, 49, 2339–2352, 2010.

46. S.P. Jang and S.U.S. Choi, Effects of various parameters on nanofluid thermal conductivity, *J. Heat Transf.*, 129, 617–623, 2007.

47. P.K. Namburu, D.P. Kulkarni, D. Misra, and D.K. Das, Viscosity of copper oxide nanoparticles dispersed in ethylene glycol and water mixture, *Exp. Therm. Fluid Sci.*, 32, 397–402, 2007.

48. J. Kim, Y.T. Kang, and C.K. Choi, Analysis of convective instability and heat transfer characteristics of nanofluids, *Phys. Fluids*, 16, 2395–2401, 2004.

49. D. Wen and Y. Ding, Formulation of nanofluids for natural convective heat transfer applications, *Int. J. Heat Fluid Flow*, 26, 855–864, 2005.

50. X.Q. Wang, A.S. Mujumdar, and C. Yap, Free convection heat transfer in horizontal and vertical rectangular cavities filled with nanofluids, *International Heat Transfer Conference IHTC-13*, Sydney, Australia, 2006.

51. J.A. Eastman, S.U.S. Choi, S. Li, L.J. Thompson, and S. Lee, Enhanced thermal conductivity through the development of nanofluids, *Materials Research Society Symposium, Pittsburgh, PA, USA*, vol. 457, pp. 3–11, 1997.

52. D. Wen and Y. Ding, Experimental investigation into convective heat transfer of nanofluids at the entrance region under laminar flow conditions, *Int. J. Heat Mass Transf.*, 47, 5181–5188, 2004.

53. Y. Xuan and Q. Li, Heat transfer enhancement of nanofluids, *Int. J. Heat Fluid Flow*, 21, 58–64, 2000.

54. K.S. Hwang, J.-H. Lee, and S.P. Jang, Buoynacy-driven heat transfer of water based Al_2O_3 nanofluids in a rectangular cavity, *Int. J. Heat Mass Transf.*, 50, 4003–4010, 2007.

55. D.Y. Tzou, Thermal instability of nanofluids in nantural convection, *Int. J. Heat Mass Transf.*, 51, 2967–2979, 2008.

56. E. Abu-Nada, Z. Masoud, and A. Hijazi, Natural convection heat transfer enhancement in horizontal concentric annuli using nanofluids, *Int. Comm. Heat Mass Transf.*, 35, 657–665, 2008.

57. E. Abu-Nada, Effects of variable viscosity and thermal conductivity of Al_2O_3-water nanofluid on heat transfer enhancement in natural convection, *Int. J. Heat Fluid Flow*, 30, 679–690, 2009.

58. E. Abu-Nada, Effects of variable viscosity and thermal conductivity of CuO-water nanofluid on heat transfer enhancement in natural convection: Mathematical model and simulation, *J. Heat Transf.*, 132, 052401, 2010.

59. G. Polidori, S. Fohanno, and C.T. Nguyen, A note on heat transfer modelling of Newtonian nanofluids in laminar free convection, *Int. J. Therm. Sci.*, 46, 739–744, 2007.

60. A.V. Kuznetsov and D.A. Nield, Natural convective boundary-layer flow of a nanofluid past a vertical plate, *Int. J. Therm. Sci.*, 49, 243–247, 2010.

61. J. Buongiorno, Convective transport in nanofluids, *J. Heat Transf.*, 128, 240–250, 2006.

62. C. Popa, S. Fohanno, C.T. Nguyen, and G. Polidori, On heat transfer in external natural convection flows using two nanofluids, *Int. J. Therm. Sci.*, 49, 901–908, 2010.

63. H.A. Mintsa, G. Roy, C.T. Nguyen, and D. Doucet, New temperature dependent thermal conductivity data for water-based nanofluids, *Int. J. Therm. Sci.*, 48, 363–371, 2009.

64. S. Lee, S.U.S. Choi, S. Li, and J.A. Eastman, Measuring thermal conductivity of fluids containing oxide nanoparticles, *J. Heat Transf.*, 121, 280–289, 1999.
65. S.U.S. Choi, Z.G. Zhang, W. Yu, F.E. Lockwood, and E.A. Grulke, Anomalous thermal conductivity enhancement in nanotube suspensions, *Appl. Phys. Lett.*, 79, 2252–2254, 2001.
66. C.-J. Yu, A.G. Richter, A. Datta, M.K. Durbin, and P. Dutta, Observation of molecular layering in thin liquid films using x-ray reflectivity, *Phys Rev. Lett.*, 82, 2326–2329, 1999.
67. C.-J. Yu, A.G. Richter, A. Datta, M.K. Durbin, P. Dutta, Molecular layering in a liquid on a solid substrate: An x-ray reflectivity study, *Physica B*, 283, 27–31, 2000.
68. W. Yu and S.U.S. Choi, The role of interfacial layers in the enhanced thermal conductivity of nanofluids: A renovated Maxwell model, *J. Nanopart. Res.*, 5, 167–171, 2003.
69. Q.-Z. Xue, Model for effective thermal conductivity of nanofluids, *Phys. Lett. A*, 307, 313–317, 2003.
70. H. Xie, M. Fujii and X. Zhang, Effect of interfacial nanolayer on the effective thermal conductivity of nanoparticle-fluid mixture, *Int. J. Heat Mass Transf.*, 48, 2926–2932, 2005.
71. K.C. Leong, C. Yang, and S.M.S. Murshed, A model for the thermal conductivity of nanofluids—the effect of interfacial layer, *J. Nanopart. Res.*, 8, 245–254, 2006.
72. D.H. Kumar, H.E. Patel, V.R.R. Kumar, T. Sundararajan, T. Pradeep, and S.K. Das, Model for heat conduction in nanofluids, *Phys. Rev. Lett.*, 93, 144301, 2004.
73. J. Koo and C. Kleinstreuer, A new thermal conductivity model for nanofluids, *J. Nanopart. Res.*, 6, 577–588, 2004.
74. Y. Ren, H. Xie, and A. Cai, Effective thermal conductivity of nanofluids containing spherical nanoparticles, *J. Phys. D: Appl. Phys.*, 38, 3958–3961, 2005.
75. R. Prasher, P. Bhattacharya, and P.E. Phelan, Thermal conductivity of nanoscale colloidal solutions (nanofluids), *Phys. Rev. Lett.*, 94, 025901, 2005.
76. R. Prasher, P. Bhattacharya, and P.E. Phelan, Brownian.motion-based convective-conductive model for the effective thermal conductivity of nanofluids, *J. Heat Transf.*, 128, 588–595, 2006.
77. Y. Xuan, Q. Li, X. Zhang, and M. Fujii, Stochastic thermal transport of nanoparticle suspensions, *J. Appl. Phys.*, 100, 043507, 2006.
78. M. Prakash and E.P. Giannelis, Mechanism of heat transport in nanofluids, *J. Comput.-Aided Mater. Des.*, 14, 109–117, 2007.
79. S.M.S. Murshed, K.C. Leong, and C. Yang, A combined model for the effective thermal conductivity of nanofluids, *Appl. Therm. Eng.*, 29, 2477–2483, 2009.
80. P.L. Kapitza, The study of heat transfer in Helium II, *J. Phys. USSR*, 4, 181–210, 1941.
81. B.-X. Wang, L.-P. Zhou, and X.-F. Peng, A fractal model for predicting the effective thermal conductivity of liquid with suspension of nanoparticles, *Int. J. Heat Mass Transf.*, 46, 2665–2672, 2003.
82. R. Prasher, W. Evans, P. Meakin, J. Fish, P. Phelan, and P. Keblinski, Effect of aggregation on thermal conduction in colloidal nanofluids, *Appl. Phys. Lett.*, 89, 143119, 2006.
83. W. Evans, R. Prasher, J. Fish, P. Meakin, P. Phelan, and P. Keblinski, Effect of aggregation and interfacial thermal resistance on thermal conductivity of nanocomposites and colloidal nanofluids. *Int. J. Heat Mass Transf.*, 51, 1431–1438, 2008.
84. Y. Xuan, Q. Li and W. Hu, Aggregation structure and thermal conductivity of nanofluids, *AIChE J.*, 49, 1038–1043, 2003.
85. R. Prasher, P.E. Phelan, and P. Bhattacharya, Effect of aggregation kinetics on the thermal conductivity of nanoscale colloidal solutions (nanofluid), *Nano Lett.*, 6, 1529–1534, 2006.
86. M. Corcione, Empirical correlating equations for predicting the effective thermal conductivity and dynamic viscosity of nanofluids, *Energy Convers. Manage.*, 52, 789–793, 2011.
87. J.A. Eastman, S.U.S. Choi, S. Li, W. Yu, and L.J. Thompson, Anomalously increased effective thermal conductivity of ethylene glycol-based nanofluids containing copper nanoparticles, *Appl. Phys. Lett.*, 78, 718–720, 2001.

88. C.H. Chon and K.D. Kihm, Thermal conductivity enhancement of nanofluids by Brownian motion, *J. Heat Transf.*, 127, 810, 2005.

89. S.M.S. Murshed, K.C. Leong, and C. Yang, Investigations of thermal conductivity and viscosity of nanofluids, *Int. J. Therm. Sci.*, 47, 560–568, 2008.

90. W. Duangthongsuk and S. Wongwises, Measurement of temperature-dependent thermal conductivity and viscosity of TiO_2-water nanofluids, *Exp. Therm. Fluid Sci.*, 33, 706–714, 2009.

91. P. Keblinski, S.R. Phillpot, S.U.S. Choi, and J.A. Eastman, Mechanisms of heat flow in suspensions of nano-sized particles (nanofluids), *Int. J. Heat Mass Transf.*, 45, 855–863, 2002.

92. J. Koo, *Computational Nanofluid Flow and Heat Transfer Analyses Applied to Micro-Systems*, Dissertation Thesis, North Carolina State University, Rayleigh, NC, USA, 2005.

93. N. Masoumi, N. Sohrabi, and A. Behzadmehr, A new model for calculating the effective viscosity of nanofluids, *J. Phys. D: Appl. Phys.*, 42, 055501, 2009.

94. S. Ganguly and S. Chakraborty, Effective viscosity of nanoscale colloidal suspensions, *J. Appl. Phys.*, 106, 124309, 2009.

95. S.K. Das, N. Putra, and W. Roetzel, Pool boiling characteristics of nano-fluids, *Int. J. Heat Mass Transf.*, 46, 851–862, 2003.

96. Y. He, Y. Jin, H. Chen, Y. Ding, D. Cang, and H. Lu, Heat transfer and flow behaviour of aqueous suspensions of TiO_2 nanoparticles (nanofluids) flowing upward through a vertical pipe, *Int. J. Heat Mass Transf.*, 50, 2272–2281, 2007.

97. J.-H. Lee, K.S. Hwang, S.P. Jang, B.H. Lee, J.H. Kim, S.U.S. Choi, and C.J. Choi, Effective viscosities and thermal conductivities of aqueous nanofluids containing low volume concentrations of Al_2O_3 nanoparticles, *Int. J. Heat Mass Transf.*, 51, 2651–2656, 2008.

98. J. Garg, B. Poudel, M. Chiesa, J.B. Gordon, J.J. Ma, J.B. Wang, Z.F. Ren, Y.T. Kang, H. Ohtani, J. Nanda, G.H. McKinley, and G. Chen, Enhanced thermal conductivity and viscosity of copper nanoparticles in ethylene glycol nanofluid, *J. Appl. Phys.*, 103, 074301, 2008.

99. B.H. Chang, A.F. Mills, and E. Hernandez, Natural convection of microparticle suspensions in thin enclosures, *Int. J. Heat Mass Transf.*, 51, 1332–1341, 2008.

100. M. Corcione, Heat transfer features of buoyancy-driven nanofluids inside rectangular enclosures differentially heated at the sidewalls, *Int. J. Therm. Sci.*, 49, 1536–1546, 2010.

101. M. Corcione, Rayleigh-Bénard convection heat transfer in nanoparticle suspensions, *Int. J. Heat Fluid Flow*, 32, 65–77, 2011.

102. J.P. Van Doormaal and G.D. Raithby, Enhancements of the simple method for predicting incompressible fluid flows, *Num. Heat Transf.*, 11, 147–163, 1984.

103. B. P. Leonard, A stable, and accurate convective modelling procedure based on quadratic upstream interpolation, *Comp. Meth. Appl. Mech. Eng.*, 19, 59–78, 1979.

104. G. de Vahl Davis, Natural convection of air in a square cavity: A bench mark numerical solution, *Int. J. Num. Meth. Fluids*, 3, 249–264, 1983.

105. H.S. Mahdi and R.B. Kinney, Time-dependent natural convection in a square cavity: Application of a new finite volume method, *Int. J. Num. Meth. Fluids*, 11, 57–86, 1990.

106. M. Hortmann, M. Peric, and G. Scheuerer, Finite volume multigrid prediction of laminar natural convection: Bench-mark solutions, *Int. J. Num. Meth. Fluids*, 11, 189–207, 1990.

107. D.C. Wan, B.S.V. Patnaik, and G.W. Wei, A new benchmark quality solution for the buoyancy-driven cavity by discrete singular convolution, *Num. Heat Transf.*, 40, 199–228, 2001.

Index

Printed and bound by CPI Group (UK) Ltd, Croydon, CR0 4YY

18/10/2024

01776261-0008